Cardiac Arrhythmia: Molecular Mechanisms and Therapeutic Strategies

Cardiac Arrhythmia: Molecular Mechanisms and Therapeutic Strategies

Guest Editors

Yosuke Okamoto
Kunichika Tsumoto

MDPI

Basel • Beijing • Wuhan • Barcelona • Belgrade • Novi Sad • Cluj • Manchester

Guest Editors
Yosuke Okamoto
Department of Cell Physiology
Akita University Graduate
School of Medicine
Akita
Japan

Kunichika Tsumoto
Department of Physiology II
Kanazawa Medical University
Uchinada
Japan

Editorial Office
MDPI AG
Grosspeteranlage 5
4052 Basel, Switzerland

This is a reprint of the Special Issue, published open access by the journal *International Journal of Molecular Sciences* (ISSN 1422-0067), freely accessible at: www.mdpi.com/journal/ijms/special_issues/OZ5L6DIQRC.

For citation purposes, cite each article independently as indicated on the article page online and using the guide below:

Lastname, A.A.; Lastname, B.B. Article Title. *Journal Name* **Year**, *Volume Number*, Page Range.

ISBN 978-3-7258-2918-7 (Hbk)
ISBN 978-3-7258-2917-0 (PDF)
https://doi.org/10.3390/books978-3-7258-2917-0

Contents

Preface

Heart disease is the leading cause of death worldwide. Arrhythmia plays an essential role in the final stages of heart disease, such as in circulatory failure due to ventricular arrhythmia and fatal cerebral infarction associated with atrial arrhythmia. For this reason, research into arrhythmia has been a priority for several years. To study the mechanisms of arrhythmia, it is first necessary to understand the complex system of cardiac excitation. Arrhythmia is a breakdown of this controlled system. Since cardiac excitation is linked to cardiac contraction, it can ultimately lead to fatal biological problems. There are several classifications of arrhythmia, such as atrial arrhythmia and ventricular arrhythmia, and in order to elucidate the mechanisms of each type of arrhythmia, there are various approaches, from basic nanoscale research to clinical and epidemiological research.

This Special Issue, titled "Cardiac Arrhythmia: Molecular Mechanisms and Therapeutic Strategies" will cover new findings on the molecular mechanisms of arrhythmias based on original research. The Special Issue includes a wide range of experimental methods, and in it, researchers will be able to discover the newest findings in the area. As Guest Editors, we thank all the authors who contributed to this Special Issue.

Yosuke Okamoto and Kunichika Tsumoto
Guest Editors

International Journal of *Molecular Sciences*

MDPI

Editorial

Cardiac Arrhythmia: Molecular Mechanisms and Therapeutic Strategies

Yosuke Okamoto [1,*,†] and Kunichika Tsumoto [2,*,†]

1 Department of Cell Physiology, Akita University Graduate School of Medicine, 1-1-1, Hondo, Akita 010-8543, Japan

2 Department of Physiology II, Kanazawa Medical University, 1-1 Daigaku, Uchinada, Ishikawa 920-0293, Japan

* Correspondence: okamoto@med.akita-u.ac.jp (Y.O.); tsumoto@kanazawa-med.ac.jp (K.T.); Tel.: +81-18-884-6074 (Y.O.); +81-76-286-2211 (ext. 33645) (K.T.); Fax: +81-18-836-2604 (Y.O.); +81-76-286-8010 (K.T.)

† These authors contributed equally to this paper.

Arrhythmias are divided into supraventricular and ventricular, depending on where they originate. Supraventricular arrhythmias are characterized by the presence of extracardiac complications, while ventricular arrhythmias have a high mortality rate.

In this Special Issue, all of the papers on supraventricular arrhythmias, with the exception of Dr. Kayser's paper, are on the theme of atrial fibrillation (AF). AF is the most frequent of all arrhythmias and can be described as the king of arrhythmias.

In a review article, Iwamiya et al. [1] explain the macroscopic anatomy, histological findings, and molecular mechanisms of AF, using "heterogeneity" as a keyword, and also explain the treatment strategies. Most cases of idiopathic AF are caused by electrical reentry between the pulmonary veins (PVs) and left atrium and are diagnosed as sustained arrhythmia [2]. The heterogeneity of thickness of the heart wall and the muscle fiber arrangement between the PVs and atrium are some of the factors that cause and sustain AF. In particular, the muscle bundles between same-side PVs are called 'carina' and are an important target for radiofrequency ablation [3]. When AF becomes persistent or chronic, atrial fibrosis becomes a major factor in the pathological condition [4]. Cultured cell study suggests that intercellular communication between myofibroblasts and cardiomyocytes is the cause of electrophysiological heterogeneity inside the atrium [5]. Recent advances in imaging technology have made it possible to visualize fibrosis using MRI non-invasively [6]. Meanwhile, there has also been progress in evaluating fibrosis using electrophysiology with catheter electrodes [7]. Interestingly, the evaluation of fibrosis using MRI and electrophysiology does not overlap. There appears to be a mismatch between structural and functional changes in the extent of fibrosis. Further research is thus necessary for future therapeutic applications in evaluating fibrosis. In particular, there is hope for developing drugs that reduce heterogeneity. Twenty-five years have passed since the discovery of the 'PV theory', and the progress of non-drug therapy, percutaneous ablation, is reaching its limits.

Perhaps it would be a good idea to introduce molecular-targeted therapy for AF, which is analogous to cancer treatment. The research by Dr. Igarashi and his colleagues [8] is a comprehensive analysis of gene expression in the pathology of non-paroxysmal AF. This research shows that the PVs lose their characteristics as cardiac tissue and instead transform into tissue similar to the cancer-related microenvironment. It is known that the microenvironment is formed prior to cancer metastasis and invasion [9]. The ectoderm, such as vascular endothelium and mucosal epithelium in the cancer-related microenvironment, is converted into mesenchymal tissue, such as fibroblasts [10]. The heart is an organ that is extremely resistant to cancer, but it is known that severe fibrosis occurs in the atria and PVs of patients with AF [11,12]. In recent years, the relationship between AF and

Citation: Okamoto, Y.; Tsumoto, K. Cardiac Arrhythmia: Molecular Mechanisms and Therapeutic Strategies. *Int. J. Mol. Sci.* **2024**, *25*, 13253. https://doi.org/10.3390/ijms252413253

Received: 2 December 2024
Accepted: 5 December 2024
Published: 10 December 2024

cancer has also been attracting attention [13]. Some reports have already suggested that endothelial-mesenchymal transition occurs in the atria of patients with AF [14,15]. The paper by Igarashi et al. was the first to suggest the possibility that a similar endothelial-mesenchymal transition occurs in the PVs of AF patients.

Dr. Kayser and colleagues [16] studied sinus node dysfunction (SND), namely that of bradycardia. A known single amino acid mutation in G protein-coupled inward rectifying potassium (GIRK) channel has been reported to cause autosomal dominant familial SND [17], and the mutation causes bradycardia. The authors used induced pluripotent stem cell (iPS) technology and gene editing techniques to confirm the phenotype. The iPS cells were differentiated using a specialized method, applying retinoic acid to induce atrial-like cells [18,19]. The cause of the bradycardia was that the GIRK channels were not closing due to a genetic mutation. This discovery is an important report suggesting that selective inhibitors against the channel could be an effective treatment strategy.

For life-threatening arrhythmias such as ventricular tachycardia (VT) and ventricular fibrillation (VF), electrical cardioversion is required for immediate life-saving intervention [20]. In the ventricles where VT (or VF) was induced, it is believed to be in a state in which the excitation wave is circling (or in which the excitation wave is splitting and fusing irregularly) within the tissue [21–23]. Electrical cardioversion is performed to reset the ventricles in the VT/VF state. Resetting the cardiac excitable state is a forced switch that accelerates repolarization from a depolarized state to a quiescent state, and the electrical stimulus delivered by the automated external defibrillator (AED) has to propagate the repolarization switch throughout the entire ventricles in an avalanche fashion. The mechanism of successful defibrillation is still unknown, and the study by Himeno et al. [24] may explain part of this defibrillation mechanism. It has been reported that the inactivation of I_{NaL} and of L-type Ca^{2+} current (I_{CaL}), the fast component of the delayed rectifier K^+ channel current (I_{Kr}), and the activation of the inward rectifier K^+ channel current (I_{K1}) were the key to repolarization propagation.

Besides, the VT/VF is inducible by the cardiotoxicity (pro-arrhythmogenicity) of drugs [25]. The human ether-à-go-go-related gene (hERG) channel screening is well known in terms of investigating cardiotoxicity in drug development. Dr. Furutani [26] reviews the pharmacological actions of hERG channel inhibitors, the underlying molecular mechanisms, and the clinical implications of these drugs. The hERG channel, which is thought to constitute $I_{Kr,}$ controls the action potential of the human heart by balancing voltage-dependent activation and inactivation. This channel is more active during the repolarization phase than the depolarization phase [27].

For this reason, suppression of this channel causes an increase in the QT interval on the electrocardiogram. Since QT prolongation increases the risk of inducing the VT/VF, unexpected suppression of the hERG channel can cause medical accidents [28]. Unfortunately, the hydrophobic space within the hERG channel is large, and many lipophilic drugs act on it [29].

On the other hand, some Class III antiarrhythmic drugs that inhibit hERG channels have a "facilitating" effect [30] and are used in clinical practice. The fact that an inhibitor is also a "facilitator" is difficult to rationalize, but Furutani's review article approaches the truth of the molecular mechanism.

As described above, research into arrhythmia is a fascinating field in which basic medicine and clinical application are closely linked. I hope that you will read the articles in this Special Issue that interest you and gain a deeper insight into the arrhythmia research.

Conflicts of Interest: The authors declare no conflicts of interest.

References

1. Iwamiya, S.; Ihara, K.; Nitta, G.; Sasano, T. Atrial Fibrillation and Underlying Structural and Electrophysiological Heterogeneity. *Int. J. Mol. Sci.* **2024**, *25*, 10193. [CrossRef] [PubMed]
2. Haïssaguerre, M.; Jaïs, P.; Shah, D.C.; Takahashi, A.; Hocini, M.; Quiniou, G.; Garrigue, S.; Le Mouroux, A.; Le Métayer, P.; Clémenty, J. Spontaneous initiation of atrial fibrillation by ectopic beats originating in the pulmonary veins. *N. Engl. J. Med.* **1998**, *339*, 659–666. [CrossRef] [PubMed]
3. Valles, E.; Fan, R.; Roux, J.F.; Liu, C.F.; Harding, J.D.; Dhruvakumar, S.; Hutchinson, M.D.; Riley, M.; Bala, R.; Garcia, F.C.; et al. Localization of Atrial Fibrillation Triggers in Patients Undergoing Pulmonary Vein Isolation. Importance of the Carina Region. *J. Am. Coll. Cardiol.* **2008**, *52*, 1413–1420. [CrossRef] [PubMed]
4. Kostin, S.; Klein, G.; Szalay, Z.; Hein, S.; Bauer, E.P.; Schaper, J. Structural correlate of atrial fibrillation in human patients. *Cardiovasc. Res.* **2002**, *54*, 361–379. [CrossRef]
5. Miragoli, M.; Salvarani, N.; Rohr, S. Myofibroblasts induce ectopic activity in cardiac tissue. *Circ. Res.* **2007**, *101*, 755–758. [CrossRef]
6. Oakes, R.S.; Badger, T.J.; Kholmovski, E.G.; Akoum, N.; Burgon, N.S.; Fish, E.N.; Blauer, J.J.E.; Rao, S.N.; Dibella, E.V.R.; Segerson, N.M.; et al. Detection and quantification of left atrial structural remodeling with delayed-enhancement magnetic resonance imaging in patients with atrial fibrillation. *Circulation* **2009**, *119*, 1758–1767. [CrossRef]
7. Sanders, P.; Morton, J.B.; Davidson, N.C.; Spence, S.J.; Vohra, J.K.; Sparks, P.B.; Kalman, J.M. Electrical remodeling of the atria in congestive heart failure: Electrophysiological and electroanatomic mapping in humans. *Circulation* **2003**, *108*, 1461–1468. [CrossRef]
8. Igarashi, W.; Takagi, D.; Okada, D.; Kobayashi, D.; Oka, M.; Io, T.; Ishii, K.; Ono, K.; Yamamoto, H.; Okamoto, Y. Bioinformatic Identification of Potential RNA Alterations on the Atrial Fibrillation Remodeling from Human Pulmonary Veins. *Int. J. Mol. Sci.* **2023**, *24*, 10501. [CrossRef]
9. Joyce, J.A.; Pollard, J.W. Microenvironmental regulation of metastasis. *Nat. Rev. Cancer* **2009**, *9*, 239–252. [CrossRef]
10. Yang, J.; Antin, P.; Berx, G.; Blanpain, C.; Brabletz, T.; Bronner, M.; Campbell, K.; Cano, A.; Casanova, J.; Christofori, G.; et al. Guidelines and definitions for research on epithelial–mesenchymal transition. *Nat. Rev. Mol. Cell Biol.* **2020**, *21*, 341–352. [CrossRef]
11. Polyakova, V.; Miyagawa, S.; Szalay, Z.; Risteli, J.; Kostin, S. Atrial extracellular matrix remodelling in patients with atrial fibrillation. *J. Cell. Mol. Med.* **2008**, *12*, 189–208. [CrossRef] [PubMed]
12. Hassink, R.J.; Aretz, H.T.; Ruskin, J.; Keane, D. Morphology of atrial myocardium in human pulmonary veins: A postmortem analysis in patients with and without atrial fibrillation. *J. Am. Coll. Cardiol.* **2003**, *42*, 1108–1114. [CrossRef] [PubMed]
13. Yun, J.P.; Choi, E.K.; Han, K.D.; Park, S.H.; Jung, J.H.; Park, S.H.; Ahn, H.J.; Lim, J.H.; Lee, S.R.; Oh, S. Risk of Atrial Fibrillation According to Cancer Type: A Nationwide Population-Based Study. *JACC CardioOncology* **2021**, *3*, 221–232. [CrossRef] [PubMed]
14. Lai, Y.J.; Tsai, F.C.; Chang, G.J.; Chang, S.H.; Huang, C.C.; Chen, W.J.; Yeh, Y.H. miR-181b targets semaphorin 3A to mediate TGF-β–induced endothelial-mesenchymal transition related to atrial fibrillation. *J. Clin. Investig.* **2022**, *132*, e142548. [CrossRef]
15. Kato, T.; Sekiguchi, A.; Sagara, K.; Tanabe, H.; Takamura, M.; Kaneko, S.; Aizawa, T.; Fu, L.T.; Yamashita, T. Endothelial–mesenchymal transition in human atrial fibrillation. *J. Cardiol.* **2017**, *69*, 706–711. [CrossRef]
16. Kayser, A.; Dittmann, S.; Šarić, T.; Mearini, G.; Verkerk, A.O.; Schulze-Bahr, E. The W101C KCNJ5 Mutation Induces Slower Pacing by Constitutively Active GIRK Channels in hiPSC-Derived Cardiomyocytes. *Int. J. Mol. Sci.* **2023**, *24*, 15290. [CrossRef]
17. Kuß, J.; Stallmeyer, B.; Goldstein, M.; Rinné, S.; Pees, C.; Zumhagen, S.; Seebohm, G.; Decher, N.; Pott, L.; Kienitz, M.C.; et al. Familial Sinus Node Disease Caused by a Gain of GIRK (G-Protein Activated Inwardly Rectifying K+Channel) Channel Function. *Circ. Genomic Precis. Med.* **2019**, *12*, E002238. [CrossRef] [PubMed]
18. Devalla, H.D.; Schwach, V.; Ford, J.W.; Milnes, J.T.; El-Haou, S.; Jackson, C.; Gkatzis, K.; Elliott, D.A.; Chuva de Sousa Lopes, S.M.; Mummery, C.L.; et al. Atrial-like cardiomyocytes from human pluripotent stem cells are a robust preclinical model for assessing atrial-selective pharmacology. *EMBO Mol. Med.* **2015**, *7*, 394–410. [CrossRef]
19. Goldfracht, I.; Protze, S.; Shiti, A.; Setter, N.; Gruber, A.; Shaheen, N.; Nartiss, Y.; Keller, G.; Gepstein, L. Generating ring-shaped engineered heart tissues from ventricular and atrial human pluripotent stem cell-derived cardiomyocytes. *Nat. Commun.* **2020**, *11*, 75. [CrossRef]
20. Szabó, Z.; Ujvárosy, D.; Ötvös, T.; Sebestyén, V.; Nánási, P.P. Handling of ventricular fibrillation in the emergency setting. *Front. Pharmacol.* **2020**, *10*, 1640. [CrossRef]
21. Winfree, A.T. Electrical turbulence in three-dimensional heart muscle. *Science* **1994**, *266*, 1003–1006. [CrossRef] [PubMed]
22. Gray, R.A.; Jalife, J.; Panfilov, A.V.; Baxter, W.T.; Cabo, C.; Davidenko, J.M.; Pertsov, A.M. Mechanisms of cardiac fibrillation. *Science* **1995**, *270*, 1222–1223. [CrossRef] [PubMed]
23. Witkowski, F.X.; Leon, L.J.; Penkoske, P.A.; Giles, W.R.; Spano, M.L.; Ditto, W.L.; Winfree, A.T. Spatiotemporal evolution of ventricular fibrillation. *Nature* **1998**, *392*, 78–82. [CrossRef]
24. Himeno, Y.; Zhang, Y.; Enomoto, S.; Nomura, H.; Yamamoto, N.; Kiyokawa, S.; Ujihara, M.; Muangkram, Y.; Noma, A.; Amano, A. Ionic Mechanisms of Propagated Repolarization in a One-Dimensional Strand of Human Ventricular Myocyte Model. *Int. J. Mol. Sci.* **2023**, *24*, 15378. [CrossRef] [PubMed]

25. For, R.; Of, R.; For, P.; Use, H.; The, V.O.F.; For, O. ICH-2005. S7B: The Non-Clinical Evaluation of the Potential for Delayed Ventricular Repolarization (QT Interval Prolongation) by Human Pharmaceuticals. 2005. Available online: https://www.ich.org/page/safety-guidelines (accessed on 3 December 2024).
26. Furutani, K. Facilitation of hERG Activation by Its Blocker: A Mechanism to Reduce Drug-Induced Proarrhythmic Risk. *Int. J. Mol. Sci.* **2023**, *24*, 16261. [CrossRef]
27. Sanguinetti, M.C.; Tristani-Firouzi, M. hERG potassium channels and cardiac arrhythmia. *Nature* **2006**, *440*, 463–469. [CrossRef]
28. Surawicz, B. Electrophysiologic substrate of torsade de pointes: Dispersion of repolarization or early afterdepolarizations? *J. Am. Coll. Cardiol.* **1989**, *14*, 172–184. [CrossRef]
29. Wang, W.; MacKinnon, R. Cryo-EM Structure of the Open Human Ether-à-go-go-Related K+ Channel hERG. *Cell* **2017**, *169*, 422–430.e10. [CrossRef]
30. Hosaka, Y.; Iwata, M.; Kamiya, N.; Yamada, M.; Kinoshita, K.; Fukunishi, Y.; Tsujimae, K.; Hibino, H.; Aizawa, Y.; Inanobe, A.; et al. Mutational analysis of block and facilitation of HERG current by A Class III anti-arrhythmic agent, nifekalant. *Channels* **2007**, *1*, 198–208. [CrossRef]

International Journal of *Molecular Sciences*

MDPI

Review

Atrial Fibrillation and Underlying Structural and Electrophysiological Heterogeneity

Satoshi Iwamiya, Kensuke Ihara, Giichi Nitta and Tetsuo Sasano *

Department of Cardiovascular Medicine, Tokyo Medical and Dental University (TMDU), 1-5-45 Yushima, Bunkyo-ku, Tokyo 113-8510, Japan; satoshi-iwamiya.cvm@tmd.ac.jp (S.I.); iharcvm@tmd.ac.jp (K.I.); giichi-nitta.cvm@tmd.ac.jp (G.N.)

* Correspondence: sasano.cvm@tmd.ac.jp; Tel.: +81-3-5803-5205

Abstract: As atrial fibrillation (AF) progresses from initial paroxysmal episodes to the persistent phase, maintaining sinus rhythm for an extended period through pharmacotherapy and catheter ablation becomes difficult. A major cause of the deteriorated treatment outcome is the atrial structural and electrophysiological heterogeneity, which AF itself can exacerbate. This heterogeneity exists or manifests in various dimensions, including anatomically segmental structural features, the distribution of histological fibrosis and the autonomic nervous system, sarcolemmal ion channels, and electrophysiological properties. All these types of heterogeneity are closely related to the development of AF. Recognizing the heterogeneity provides a valuable approach to comprehending the underlying mechanisms in the complex excitatory patterns of AF and the determining factors that govern the seemingly chaotic propagation. Furthermore, substrate modification based on heterogeneity is a potential therapeutic strategy. This review aims to consolidate the current knowledge on structural and electrophysiological atrial heterogeneity and its relation to the pathogenesis of AF, drawing insights from clinical studies, animal and cell experiments, molecular basis, and computer-based approaches, to advance our understanding of the pathophysiology and management of AF.

Keywords: atrial fibrillation; heterogeneity; arrhythmogenicity; electrophysiology; atrial remodeling

Citation: Iwamiya, S.; Ihara, K.; Nitta, G.; Sasano, T. Atrial Fibrillation and Underlying Structural and Electrophysiological Heterogeneity. *Int. J. Mol. Sci.* **2024**, *25*, 10193. https://doi.org/10.3390/ijms251810193

Academic Editors: Yosuke Okamoto and Kunichika Tsumoto

Received: 19 August 2024
Revised: 16 September 2024
Accepted: 19 September 2024
Published: 23 September 2024

1. Introduction

Atrial fibrillation (AF), a common arrhythmia, has been associated with increased risks of heart failure and stroke, consequently exacerbating patient prognoses. Despite its prevalence, AF treatments have not yet been optimized. Pharmacotherapy for AF using antiarrhythmic drugs is characterized by limited efficacy, with no substantial evidence indicating improvements in long-term outcomes [1]. Upstream therapies, such as angiotensin-converting enzyme inhibitors (ACEis) and angiotensin-receptor blockers, have not been proven to prevent the occurrence of AF [2,3]. Despite the effectiveness of catheter ablation for paroxysmal AF, it has therapeutic limitations for persistent AF. As AF progresses from the paroxysmal to the persistent phase, it becomes necessary to focus not only on the triggers but also the vulnerable substrates that drive fibrillatory activity.

The detailed pathophysiological mechanisms underlying AF remain largely elusive. In particular, the processes by which the triggers induce electrical fibrillatory activity and the factors that sustain this abnormal activity remain poorly understood. Furthermore, the determining factors of the seemingly disorganized propagation pattern during AF remain unclear. These mechanisms have been investigated from various perspectives, such as anatomical structure, tissue, sarcolemmal ion currents, and electrophysiology. One of the contributing factors to the vulnerable substrates for AF, namely arrhythmogenic substrate, is heterogeneity. The electrical activity in an atrium is believed to repeat even and synchronous excitement unless heterogeneity exists. When a normal atrial excitation wave advances evenly and synchronously on a homogeneous substrate but encounters a heterogeneous point, it breaks up (wavebreak) and occasionally rotates. If a cascade of wavebreaks is induced, the wave begins to exhibit chaotic propagation. As the propagating wave rotates consistently around the point of the wavebreak, if sustained, it is termed a rotor, which typically has a short cycle length and produces rapid tachycardia. A rotor can be a driver of AF, which represents a focal source demonstrating repetitive and fast activity that propagates outward from the source. The process from the inception of a wavebreak to the sustenance of fibrillation can be attributed to heterogeneity [4]. Many studies have reported a correlation between atrial heterogeneity and susceptibility to AF; however, how these atrial heterogeneities correlate with AF remains unclear, resulting in the unawareness that the underlying heterogeneities are a causative factor of AF. Two main types of heterogeneity exist: structural and electrophysiological. Structural heterogeneity can be further viewed based on three subcategories: macroscopic (segmental), mesoscopic (tissue), and microscopic (cellular and molecular) structure. Each level of structural heterogeneity can affect electrophysiological heterogeneity, contributing to the substrate for AF [5–7] (Figure 1).

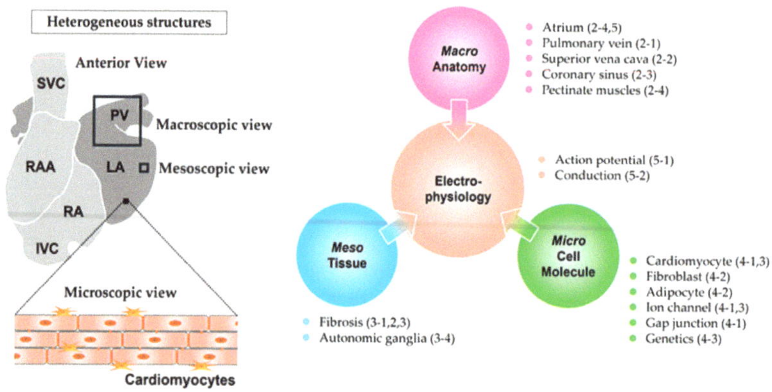

Figure 1. Macro-, meso-, microscopic structural and electrophysiological heterogeneities. RA: right atrium, RAA: right atrial appendage, LA: left atrium, PV: pulmonary vein, SVC: superior vena cava, IVC: inferior vena cava, Macro: macroscopic, Meso: mesoscopic, and Micro: microscopic. Numbers in parentheses indicate the corresponding section numbers in the text.

Highlighting atrial heterogeneity will help us understand its contribution to AF and investigate new therapeutic strategies for AF [8]. Herein, we will summarize the atrial structural and electrophysiological heterogeneities which are incorporated into the pathophysiology of AF, drawing insights from human clinical studies, animal experiments, cell experiments, and theoretical models.

2. Macroscopic Heterogeneity

The atria are complex structures composed of various components, including chambers divided by the atrial septum, venous elements, appendages, and vestibules, rather than a single uniform structure (Figure 2). At the boundaries of these segments or within each segment, inherent heterogeneities exist, even in the absence of pathophysiological conditions. In particular, venous components, such as the pulmonary vein (PV), vena cava, and coronary sinus (CS), have common peculiar features. They have in common myocardial tissue stretching from the atrium, which is referred to as a myocardial sleeve. The features of the myocardial sleeve differ from those of the atrial myocardium in terms of myocardial thickness and arrangement. Around the boundary between the veins and atrium, abrupt changes in musculature structure may cause electrical divergence or convergence, possibly resulting in an electrical source–sink mismatch. Meanwhile, the right atrial (RA) appendage is composed of pectinate muscles, which have an uneven and rough surface characterized by the branching and overlapping arrangement of trabeculae. Moreover, the crista terminalis, the connection between pectinate muscles and the relatively smooth muscle region derived from the sinus venosus, generates an electrical boundary. The complex structure of pectinate muscles may contribute to the arrhythmogenic substrates [9]. Although these distinct structural heterogeneities are not arrhythmogenic in most healthy individuals, they become arrhythmogenic and evoke AF when certain pathophysiological conditions arise. The causative factors of AF based on anatomical heterogeneity are listed in Table 1.

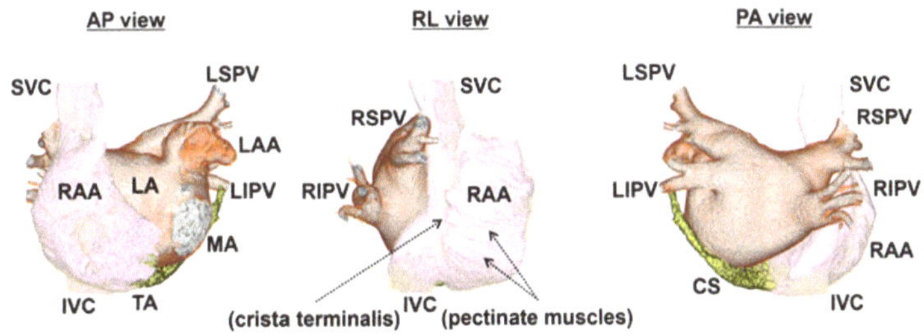

Figure 2. Anatomy of human atria in 3D computed tomography. The endocardial aspects of the atria and surrounding structures above the mitral and tricuspid annulus are cropped using 3D computed tomography. With the 3D images of contrast-enhanced computed tomography reflecting the endocardial aspects of the atria, the endocardial structures, crista terminalis, and pectinate muscles, in particular, are recognizable. Of note, the endocardial aspects of an RAA demonstrate a rough surface due to pectinate muscles, while the body of an RA has a smooth surface. The marks of the crista terminalis and pectinate muscles from the endocardial aspect are shown by the arrows. The images are separated by colors: RA, SVC, and IVC (white); LA, LAA, and PVs (brown); and CS (yellow). AP view: anterior–posterior view, RL view: right lateral view, PA view: posterior–anterior view, RA: right atrium, RAA: right atrial appendage, LA: left atrium, LAA: left atrial appendage, LSPV: left superior pulmonary vein, LIPV: left inferior pulmonary vein, RSPV: right superior pulmonary vein, RIPV: right inferior pulmonary vein, SVC: superior vena cava, IVC: inferior vena cava, CS: coronary sinus, MA: mitral annulus, and TA: tricuspid annulus.

2.1. Pulmonary Vein

PVs play prominent roles in AF, and catheter ablation for the electrical isolation of PVs has become an established treatment because the majority of AF is initiated by ectopic discharges mostly within the PVs. Abnormal automaticity and triggered activity have been suggested as the triggers of AF. The architecture of the PV myocardial sleeve can be a trigger source for AF, and a few mechanisms have been advocated in canine experiments. Previous anatomic-electrophysiological studies in isolated PV specimens have demonstrated that PVs contain a mixture of pacemakers and myocardial cells [10]. These pacemaker cells with spontaneous activity have a significantly lower density of inward rectifier potassium current, leading to less negative diastolic potential, with which increased automaticity is generally associated [11]. Thus, pacemaker activity from these cells is thought to result in the formation of the ectopic beats that initiate AF. Several studies using the atria of autopsy patients support these findings in canine experiments. Light microscopy with periodic acid–Schiff staining and electron microscopy identified that the PV has cells that are histologically and morphologically similar to node, transitional, and Purkinje cells [12]. Not only enhanced automaticity, but also triggered activity is referred to as impulse initiation in cardiac fibers which is dependent on afterdepolarizations [13]. Animal experiments showed that isolated PV cardiomyocytes (CMs) from healthy dogs and rabbits manifested arrhythmogenic afterdepolarizations, suggesting that triggered activity may account for the trigger of AF within the PV [11,14]. In addition, PVs contain enriched autonomic innervation, and the autonomic nervous system is suggested to enhance ectopic triggers via sympathetic or vagal activation [15,16]. These findings indicate that the PV myocardial sleeve may act as a focal trigger. Furthermore, not only do PVs act as a trigger of AF, but they also contribute to the initiating and sustaining mechanism of AF [17]. The arrhythmogenicity of PVs can be evidenced in humans by persisting tachyarrhythmia confined to the PV during sinus rhythm after its isolation, suggesting that the PV myocardial sleeve is involved in the maintenance mechanism of AF. Furthermore, many of the focal left atrial (LA) tachycardias after segmental PV isolation have been reported to be caused by a focal reentrant circuit located at the PV ostium [18]. From a structural perspective, this is partly attributed to the arrangement and thickness of the myocyte bundles within the PV myocardial sleeves and at the PV-LA junction. The sleeves are composed of circularly or spirally oriented bundles of myocytes with additional bundles that are longitudinally or obliquely oriented [19]. These intricate arrangements of myocardial fibers in the PVs, which are often diagonal or perpendicular to LA CMs near the PV-LA junction, contribute to conduction slowing because of the conduction anisotropy, serving as a possible reason for their complexity within the PV and even at the PV-LA junction [19,20]. According to a histological study performed on human hearts, the thickness of the sleeves is not uniform. The myocardial thickness of the PV is almost uniformly thin (0.3–0.8 mm) in contrast to that of LA (1.1–2.6 mm). An abrupt change in myocardial thickness exists around the PV-LA junction (mean 1.1 mm), and the sleeve tapers distally [21]. The differential muscle narrowing and complex conduction patterns at the PV-LA junction can provide a robust anatomical basis for source–sink mismatch and conduction anisotropy. The arrangement of muscle bundles between different ipsilateral PVs, the carina, is also convoluted and interwoven, which may produce non-uniform anisotropic properties [22,23]. The catheter ablation strategy of PV antrum isolation to encircle ipsilateral PVs by ablating on the atrial side of the PV-LA junction has been widely accepted. PV antrum isolation has achieved a higher clinical success rate and a lower incidence of postprocedural atrial arrhythmias [24], since it can be attributed not only to the isolation of PVs but also to the isolation of the PV-LA junction and the ipsilateral carina. These structural characteristics of the PV myocardium potentially represent a primary substrate for AF.

2.2. *Vena Cava*

Myocardial sleeves within the superior vena cava (SVC) can also serve as the origins of AF [25], and the SVC accounts for the major portion of non-PV foci [26–28]. Unlike PVs, limited data regarding the morphology of SVC myocardial sleeves have been published. However, a previous canine study revealed that half of the isolated CMs from SVC myocardial sleeves have pacemaker activity, showing spontaneous depolarization and less negative resting membrane potentials [29]. The presence of CMs with pacemaker activity suggests that automaticity plays a role in the arrhythmogenicity of the SVC. This study also demonstrated that the infusion of autonomic agents, such as isoproterenol, atropine, and phenylephrine, into SVC CMs accelerated the spontaneous activity and induced afterdepolarizations, indicating that the enhanced automaticity and afterdepolarization are involved in the arrhythmogenic activity of the SVC [29]. Additionally, a recent study showed that the muscle fibers of SVC myocardial sleeves exhibit a morphology similar to that of Purkinje fibers [30], supporting the indication of possible arrhythmogenicity in the SVC. In addition to the cell types in SVC myocardial sleeves, the geometric size of SVC myocardial sleeves can be an important element of SVC arrhythmogenicity. In a clinical study, an SVC myocardial sleeve longer than 30 mm has been reported as an independent risk factor for arrhythmogenic trigger sources from the SVC [31]. Long SVC myocardial sleeves were also reported to be correlated with the SVC potential amplitude, reflecting greater myocardial volume. Thus, these findings suggest that long myocardial extension and large amounts of CMs in the SVC can be a source of arrhythmogenicity. SVC myocardial sleeves may serve not only as a trigger of AF, but also as a perpetuator that initiates and sustains AF. Similar to PV myocardial sleeves, the arrangement of musculature in myocardial sleeves within the SVC is intricate. The cardiac fibers are bundled and disposed in longitudinal, oblique, or circumferential directions in the venous wall, which may cause anisotropic conduction within the SVC. Based on high-resolution electroanatomic mapping, slow and heterogeneous conduction has been observed during sinus rhythm and is exaggerated by premature stimulation, indicating that SVC myocardial sleeves have the potential for a reentrant mechanism [32]. Several observations that the SVC acts not only as an AF trigger but also as a driver have been reported [33,34]. Miyazaki et al. reported that SVC fibrillation was confined to the SVC after its electrical isolation, indicating the existence of the driver within the SVC [34]. Therefore, the intricate structure and heterogeneity of the SVC may be involved in the AF substrate, which could potentially facilitate the heterogeneity of the underlying electrophysiological property. For non-responders to PV isolation and patients undergoing repeated ablation, SVC isolation should be considered, as it can separate a highly heterogeneous region with a possible trigger or an AF perpetuator from the RA [35].

The musculature extension into the inferior vena cava (IVC) is shorter than that into the SVC, rendering a lower possibility of an AF source [36,37]. However, a few reports have shown that the IVC serves as a driver for the maintenance of AF in addition to being a trigger [38,39]. The optimal IVC ablation strategy for AF remains unknown because of the limited studies, although focal ablation and IVC isolation have been reported [39].

2.3. Coronary Sinus

The CS occupies the atrioventricular groove, and its wall is covered with a myocardial sleeve. The CS is a remnant of the sinus venosus, and its muscle sleeve can be an extension of the RA myocardium over the CS. Histological studies have shown that the CS myocardial sleeve has muscular connections with both RA and LA fibers [40,41]. Muscular fibers of varying thickness arise from the myocardial sleeve along the inferior mitral annulus, providing electrical continuity with the LA. Striated myocardial connections exist between the CS and LA ranging from one to two fascicles surrounded by insulating compartments of adipose tissue [42]. The CS muscle sleeve and its branching connections to the atrium have been implicated in the genesis of atrial tachyarrhythmias. Non-PV ectopic beats arising from the CS have also been reported in clinical studies, suggesting that AF may originate within the CS [27,43]. Little evidence regarding the abnormal automaticity within the CS has been reported; however, some previous reports on CMs isolated from canine CS have confirmed the triggered activity [44,45]. Conversely, several studies have demonstrated that the CS myocardial sleeve may serve not only as a source generating a focal trigger but also as a part of a reentrant circuit [46,47]. The complex orientation of muscular fibers around the CS and their discrete insertion sites within the LA construct a heterogeneous structure and may facilitate the source–sink mismatch, leading to conduction disturbance within the connections [48]. These electrical connections between the CS and both atria may be involved in arrhythmogenesis by forming a part of the reentrant circuit [49]. Optical mapping in a canine model showed a functional conduction block at the RA-CS junction, and this site was involved in the rapid pacing-induced reentrant circuit [50]. Activation maps derived from another canine model demonstrated that the CS musculature developed unstable reentry and AF, which were prevented by the isolation of CS musculature from LA tissue [51]. Similarly, in a clinical study, electrical dissociation of the CS from the LA led to less inducibility of sustained AF after PV isolation, which suggests that the CS is involved in perpetuating AF [46].

2.4. Crista Terminalis and Pectinate Muscles

The RA appendage is readily distinguished from the RA because of the crista terminalis and pectinate muscles [52]. The crista terminalis has been demonstrated as a conduction barrier during typical atrial flutter. The investigations of the conduction properties within the crista terminalis demonstrated that it has marked anisotropic conduction with enhanced conduction velocity in the longitudinal direction compared to that in the transverse direction [53]. Meanwhile, in terms of the genesis of AF, animal experiments reported that isolated cells from the crista terminalis in rabbits have spontaneous pacemaker activity, whereas those in dogs showed delayed afterdepolarization induced by norepinephrine [53–55]. In a clinical study, the non-PV ectopic triggers of paroxysmal AF originating from the crista terminalis were observed [27]. These findings endorse the view that spontaneous activation may be initiated at the crista terminalis. Moreover, the complex branching structure of the pectinate muscle network has been proposed to provide a substrate for complex patterns of propagation during AF. Pectinate muscles spread throughout the wall of the atrial appendage with a branching and overlapping arrangement. The pectinate muscles are composed of ridges; however, the wall in the groove between the ridges is very thin [36]. This complex structure results in a source–sink mismatch at the branching site of bundles. An electrophysiological study of canine RA tissue showed that a premature stimulus induced conduction delay along the ridge of large pectinate muscles, leading to a wavebreak and resulting in the initiation of reentry [56]. The direction of the propagation in the RA appendage was also demonstrated to be rate-dependent with variability from beat to beat at the site of pectinate bundles, leading to a transformation of fibrillatory conduction [9]. Thus, pectinate muscles may contribute to the vulnerable substrate, allowing the sustenance of AF. Owing to the recent progress in electrode mapping techniques, AF rotational drivers were identified and described within the RA appendage [57,58].

Pectinate muscles are much less extensive in the LA appendage, and the LA appendage lacks a muscular bundle equivalent to the crista terminalis [36]. Despite these anatomical characteristics of the LA appendage being distinct from the RA appendage, the LA appendage has been identified as a possible trigger origin and substrate for AF maintenance after PV isolation [59,60]. Previous studies have demonstrated that LA appendage isolation in addition to PV isolation by catheter ablation reduced AF recurrence compared with PV isolation alone [61,62]. However, procedures with LA appendage isolation had higher rates of thromboembolism compared to those without it [63].

2.5. *Left Atrial Posterior Wall*

The LA posterior wall is a major non-PV trigger that initiates paroxysmal AF [27]. Previous studies have shown that the shortest AF cycle length and highest dominant frequency during AF are found in the LA posterior wall in dogs, sheep, and patients with chronic AF, implying that the LA posterior wall can be a potential source of AF [64–67]. However, to date, the reason why the LA posterior wall is so arrhythmogenic remains unclear. Of note, the LA posterior wall has a common embryologic origin with PVs, which may contribute to the factors that make the LA posterior wall distinct from other areas of the atrial wall [68–72]. However, the way in which the embryologic origin of the LA posterior wall is associated with the sources of AF remains elusive. On the other hand, an anatomical factor potentially characterizes the arrhythmogenicity in the LA posterior wall, despite the lack of evidence regarding the involvement of abnormal automaticity and triggered activity. The fiber orientation and myocardial thickness of the LA posterior wall abruptly change with partial involvement of the septopulmonary bundle. This bundle arises from the anterior interatrial raphe, ascends obliquely, and combines with longitudinal fibers from the anterior vestibule to run between the left and right PVs on the LA posterior wall, forming heterogeneous fiber orientation [73]. Furthermore, the LA posterior wall exhibits varying degrees of thickness, with the largest increase in myocardial thickness occurring at the border of the septopulmonary bundle [73]. These abrupt changes in fiber orientation and thickness may characterize the electrophysiological properties within the LA posterior wall. Markides et al. demonstrated that a line of functional conduction block in the LA posterior wall exists, running craniocaudally between the PVs during sinus rhythm and PV ectopy in humans [74]. They also demonstrated that patients with AF developed this line of conduction delay, corresponding to a change in fiber orientation in this region. Wavefronts entering the atrium from the PV interact with this functional line of conduction block, resulting in the reentry formation of daughter wavefronts. In an ovine study, electric source–sink mismatch and the subsequent wavebreaks mostly appeared at the septopulmonary bundle near the right superior PV, where the myocardial thickness dramatically expanded [73]. These findings indicate that the structural peculiarity may promote arrhythmogenicity in the LA posterior wall.

To summarize this section, macroscopic views show several common features throughout the atrial region. The pacemaker activity underlying myocardial sleeves within PV and SVC, and the arrhythmogenic afterdepolarization in the myocytes of PV, SVC, and CS, are potential triggers of AF. The intricate and anisotropic arrangement, and varying thickness of muscle fibers can provide an anatomical basis for source–sink mismatch and anisotropic conduction, which are causative mechanisms of the reentrant circuit, located in the PV (including PV-LA junction), SVC, CS-LA junction, crista terminalis, pectinate muscles, and septopulmonary bundle of the LA posterior wall.

Table 1. Macroscopic heterogeneities predisposing to AF. RA: right atrium, RAA: right atrial appendage, LA: left atrium, and LAA: left atrial appendage.

Macroscopic Structure					
Region	**Structure**	**Causative Factor**	**Species and Reference**	**Mechanism for Arrhythmia**	**Section**
Pulmonary vein (PV)	PV myocardial sleeve	pacemaker activity	dog [10,11], human [12]	arrhythmogenic trigger	2.1
		afterdepolarization	dog, rabbit [11,14]	arrhythmogenic trigger	
		intricate muscular bundle arrangement	human [19]	reentry	
		autonomic innervation	human [15,16]	arrhythmogenic trigger	
	PV-LA junction	varying thickness of muscular fiber	human [19], dog [20]	reentry	
		intricate muscular bundle arrangement	human [21]	reentry	
	carina	intricate muscular bundle arrangement	human [22,23]	reentry	
Superior vena cava (SVC)	SVC myocardial sleeve	pacemaker activity	dog [29], human [30]	arrhythmogenic trigger	2.2
		afterdepolarization	dog [29]	arrhythmogenic trigger	
		intricate muscular bundle arrangement	human [33,34]	reentry	
Coronary sinus (CS)	CS myocardial sleeve	afterdepolarization	dog [44,45]	arrhythmogenic trigger	2.3
	RA-CS connection	functional conduction block	dog [50]	reentry	
	CS-LA connection	intricate muscular bundle arrangement	human [46–48]	reentry	
		varying thickness of muscular fiber		reentry	
Crista terminalis (CT)	CT	pacemaker activity	rabbit [54,55]	arrhythmogenic trigger	2.4
		triggered activity	dog [53]	arrhythmogenic trigger	
		anisotropic conduction	dog [53]	reentry	
Pectinate muscle	RAA (LAA)	branching structure (ridge and groove)	human [36], dog [56]	reentry	
Left atrial posterior wall	septopulmonary bundle	intricate muscular bundle arrangement	sheep [73], human [74]	reentry	2.5
		varying thickness of muscular fiber		reentry	

3. Mesoscopic Heterogeneity

Anatomical structural heterogeneity at the segmental level, as described above, is inherently present and may become a potential factor of triggers and substrates for AF. However, most AFs are believed to occur when combined with other atrial pathophysiological conditions. Particularly in heart failure, increased atrial volume overload leads to the elevation of atrial pressure and stretching of atrial muscle, resulting in the reconstruction of tissue and the alteration of electrophysiological properties, known as atrial remodeling [6]. Atrial remodeling histopathologically involves the activation and proliferation of myofibroblasts (MFs), which are responsible for the uncontrolled deposition of the extracellular matrix. The enhanced fibrosis leads to tissue anisotropy, and obstructs the electrical wave propagation [75,76]. The pathological sources of AF have been proposed from the view of tissue-level heterogeneity.

3.1. *Myofibroblasts and Fibrosis*

Fibrosis refers to an increased deposition of collagen and other extracellular matrix proteins in the interstitial space. The excessive deposition of collagen enlarges interstitial spaces and decouples electrical cell-to-cell communication at gap junctions, resulting in the reduction in inter-myocyte electrical coupling. As described above, PVs are a major segment that proposes high arrhythmogenicity. An important tissue feature of PVs is the patchy areas of fibrosis, which can be detected even within healthy PVs [19]. In addition, a higher degree of fibrosis in the atrial myocardium extends to the PVs in patients with AF [21]. The inherent and acquired fibrosis within PVs is a possible source of arrhythmogenicity. Furthermore, fibroblast proliferation not only promotes collagen production which serves as an electrical insulator, but also mediates the electrical coupling with CMs. MFs are basically unexcitable cells, but they can form hetero-cellular gap junctions with CMs [77]. Some in vitro experiments using CMs isolated from animal hearts or in silico experiments demonstrated that the electrical coupling with MFs depolarizes the resting membrane potential of CMs [77–79]. According to a numerical study, these actions of MFs alter the electrophysiological characteristics of CMs, such as the shortening of myocyte action potential duration (APD) and a decrease in myocyte maximum depolarization velocity, leading to slow conduction velocity [80]. As a result, both the extracellular matrix and MFs bring about further complex electrophysiological changes, potentially causing arrhythmogenicity.

3.2. *Heterogeneous Distribution of Fibrosis*

As atrial remodeling progresses, the heterogeneous distribution of fibrosis is exacerbated, which is one of the factors promoting the pathophysiology of AF. To assess the tissue heterogeneity caused by fibrosis in an entire atrium, one non-invasive method for visualizing atrial fibrosis is cardiac magnetic resonance (CMR) imaging. Late gadolinium-enhanced CMR (LGE-CMR) has been recognized as an indicator to quantify the scar areas in an atrium [81,82], making LGE-CMR a promising tool for visualizing atrial fibrosis. LGE-CMR in patients with AF showed that the fibrotic area is heterogeneously distributed, and preferentially located at the LA posterior wall and around the antrum of the left inferior PV [83–85]. Previous studies have demonstrated that more advanced fibrosis detected by LGE-CMR is related to the recurrence of AF after PV isolation [86,87]. Therefore, to investigate the efficacy of catheter ablation targeting atrial fibrosis detected by LGE-CMR, the DECAAF II trial executed the LGE-CMR-guided strategy of identifying fibrotic regions compared with conventional methods for persistent AF [88]. However, no significant difference in the recurrence rate of AF was observed between the two groups. Much debate remains regarding the therapeutic applications based on the LGE-CMR. Another method currently used to estimate fibrotic tissue is electroanatomic voltage mapping. Low-voltage zone (LVZ), defined as an area with less than a certain cutoff using bipolar voltage mapping, has been widely used as a surrogate for local fibrotic tissue [89]. Several previous studies have shown that the most frequent localization of the LVZ tends to be within the LA anterior wall, followed by the septum and LA posterior wall [85,90]. Another observational study suggested that this segmental LVZ distribution pattern advances as AF perpetuates; initially in the LA anterior wall, followed by the LA posterior wall [91]. The severity of fibrosis estimated by the LVZ during sinus rhythm is associated with poor outcomes after PV isolation [92,93]. Several meta-analyses have reported that LVZ-guided ablation for persistent AF provides a significant reduction in recurrence rates [94–96]. In a study comparing the fibrotic areas shown by LGE-CMR and LVZ in the LA, LGE-CMR and LVZ covered 55% and 24% of the LA, respectively, while 61% of the LVZ was co-located with the LGE area, and only 28% of the LGE area displayed LVZ [85]. A high mismatch was observed in the distribution and volume of fibrosis detected by the LGE-CMR compared with that detected by the LVZ. The cause of this significant spatial mismatch is still under debate. Considering the association of the fibrotic area detected by these modalities and the electrophysiological properties, both of them correlate with a decrease in atrial conduction velocity [97–99]. On the other hand, the relationship between the LGE area and local reentry or rotor remains debated at present. While some reports suggest a higher prevalence of reentry in areas where enhancement is observed on LGE-CMR, other reports do not support this correlation [100–103]. Similarly, a wide discrepancy exists in the distribution of the LVZ and sites of identified rotors [104]. In fact, several studies have reported that reentrant AF sources were identified at only 8–37% of the LVZ [85,104].

3.3. Fibrosis and Arrhythmogenicity

Although the heterogeneous distribution of fibrosis does not necessarily allow for identifying the site of AF sources, the factors causing local conduction disturbances are believed to be partly ascribed to increased fibrosis [105]. Fibrosis has been indicated as the primary factor causing AF. However, among the fibrotic volume, distribution, or both, the factor that contributes the most to arrhythmogenesis remains unclear. Previous studies have implied that the heterogeneity of fibrotic distribution is a more contributive factor to the development of AF [106,107]. In a recent study with a porcine ischemic model, heterogeneous fibrosis in the LA rather than the overall level of fibrosis was associated with an increased AF susceptibility [108]. Several studies have demonstrated that the areas adjacent to dense fibrosis, in relatively healthy zones, show high levels of arrhythmogenic activity [109,110]. Another study also reported that the patchy and heterogeneous distribution of fibrous tissue can lead to conduction delays and the fragmentation of signals [111]. Jadidi et al. reported fractionated continuous activity in the vicinity of the border zone within the patchy fibrotic area, rather than within the dense fibrotic area [109]. However, determining the exact role of fibrotic tissue in arrhythmogenesis experimentally is difficult; therefore, experimental studies have been substituted with some numerical studies. A simulation study demonstrated that an important factor determining the formation and dynamics of arrhythmia in heterogeneous fibrosis stems from the maximum local fibrosis density occurring within the heterogeneous tissue [112]. A highly heterogeneous distribution can induce wavebreaks resulting in the formation of meandering rotors [113]. Other theoretical studies demonstrated that the degree and distribution of fibrosis had a large effect on rotor locations [114], and AF can be perpetuated by rotors meandering in the border zones of patchy fibrosis [115,116]. These numerical findings support the experimental results that the heterogeneous fibrotic distribution provides a certain mechanism of arrhythmogenesis and plays a key role in local fibrillatory activity.

3.4. Autonomic Ganglion Plexus

Electrophysiological properties in atria are modified by the autonomic nervous system, in which AF is often mediated by adrenergic or vagal activation [117]. Autonomic ganglia in the atrium are heterogeneously distributed, and ganglion plexuses (GPs) are concentrated at the SVC, CS, PV, and LA posterior wall in humans [118]. Preliminary studies have shown that GP stimulation induces the shortening of APD observed within PV sleeves in dogs [119,120]. Additionally, they demonstrated that injecting acetylcholine into the GP in adipose tissue next to a PV leads to rapid firing from the ipsilateral PV, transitioning to AF, followed by the cessation of AF by ablating the GP at the same site [119,121]. Excessive activity of the autonomic ganglia is now believed to be partially involved in the generation of AF, at least in cases showing focal firing. In clinical practice, attempts have been made to achieve catheter ablation for GP denervation [122]. Recent meta-analyses have shown a reduction in arrhythmia recurrence by adjunctive GP ablation plus PV isolation compared with PV isolation alone [123,124]. When it comes to autonomic nervous activity and AF, the oblique ligament of Marshall (LOM) also plays a key role. It is a remnant of the left SVC and is located between the LA appendage and the left superior/inferior PVs. It runs inferiorly along the inferior atrial wall, while only its intracardiac portion remains patent as the vein of Marshall which drains into the CS. The LOM contains fat and fibrotic tissues, vessels, muscle bundles, nerve fibers, and ganglia. Not only can the LOM's complex structure serve as an arrhythmogenic substrate, but also its rich innervation by sympathetic nerves may serve as a source of catecholamine-sensitive focal automaticity. Ectopic activity and focal automaticity in the LOM induced by isoproterenol have been demonstrated in dogs, indicating that it contributes to the breakout of AF [125]. The effectiveness of LOM ablation has been reported in patients with persistent AF after PV isolation, implying that enhanced atrial denervation, elimination of AF triggers, or conduction block correlate with a reduction in the vulnerability to AF [126,127].

4. Microscopic Heterogeneity

Since cardiac action potentials are governed mainly by sarcolemmal ion currents, the alteration of these ion currents can help understand the arrhythmogenic mechanism. Once arrhythmogenic substrates result in AF, AF further adversely progresses the arrhythmogenic substrates, which is termed "AF begets AF". In some reports, electrical changes in sarcolemmal ion currents resulting from AF were investigated using a whole-cell patch-clamp system with CMs isolated from the appendages of patients with persistent AF [128–132]. Despite the overall changes in ion currents in AF, to the best of our knowledge, the localized inter- or intra-cellular heterogeneous distribution of ion channels has not been discussed so far. However, a better understanding of the structural and electrophysiological heterogeneity underlying AF at a microscopic level will lead to advancements in treatment strategies. Regardless of the underlying diverse causes, abnormal discharge and electrical reentry are the two main determinants of ectopic firing, initiation, and maintenance of AF. Herein, we will review the microscopic heterogeneities at a cellular level and the molecular basis of the connexins, the profibrotic process, and genetics in association with their predisposition to AF.

4.1. Connexin Remodeling and Regulation

As regards the conduction heterogenicity exacerbated by AF, the redistribution of connexin has been mainly argued. Electrical intercellular conduction is highly dependent on connexins, which facilitate cell-to-cell connections (gap junctions) and are known to be involved in the electrochemical coupling to adjacent cells at the intercalated disk [133]. An action potential in a CM propagates longitudinally into the adjacent CM via gap junctions localized normally in the intercalated disk. The heterogeneous distribution in a tissue or lateralization of connexin distribution in a CM has often been discussed. Of note, heterogeneous distribution is termed a patchy distribution in a tissue, and lateralization is termed a lateralized intracellular distribution. Connexin remodeling in a CM, such as reduced expression and lateralization, can decrease longitudinal cell-to-cell electrical coupling resulting in a change in anisotropic conduction. van der Velden et al. investigated the role of gap junctions in a goat AF model [134,135]. They found no changes in the overall gene expression and protein levels of Cx40 and Cx43; however, according to immunohistochemical studies and confocal laser scanning microscopy, a heterogeneous distribution of Cx40 in tissue was observed in AF. Subsequently, Polontchouk et al. demonstrated that in atrial CMs obtained from patients with AF, while the protein level of Cx40 increased, its lateralization in the CMs also increased [136,137]. In patients with AF undergoing a biopsy procedure, the reduced expression and lateralized distribution of Cx40 were observed [138]. Overall, these results showed inconsistent expression levels of Cx40, but the intracellular lateralization of Cx40 was commonly observed. Previous findings on the expression levels and intercellular lateralization of Cx43 in AF were similar to those of Cx40 [133,139]. van der Velden et al. also demonstrated that the total levels and localization of Cx43 remained unchanged in a goat AF model; however, several studies with canine AF models reported increased Cx43 expression together with increased lateralization [140,141]. In humans, a study showed that the LA in patients with lone AF exhibited an increase in the protein expression of Cx43, while another study demonstrated the lateralization of Cx43 in patients with persistent AF [136]. In summary, the heterogeneous expression level and the heterogeneous distribution, including lateralization, of Cx40 and Cx43 may result in heterogeneous intercellular coupling, leading to conduction defects and anisotropy, which can facilitate wavebreaks and pathophysiological substrates of AF [142].

Abnormalities in connexin assembly, permeability, or localization impair electrical propagation and can lead to conduction disturbance which contributes to the development of AF [143]. Those abnormalities manifest in the form of phosphorylation, Cx40/Cx43 protein ratios, and lateralization to the surface membrane [143]. Connexin phosphorylation has been thought to switch on several types of functions. For instance, it regulates gap junctional protein trafficking and assembly. The increase in lateralized connexins can lead to a reduction in gap junctional conductance. Cx43 is phosphorylated specifically by various kinases including protein kinase A (PKA), protein kinase C (PKC), Ca^{2+}/calmodulin-dependent kinase II (CaMKII), and mitogen-activated protein kinases (MAPKs) [144]. The sympathetic and β-adrenergic induction of activated cyclic adenosine monophosphate/PKA promote the synthesis and the phosphorylation of Cx43 (Ser364) [145]. PKA enhances the assembly of connexins into the gap junction and suppresses proteolytic degradation. As a result, PKA increases channel conductance. On the other hand, angiotensin II (Ang II) induces PKC activation, which augments the phosphorylation of Cx43 (Ser368) [146], inhibits the assembly of the connexins into the gap junction, and accelerates proteolytic degradation, followed by the impairment of the gap junctional communication. A remarkable reduction in connexins in the intercalated disk, together with an increase in the lateralized delocalization, attenuates cell-to-cell coupling. Ca^{2+} is also an important factor that decreases the gap junctional conductance. Ca^{2+} overload elevated the resistance of gap junctions by hindering the PKA-mediated phosphorylation of Cx43. In addition, Ca^{2+}/calmodulin binding activates CaMKII. Not only does CaMKII phosphorylate the ryanodine receptor 2 and increase its Ca^{2+} sensitivity to exhibit Ca^{2+} leak and triggered activity via an increase in channel open probability, but it also exerts an indirect effect by regulating Cx43 expression and subcellular localization in the intercalated disk [147]. Furthermore, inflammatory cytokines activate MAPK to upregulate phosphorylated Cx43 and impair cell-to-cell communication because Cx43 becomes extensively dispersed at the intercalated disks [148,149]. Taken together, the phosphorylation of connexins causes the abnormal distribution and expression of the gap junction, which promotes conduction heterogeneity. Remodeling of the gap junction can be an arrhythmogenic substrate.

4.2. Molecular Mechanism of Atrial Fibrosis

Atrial fibrosis is a major factor contributing to atrial structural and electrophysiological heterogeneity. The proliferation of the extracellular matrix (ECM) is promoted by the neurohormonal dysregulation of the renin–angiotensin–aldosterone system (RAAS) and the activation of fibrotic pathways, mainly initiated by the transforming growth factor beta (TGFβ), connective tissue growth factor, and platelet-derived growth factor [150]. These profibrotic signaling molecules activate fibroblasts, resulting in proliferation and differentiation into secretory myofibroblasts, often accompanied by the upregulation of matrix metalloproteinases (MMPs) and the downregulation of the tissue inhibitors of metalloproteinases (TIMPs). AF can activate inflammatory cells, such as macrophages, leading to oxidative stress and the production of reactive oxygen species (ROS). Consequently, the RAAS is activated, and Ang II contributes to a profibrotic process by binding to Ang II type 1 receptor [151], followed by the further stimulation of phospholipase C, which acts through inositol 1,4,5-trisphosphate (IP3) and diacylglycerol (DAG). IP3 mediates an increase in Ca^{2+} levels in the cytoplasm. The intracellular Ca^{2+} overload promotes fibroblast proliferation and differentiation. DAG activates PKC, which activates the downstream signaling of MAPK. Ang II serves as a potent nicotinamide adenine dinucleotide phosphate oxidase activator, leading to ROS overproduction, which activates MAPK as well. The activation of the MAPK signaling pathway promotes the secretion of various transcription factors such as TGFβ and MMP. TGFβ binds to serine/threonine kinase receptors, accompanied by the suppressor of mother against decapentaplegic protein-mediated signal transduction, thereby further promoting the profibrotic process. These processes in cooperation accelerate the formation of vulnerable substrates for AF. In terms of the association between the profibrotic molecular process and heterogeneity, AF vulnerability has been demonstrated to be associated with a segmental heterogeneous pattern of atrial ECM remodeling (molecular MMP/TIMP profile) in a porcine AF model with myocardial infarction [152]. This result implied that a substantial regional heterogeneity exists at the molecular level. Furthermore, atrial inflammation and adipose tissue depots play an important role in AF substrate formation [153,154]. In particular, many studies have suggested an association between epicardial adipose tissue (EAT) and the fibrosis of the underlying atrial myocardium [155–157]. EAT contains various cell types, especially adipocytes and immune cells such as macrophages and lymphocytes. Both the immune cells and adipocytes within EAT can release various cytokines, contributing to pathogenic inflammation. In patients with coronary artery disease, EAT has increased levels of inflammatory cytokines [158]. Another study with resected LA appendage and associated EAT from cardiac surgery patients with AF demonstrated that the fibrosis of EAT was associated with LA fibrosis [157]. The fibrosis and inflammation of EAT can predispose to the fibrosis of the adjacent atrial myocardium through the proinflammatory and profibrotic bioactive secretome from EAT [154]. Adipocytes can also have a remote effect on the myocardium by the secretion of adipokines. Profibrotic secretome analysis identified an adipokine, activin A, which is a member of the TGFβ superfamily. Ventedlef et al. demonstrated that atrial fibrosis was promoted with a high concentration of activin A secreted from adipocytes [155]. Taken together, a great amount of EAT correlates with the development of AF through the profibrotic secretome. In terms of the distribution of EAT in the LA, a study evaluating the location of EAT using computed tomography in patients with AF showed that large clusters of EAT are observed adjacent to the anterior roof, LA appendage, and lateral mitral isthmus [159]. This result implies that the heterogeneous distribution of EAT can be a causative factor of the regionally dependent severity of fibrosis in atria.

4.3. Genetics of Atrial Fibrillation

Many studies have demonstrated that AF has substantial genetic factors predisposing to the development of AF [160]. The linkage analyses of AF have shown various causative genes for familial AF. Mutations in the potassium channels were discovered early as causes of familial AF: *KCNQ1*, which encodes a subunit of KvLQT1; *KCNH2* which encodes HERG; and *KCNJ2* which encodes Kir2.1 [161–163]. An increase in potassium currents due to a gain of function, which shortens the action potential and refractory period of atrial CMs, forms a reentry substrate that promotes the occurrence and maintenance of AF. However, AF is rarely a monogenic disorder; AF is often caused by environmental factors and multiple common genetic predispositions that affect AF susceptibility. Genome-wide association studies have clarified the genetic polymorphisms strongly associated with AF. Multiple variants were found in genes responsible for cardiac development (*NKX2-5*, *TBX5*, and *PITX2*) and electrophysiology, including the coding genes of potassium ion channels (*KCNH2* and *KCNJ2*) and connexins (*GJA1* and *GJA5*) [164]. Although the susceptibility to AF is different across ancestries, some of the variants found across different ancestries are concentrated in the 4q25 region of the long arm of chromosome 4, in which the transcription factor *PITX2* is closest [165]. *PITX2* is expressed exclusively in the LA and has a critical function of regulating the left–right differentiation of the embryonic heart [166,167]. The asymmetrical organ morphogenesis by *PITX2* confines the sinoatrial node, pacemaker cells, to the RA. *PITX2C* is the major isoform expressed in the heart, particularly in the LA, and *PITX2C* expression was significantly downregulated in human patients with AF [168]. Thus, it is probable that a deficiency in *PITX2* causes the incomplete suppression of pacemaker activity in the LA, resulting in enhanced pacemaker activity in the LA. Furthermore, a transgenic murine model with a *Pitx2c* deficiency, which exhibited a decrease in *Pitx2c* expression in only the LA compared to the wild-type, showed the shortening of the LA action potential, and more susceptibility to AF compared to the wild-type [169]. Similarly, the chamber-specific *Pitx2* conditional mutant mice revealed that the LA displayed a more depolarized resting membrane potential and a smaller action potential amplitude [168]. *PITX2* has also been reported to regulate the extension of myocardial sleeves into PVs from the LA during development [170]. The differentiation of dorsal mesenchymal cells into the pulmonary myocardium requires *PITX2*, and transgenic *Pitx2*-deficient mice were found to have normally developed PVs, but absent PV myocardial sleeves [170]. In summary, these findings indicate that genetic predisposition increases susceptibility to AF, either electrophysiologically or anatomically. The regionally heterogeneous expression patterns of these genes can facilitate the basis of electrophysiological heterogeneity.

5. Electrophysiological Heterogeneity

The progression of AF is believed to require an ectopic or reentrant basis. When a focal discharge exceeds the threshold to excite the surrounding muscle, the propagation from this ectopic focus possibly acts as a trigger. If the electrophysiological heterogeneity meets the sufficient need for the AF substrate, the wavefront of the propagation wave can trigger a cascade of wavebreaks, leading to continuous reentry and the initiation of AF. The maintenance of continuous activity depends on the existing electrophysiological properties with refractory and excitability determinants; shortened refractory period and reduced conduction velocity (CV). As described above, segmentally differentiated tissue properties, fiber arrangements, and redistributed intra- and inter-cellular connexins facilitate the structural heterogeneities, which in turn connect with electrophysiological heterogeneities in the refractoriness and conduction of action potential.

5.1. Repolarization Heterogeneity

The heterogeneity of repolarization in the atria has been reported as a source of AF [171], similar to that in the ventricles where repolarization heterogeneity has been widely reported to cause ventricular tachycardia and fibrillation [172,173]. The analyses of atrial action potentials derived from the patch-clamp technique, monophasic action potential (MAP) method [174], endo- or epicardial electrophysiological studies, or optical mapping [175] are helpful to assess atrial repolarization. According to studies that used these methods, the heterogeneity in the APD or effective refractory period (ERP) serves as a substrate for AF [176–178]. In a study comparing action potential morphology using isolated CMs from various atrial segments in adult or aged dogs, and with sinus rhythm or chronic AF, patch-clamp recordings showed a tendency for APD shortening and increased APD heterogeneity in aged dogs with chronic AF compared to that in adult dogs with sinus rhythm [179]. Another canine study investigating the specific cellular electrophysiological properties using a patch-clamp system showed that resting membrane potential was more depolarized, and the APD was shorter in PVs than in the LA [180]. MAP studies in patients with paroxysmal AF showed a significant increase in the segmental dispersion of APD and ERP compared to those in patients with sinus rhythm [171,181,182]. In a canine model, an electrophysiological study demonstrated that the regional heterogeneity of refractory periods increases in AF [176]. Optical mapping with isolated hearts of diabetic rats demonstrated that an increase in the spatial dispersion of APD corresponds to the vulnerability of atrial arrhythmia [183]. Another optical mapping study in canine models showed the progressive shortening of APD and the depolarization of the resting membrane from the LA to the distal PVs [120]. This study also substantiated that the heterogeneity of the APD in PVs provided a favorable substrate for reentry formation. Simulation studies have supported these findings that the heterogeneity in refractoriness can account for the initiation of AF [176,184]. They revealed the presence of segmental dispersion or heterogeneity of APD in chronic AF and even paroxysmal AF. Further studies numerically showed that wavebreaks occur at sites with the steepest gradients of APD dispersion [185–187]. Another computer-based study with a 3D human model, which incorporated the regional distribution of APD from previous experimental data, showed that marked regional differences in the APD co-exist with shortened APD, facilitating the initiation and maintenance of reentrant waves. In particular, reentrant excitation is stabilized in the presence of pronounced regional differences in APD at the PV-LA junction and the crista terminalis/pectinate muscle junction in the RA appendage [188].

Many reports have suggested that an increase in inward rectifier potassium current is crucial in AF by shortening repolarization and causing hyperpolarization [128–132]. Consistent with the dispersion of APD, an in vitro assay with isolated CMs from different segments in an atrium showed the heterogeneous expression of the inward rectifier potassium channel [189,190]. Since the functional effects of structural remodeling induced by AF on the regional heterogeneity of APD have not yet been elucidated, computational models have been proposed to characterize these relationships. Several theoretical studies have compared the remodeled ionic currents with AF-induced reduction in APD in a human atrium, demonstrating that the up-regulation of inward rectifier potassium current has a greater influence on APD than any changes in other ion currents [191]. Moreover, Berenfeld et al. introduced the ionically regional gradient model to investigate the relation between the distribution of various types of ion channels and rotor stability. Among them, the gradient of the inward rectifier potassium current had the most significant impact on rotor stability [192,193]. The variation in the conductance of the inward rectifier potassium current resulted in gradients of APD, leading to the trapping of rotors at sites with the steepest gradients of APD [114].

Of note, what complicates the understanding of the reentrant mechanism is that the determinants of electrophysiological properties may change, not only spatially but also temporally [194–196]. It is not easy to thoroughly visualize spatio-temporally heterogeneous electrophysiological factors. APD is not just a static electrophysiological parameter but a dynamically changing one that can be influenced by the preceding diastolic interval (DI). If the preceding DI is short, the following APD tends to shorten, whereas if it is long, the following APD tends to be prolonged. The relationship between APD and DI is defined as APD restitution [197]. A steep curve in the APD restitution slope can cause oscillations in APD alternans, a phenomenon in which APD (or repolarization time) alternates between longer and shorter durations over time. When APD alternans occurs in a spatially inconsistent pattern, it is termed discordant APD alternans. Discordant alternans is a phenomenon in which two spatially distinct regions exhibit APD alternance of opposite phases. At the boundaries between the regions in which APD oscillates discordantly, there turns out to be an increased APD gap. The APD gap can induce the functional conduction block at the border, and thus yield heterogeneous conduction patterns, resulting in a possible source of fibrillatory conduction. Previous studies have demonstrated that discordant alternance is associated with reentrant activation and susceptibility to AF [198–200]. This substantiates that APD possesses the property of spatio-temporal variability, which can be a cause of electrophysiological heterogeneity and serve as the substrate for arrhythmogenicity. Taken together, repolarization heterogeneity can be a strong determinant of arrhythmogenic substrate.

5.2. Conduction Heterogeneity

Atrial conduction disorder can be decomposed into two elements: altered conduction propagation (direction) and reduced magnitude (velocity). When longitudinal propagation along the muscular fiber is disturbed because of the reduced electrical coupling between myocytes or collagenous obstacles, conduction in the transverse direction through lateral cell connections gets involved in the propagation. Additionally, tissue discontinuities due to fibrosis or anisotropic fiber orientation can cause electrical source–sink mismatch along these pathways, resulting in the local slowing of conduction or conduction block, and the tortuous conduction pathway. Furthermore, the hetero-cellular electrical coupling of myofibroblasts and myocytes can increase heterogeneity in excitability, refractoriness, and electrical load, potentially inducing conduction slowing [201]. This can lead to a pivoting or zigzag trajectory of activation. Electroanatomic maps during sinus rhythm of patients with paroxysmal or persistent AF showed that the atrial conduction properties in persistent AF were characterized by a higher wavefront curvature and a larger number of pivoting points than those in paroxysmal AF [202]. Several animal studies with optical mapping recognized zigzag conduction in pathological states such as stretch-induced and age-related atria [203,204]. Slow conduction with pivoting or zigzag pathways has become a factor of localized reentrant circuits [204]. Slow conduction or pivoting points can be viewed as fragmented potential or complex fractionated atrial electrograms (CFAEs) in the local potentials obtained from the electrode catheter during catheter ablation procedures [205,206]. Ablating the localized regions that represent CFAE in addition to PV isolation has been advocated to increase the procedural success rate. A meta-analysis demonstrated that the adjunctive CFAE ablation could provide additional benefits, such as reducing the recurrence of AF for patients with persistent AF [207]. Of note, a CFAE during AF is not always an indicator of an abnormal substrate; however, CFAE during sinus rhythm is a preferential marker of slow conduction, wavefront collision, or pivoting sites. Furthermore, to understand atrial arrhythmogenicity, transmural conduction should also be considered. de Groot et al. demonstrated the asynchronous activation of the endo-epicardial wall during AF in humans using simultaneous endo-epicardial mapping, which was not apparently observed during sinus rhythm [208,209]. These results implied that the dissociation of endo-epicardial activation may be important in the maintenance of AF. Optical mapping studies are helpful to assess the localized CV and visualize the heterogeneity. In animal experiments, atrial heterogeneous CV has been reported using isolated hearts in models such as pressure-overload mice [210], diabetic rats [183], and rabbits with atrial stretch stress [211]. These studies substantiated that incremental atrial heterogeneity in CV is closely related to the increased occurrence of AF. In addition, just as APD depends on the preceding DI, CV depends on it as well [212]. This dynamic is referred to as CV restitution, similar to APD restitution. When static conduction abnormalities are combined with dynamic conduction abnormalities due to CV restitution, conduction becomes more heterogeneous. In other words, the source of AF is present in regions where the decrease in atrial CV and the increase in the heterogeneity of CV with rate dependence are prominent [210,213,214] (Figure 3). To date, while a positive relation between the severity of the conduction heterogeneity and the susceptibility of AF has been suggested, how the conduction heterogeneity is linked to the initiation and sustenance of AF and the propagation pattern during AF remains unclear.

5.3. Ectopic Excitement

Ectopic premature contractions are widely acknowledged to serve as critical triggers for AF. The underlying mechanism by which these focal discharges act as triggers is their ability to induce conduction disturbances, thereby manifesting the AF substrate. Mapping studies have demonstrated that an atrial extrasystole can cause a functional conduction block [215,216]. Not only can a premature firing with a short-coupling interval trigger the initiation of AF via heterogenous APD or CV restitution as described in the previous sections, but conduction-directional effects can also predispose to functional conduction block and conduction slowing. In terms of the directional dependence of conduction, Kumagai et al. showed that the conduction delay from the distal PV to the PV-LA junction was significantly longer than that from the PV-LA junction to the distal PV, and a short-coupled extra-stimulus from the PV formed a PV-LA reciprocating reentrant circuit involving exit and entrance breakthrough points at the PV-LA junction [217]. Spach et al. demonstrated with human and canine models that premature stimuli resulted in slower conduction in the transverse direction, leading to anisotropic conduction [218–220]. They showed that ectopic premature excitation provokes more conduction disorders and reentries. An intra-operative epicardial mapping during 503 premature atrial contractions revealed that the CV decreases within the PV area, and the local directional conduction heterogeneity increases [216]. Furthermore, they demonstrated that patients with AF have a slower CV and more conduction heterogeneity during sinus rhythm, which becomes more pronounced during premature atrial contractions than in patients without AF.

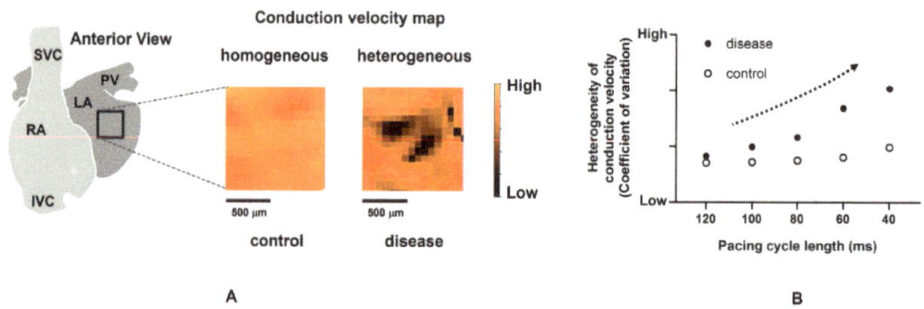

Figure 3. Increased heterogeneity with rate dependence in a diseased atrium. (**A**) Conduction velocity maps (during sinus rhythm) in a control and a diseased murine atrium with heart failure. (**B**) Increase in heterogeneity of conduction velocity dependent on pacing cycle length in a diseased atrium. The dotted arrow indicates an increase in the heterogeneity of conduction velocity. Taken from the reference [210] with partial modification.

6. Clinical Applications for Reducing Heterogeneity

Three therapeutic approaches can be considered to alleviate the heterogeneity. The first approach is to electrically separate the area with severe heterogeneity from the healthy area (Figure 4A). Isolating not only the trigger source but also regions with significant heterogeneity that deeply contribute to the initiation and maintenance of AF can be considered beneficial. In this context, PV isolation, PV antrum isolation, SVC isolation, and LA posterior wall isolation have been performed, with demonstrated efficacy. To further improve this approach, developing novel procedural techniques and devices for isolating anatomically challenging regions and advancing imaging and mapping technologies for more accurate identification of areas that need isolation is essential. While isolating a larger area can potentially be more effective by surrounding a greater amount of AF substrate with a conduction block, the isolated atrial regions lose their physiological contractile function, which can lead to thrombus formation. Therefore, the isolation should be limited to a necessary and sufficient extent. The second way is to ablate the individual distinctive features contributing to triggers, drivers, or other factors that are considered strongly correlated to AF sources using catheter interventions (Figure 4B). To this end, catheter ablation targeting sites with distinctive local features, such as reentrant circuit with conduction delay [221,222], fragmented potential [206,207], ganglionated plexi [122], local driver [206], and rotor [223], have been developed. However, the indicators to be targeted to achieve optimal electrical modification remain unclear. Particularly in the cases of persistent AF, where target regions are widely distributed throughout the atria, the complete ablation of all the target areas is difficult. Furthermore, the ablation process itself can contribute to myocardial damage and the formation of new conduction disturbances, presenting numerous challenges that need to be addressed. In this approach, developing new methods to identify the required and sufficient targeted areas and innovative interventions to modify myocardial conduction properties without proposing another source of AF is highly desired. The third approach is to decrease the heterogeneity or restore it to the original state with pharmacotherapy (Figure 4C). The development of new pharmacotherapy to reduce heterogeneity, including the profibrotic process or connexin remodeling, which is acquired under pathological conditions, is anticipated. To date, several drugs for atrial arrhythmia and heart failure have been proven to reduce the atrial heterogeneity in animal experiments: sodium channel blockers [217], calcium channel blockers [224], ACEis [225], and neprilysin and angiotensin receptor blockers [210]. However, these drugs have not been proven to reduce the occurrence of AF in clinical settings. While a new treatment is being strongly anticipated, sodium–glucose cotransporter-2 (SGLT-2) inhibitors are currently used for heart failure, and their effectiveness on AF has been gradually uncovered [226]. In short, therapeutic interventions for advanced AF are not established yet since the treatment outcomes decline as heterogeneity is exacerbated. Thus, finding strategies through catheter-based procedures or pharmacotherapy to reduce the heterogeneity could potentially lead to improved treatment outcomes. In addition to elucidating the underlying cause, a new approach to heterogeneity is anticipated.

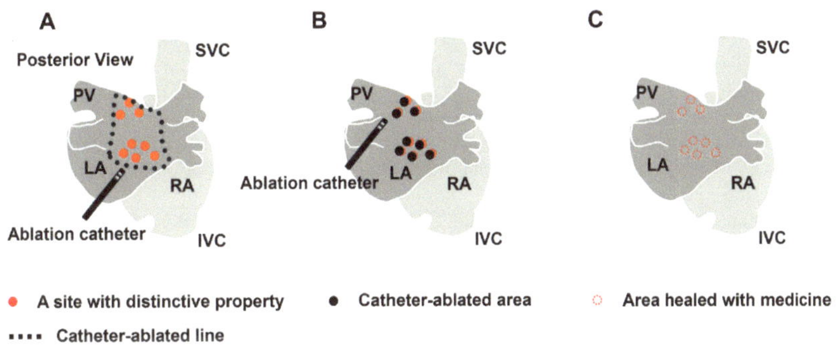

Figure 4. Therapeutic approach of ablation or pharmacotherapy for the heterogeneity. (**A**) Isolation strategy by ablation. (**B**) Ablation strategy for the individual structural and electrophysiological features. (**C**) Pharmacotherapy to reduce the overall heterogeneity by inhibiting the progression of pathophysiology such as profibrotic process or connexin remodeling. RA: right atrium, LA: left atrium, PV: pulmonary vein, SVC: superior vena cava, and IVC: inferior vena cava.

7. Conclusions

The experimental and clinical evidence that the structural and electrophysiological heterogeneity contributes to the substrate for AF is accumulating. Advances in the recognition of the underlying heterogeneity with mapping technologies have elucidated the substrates incorporated into the pathophysiology of AF. Further studies on the underlying heterogeneity would provide more profound insights into the mechanisms of the sustenance and dynamics of AF.

Author Contributions: Conceptualization, S.I., K.I. and G.N.; methodology, S.I.; software, S.I.; validation, S.I. and K.I.; formal analysis, S.I.; investigation, S.I.; resources, S.I.; data curation, S.I.; writing—original draft preparation, S.I., K.I. and G.N.; writing—review and editing, K.I. and T.S.; visualization, S.I.; supervision, T.S.; project administration, T.S. All authors have read and agreed to the published version of the manuscript.

Funding: This research received no external funding.

Institutional Review Board Statement: Not applicable.

Informed Consent Statement: Not applicable.

Conflicts of Interest: The authors declare no conflicts of interest.

References

1. Wyse, D.G.; Waldo, A.L.; DiMarco, J.P.; Domanski, M.J.; Rosenberg, Y.; Schron, E.B.; Kellen, J.C.; Greene, H.L.; Mickel, M.C.; Dalquist, J.E.; et al. A comparison of rate control and rhythm control in patients with atrial fibrillation. *N. Engl. J. Med.* **2002**, *347*, 1825–1833. [CrossRef] [PubMed]
2. Disertori, M.; Latini, R.; Barlera, S.; Franzosi, M.G.; Staszewsky, L.; Maggioni, A.P.; Lucci, D.; Di Pasquale, G.; Tognoni, G. Valsartan for prevention of recurrent atrial fibrillation. *N. Engl. J. Med.* **2009**, *360*, 1606–1617. [CrossRef] [PubMed]
3. Mohammad, Z.; Ahmad, J.; Sultan, A.; Penagaluri, A.; Morin, D.; Dominic, P. Effect of sacubitril-valsartan on the incidence of atrial fibrillation: A meta-analysis. *J. Cardiovasc. Electrophysiol.* **2023**, *34*, 1037–1042. [CrossRef] [PubMed]
4. Rabinovitch, A.; Rabinovitch, R.; Biton, Y.; Braunstein, D.; Thieberger, R. A possible new cardiac heterogeneity as an arrhythmogenic driver. *Sci. Rep.* **2023**, *13*, 7571. [CrossRef]
5. Allessie, M.A.; Boyden, P.A.; Camm, A.J.; Kléber, A.G.; Lab, M.J.; Legato, M.J.; Rosen, M.R.; Schwartz, P.J.; Spooner, P.M.; Van Wagoner, D.R.; et al. Pathophysiology and prevention of atrial fibrillation. *Circulation* **2001**, *103*, 769–777. [CrossRef]
6. Nattel, S.; Burstein, B.; Dobrev, D. Atrial remodeling and atrial fibrillation: Mechanisms and implications. *Circ. Arrhythm. Electrophysiol.* **2008**, *1*, 62–73. [CrossRef]
7. Dobrev, D.; Carlsson, L.; Nattel, S. Novel molecular targets for atrial fibrillation therapy. *Nat. Rev. Drug Discov.* **2012**, *11*, 275–291. [CrossRef]
8. Avula, U.M.R.; Abrams, J.; Katchman, A.; Zakharov, S.; Mironov, S.; Bayne, J.; Roybal, D.; Gorti, A.; Yang, L.; Iyer, V.; et al. Heterogeneity of the action potential duration is required for sustained atrial fibrillation. *JCI Insight* **2019**, *4*, e128765. [CrossRef]
9. Berenfeld, O.; Zaitsev, A.V.; Mironov, S.F.; Pertsov, A.M.; Jalife, J. Frequency-dependent breakdown of wave propagation into fibrillatory conduction across the pectinate muscle network in the isolated sheep right atrium. *Circ. Res.* **2002**, *90*, 1173–1180. [CrossRef]
10. Chen, Y.J.; Chen, S.A.; Chang, M.S.; Lin, C.I. Arrhythmogenic activity of cardiac muscle in pulmonary veins of the dog: Implication for the genesis of atrial fibrillation. *Cardiovasc. Res.* **2000**, *48*, 265–273. [CrossRef]
11. Chen, Y.J.; Chen, S.A.; Chen, Y.C.; Yeh, H.I.; Chan, P.; Chang, M.S.; Lin, C.I. Effects of rapid atrial pacing on the arrhythmogenic activity of single cardiomyocytes from pulmonary veins: Implication in initiation of atrial fibrillation. *Circulation* **2001**, *104*, 2849–2854. [CrossRef] [PubMed]
12. Perez-Lugones, A.; McMahon, J.T.; Ratliff, N.B.; Saliba, W.I.; Schweikert, R.A.; Marrouche, N.F.; Saad, E.B.; Navia, J.L.; McCarthy, P.M.; Tchou, P.; et al. Evidence of specialized conduction cells in human pulmonary veins of patients with atrial fibrillation. *J. Cardiovasc. Electrophysiol.* **2003**, *14*, 803–809. [CrossRef] [PubMed]
13. Wit, A.L.; Boyden, P.A. Triggered activity and atrial fibrillation. *Heart Rhythm* **2007**, *4*, S17–S23. [CrossRef] [PubMed]
14. Chen, Y.J.; Chen, S.A.; Chen, Y.C.; Yeh, H.I.; Chang, M.S.; Lin, C.I. Electrophysiology of single cardiomyocytes isolated from rabbit pulmonary veins: Implication in initiation of focal atrial fibrillation. *Basic Res. Cardiol.* **2002**, *97*, 26–34. [CrossRef] [PubMed]
15. Vaitkevicius, R.; Saburkina, I.; Rysevaite, K.; Vaitkeviciene, I.; Pauziene, N.; Zaliunas, R.; Schauerte, P.; Jalife, J.; Pauza, D.H. Nerve supply of the human pulmonary veins: An anatomical study. *Heart Rhythm* **2009**, *6*, 221–228. [CrossRef]
16. Nguyen, B.L.; Fishbein, M.C.; Chen, L.S.; Chen, P.S.; Masroor, S. Histopathological substrate for chronic atrial fibrillation in humans. *Heart Rhythm* **2009**, *6*, 454–460. [CrossRef]
17. Iwasaki, Y.K.; Nishida, K.; Kato, T.; Nattel, S. Atrial fibrillation pathophysiology: Implications for management. *Circulation* **2011**, *124*, 2264–2274. [CrossRef]
18. Gerstenfeld, E.P.; Callans, D.J.; Sauer, W.; Jacobson, J.; Marchlinski, F.E. Reentrant and nonreentrant focal left atrial tachycardias occur after pulmonary vein isolation. *Heart Rhythm* **2005**, *2*, 1195–1202. [CrossRef]
19. Ho, S.Y.; Cabrera, J.A.; Tran, V.H.; Farré, J.; Anderson, R.H.; Sánchez-Quintana, D. Architecture of the pulmonary veins: Relevance to radiofrequency ablation. *Heart* **2001**, *86*, 265–270. [CrossRef]
20. Hamabe, A.; Okuyama, Y.; Miyauchi, Y.; Zhou, S.; Pak, H.N.; Karagueuzian, H.S.; Fishbein, M.C.; Chen, P.S. Correlation between anatomy and electrical activation in canine pulmonary veins. *Circulation* **2003**, *107*, 1550–1555. [CrossRef]
21. Hassink, R.J.; Aretz, H.T.; Ruskin, J.; Keane, D. Morphology of atrial myocardium in human pulmonary veins: A postmortem analysis in patients with and without atrial fibrillation. *J. Am. Coll. Cardiol.* **2003**, *42*, 1108–1114. [CrossRef] [PubMed]
22. Valles, E.; Fan, R.; Roux, J.F.; Liu, C.F.; Harding, J.D.; Dhruvakumar, S.; Hutchinson, M.D.; Riley, M.; Bala, R.; Garcia, F.C.; et al. Localization of atrial fibrillation triggers in patients undergoing pulmonary vein isolation: Importance of the carina region. *J. Am. Coll. Cardiol.* **2008**, *52*, 1413–1420. [CrossRef] [PubMed]
23. Cabrera, J.A.; Ho, S.Y.; Climent, V.; Fuertes, B.; Murillo, M.; Sánchez-Quintana, D. Morphological evidence of muscular connections between contiguous pulmonary venous orifices: Relevance of the interpulmonary isthmus for catheter ablation in atrial fibrillation. *Heart Rhythm* **2009**, *6*, 1192–1198. [CrossRef] [PubMed]
24. Arentz, T.; Weber, R.; Bürkle, G.; Herrera, C.; Blum, T.; Stockinger, J.; Minners, J.; Neumann, F.J.; Kalusche, D. Small or large isolation areas around the pulmonary veins for the treatment of atrial fibrillation? Results from a prospective randomized study. *Circulation* **2007**, *115*, 3057–3063. [CrossRef]
25. Yeh, H.I.; Lai, Y.J.; Lee, S.H.; Lee, Y.N.; Ko, Y.S.; Chen, S.A.; Severs, N.J.; Tsai, C.H. Heterogeneity of myocardial sleeve morphology and gap junctions in canine superior vena cava. *Circulation* **2001**, *104*, 3152–3157. [CrossRef]
26. Watanabe, K.; Nitta, J.; Inaba, O.; Sato, A.; Inamura, Y.; Kato, N.; Suzuki, M.; Goya, M.; Hirao, K.; Sasano, T. Predictors of non-pulmonary vein foci in paroxysmal atrial fibrillation. *J. Interv. Card. Electrophysiol.* **2021**, *61*, 71–78. [CrossRef]

27. Lin, W.S.; Tai, C.T.; Hsieh, M.H.; Tsai, C.F.; Lin, Y.K.; Tsao, H.M.; Huang, J.L.; Yu, W.C.; Yang, S.P.; Ding, Y.A.; et al. Catheter ablation of paroxysmal atrial fibrillation initiated by non-pulmonary vein ectopy. *Circulation* **2003**, *107*, 3176–3183. [CrossRef]
28. Yamaguchi, T.; Tsuchiya, T.; Miyamoto, K.; Nagamoto, Y.; Takahashi, N. Characterization of non-pulmonary vein foci with an EnSite array in patients with paroxysmal atrial fibrillation. *Europace* **2010**, *12*, 1698–1706. [CrossRef]
29. Chen, Y.J.; Chen, Y.C.; Yeh, H.I.; Lin, C.I.; Chen, S.A. Electrophysiology and arrhythmogenic activity of single cardiomyocytes from canine superior vena cava. *Circulation* **2002**, *105*, 2679–2685. [CrossRef]
30. Kugler, S.; Nagy, N.; Rácz, G.; Tőkés, A.M.; Dorogi, B.; Nemeskéri, Á. Presence of cardiomyocytes exhibiting Purkinje-type morphology and prominent connexin45 immunoreactivity in the myocardial sleeves of cardiac veins. *Heart Rhythm* **2018**, *15*, 258–264. [CrossRef]
31. Higuchi, K.; Yamauchi, Y.; Hirao, K.; Sasaki, T.; Hachiya, H.; Sekiguchi, Y.; Nitta, J.; Isobe, M. Superior vena cava as initiator of atrial fibrillation: Factors related to its arrhythmogenicity. *Heart Rhythm* **2010**, *7*, 1186–1191. [CrossRef] [PubMed]
32. Shah, D.C.; Haïssaguerre, M.; Jaïs, P.; Clémenty, J. High-resolution mapping of tachycardia originating from the superior vena cava: Evidence of electrical heterogeneity, slow conduction, and possible circus movement reentry. *J. Cardiovasc. Electrophysiol.* **2002**, *13*, 388–392. [CrossRef] [PubMed]
33. Miyazaki, S.; Kuwahara, T.; Takahashi, A. Confined driver of atrial fibrillation in the superior vena cava. *J. Cardiovasc. Electrophysiol.* **2012**, *23*, 440. [CrossRef] [PubMed]
34. Miyazaki, S.; Takigawa, M.; Kusa, S.; Kuwahara, T.; Taniguchi, H.; Okubo, K.; Nakamura, H.; Hachiya, H.; Hirao, K.; Takahashi, A.; et al. Role of arrhythmogenic superior vena cava on atrial fibrillation. *J. Cardiovasc. Electrophysiol.* **2014**, *25*, 380–386. [CrossRef] [PubMed]
35. Simu, G.; Deneke, T.; Ene, E.; Nentwich, K.; Berkovitz, A.; Sonne, K.; Halbfass, P.; Arvaniti, E.; Waechter, C.; Müller, J. Empirical superior vena cava isolation in patients undergoing repeat catheter ablation procedure after recurrence of atrial fibrillation. *J. Interv. Card. Electrophysiol.* **2022**, *65*, 551–558. [CrossRef]
36. Ho, S.Y.; Sánchez-Quintana, D. The importance of atrial structure and fibers. *Clin. Anat.* **2009**, *22*, 52–63. [CrossRef]
37. Hashizume, H.; Ushiki, T.; Abe, K. A histological study of the cardiac muscle of the human superior and inferior venae cavae. *Arch. Histol. Cytol.* **1995**, *58*, 457–464. [CrossRef]
38. Alonso-Martín, C.; García Mancebo, S.; Campos García, B.; Guerra Ramos, J.; Moreno Weidman, Z.; Méndez Zurita, F.; Montiel Quintero, R.; Betancur Gutiérrez, A.; Viñolas, X.; Rodríguez Font, E. Atrial Fibrillation Originating in the Inferior Vena Cava: A Typical Presentation of an Atypical Location. *JACC Case Rep.* **2021**, *3*, 1918–1923. [CrossRef]
39. Tao, Y.; Yang, D.; Chen, L. Paroxysmal Atrial Fibrillation Originating from the Inferior Vena Cava: A Case Report and Literature Review. *Front. Cardiovasc. Med.* **2022**, *9*, 935524. [CrossRef]
40. Habib, A.; Lachman, N.; Christensen, K.N.; Asirvatham, S.J. The anatomy of the coronary sinus venous system for the cardiac electrophysiologist. *Europace* **2009**, *11* (Suppl. S5), v15–v21. [CrossRef]
41. Antz, M.; Otomo, K.; Arruda, M.; Scherlag, B.J.; Pitha, J.; Tondo, C.; Lazzara, R.; Jackman, W.M. Electrical conduction between the right atrium and the left atrium via the musculature of the coronary sinus. *Circulation* **1998**, *98*, 1790–1795. [CrossRef] [PubMed]
42. Chauvin, M.; Shah, D.C.; Haïssaguerre, M.; Marcellin, L.; Brechenmacher, C. The anatomic basis of connections between the coronary sinus musculature and the left atrium in humans. *Circulation* **2000**, *101*, 647–652. [CrossRef] [PubMed]
43. Volkmer, M.; Antz, M.; Hebe, J.; Kuck, K.H. Focal atrial tachycardia originating from the musculature of the coronary sinus. *J. Cardiovasc. Electrophysiol.* **2002**, *13*, 68–71. [CrossRef] [PubMed]
44. Johnson, N.; Danilo, P., Jr.; Wit, A.L.; Rosen, M.R. Characteristics of initiation and termination of catecholamine-induced triggered activity in atrial fibers of the coronary sinus. *Circulation* **1986**, *74*, 1168–1179. [CrossRef]
45. Tseng, G.N.; Wit, A.L. Characteristics of a transient inward current that causes delayed afterdepolarizations in atrial cells of the canine coronary sinus. *J. Mol. Cell Cardiol.* **1987**, *19*, 1105–1119. [CrossRef]
46. Oral, H.; Ozaydin, M.; Chugh, A.; Scharf, C.; Tada, H.; Hall, B.; Cheung, P.; Pelosi, F.; Knight, B.P.; Morady, F. Role of the coronary sinus in maintenance of atrial fibrillation. *J. Cardiovasc. Electrophysiol.* **2003**, *14*, 1329–1336. [CrossRef]
47. Katritsis, D.; Ioannidis, J.P.; Giazitzoglou, E.; Korovesis, S.; Anagnostopoulos, C.E.; Camm, A.J. Conduction delay within the coronary sinus in humans: Implications for atrial arrhythmias. *J. Cardiovasc. Electrophysiol.* **2002**, *13*, 859–862. [CrossRef]
48. Ho, S.Y.; Sanchez-Quintana, D.; Cabrera, J.A.; Anderson, R.H. Anatomy of the left atrium: Implications for radiofrequency ablation of atrial fibrillation. *J. Cardiovasc. Electrophysiol.* **1999**, *10*, 1525–1533. [CrossRef]
49. Ahmed, N.; Perveen, S.; Mehmood, A.; Rani, G.F.; Molon, G. Coronary Sinus Ablation Is a Key Player Substrate in Recurrence of Persistent Atrial Fibrillation. *Cardiology* **2019**, *143*, 107–113. [CrossRef]
50. Morita, H.; Zipes, D.P.; Morita, S.T.; Wu, J. The role of coronary sinus musculature in the induction of atrial fibrillation. *Heart Rhythm* **2012**, *9*, 581–589. [CrossRef]
51. Morita, H.; Zipes, D.P.; Morita, S.T.; Wu, J. Isolation of canine coronary sinus musculature from the atria by radiofrequency catheter ablation prevents induction of atrial fibrillation. *Circ. Arrhythm. Electrophysiol.* **2014**, *7*, 1181–1188. [CrossRef] [PubMed]
52. Zhao, J.; Trew, M.L.; Legrice, I.J.; Smaill, B.H.; Pullan, A.J. A tissue-specific model of reentry in the right atrial appendage. *J. Cardiovasc. Electrophysiol.* **2009**, *20*, 675–684. [CrossRef] [PubMed]
53. Zhao, Q.Y.; Huang, H.; Tang, Y.H.; Wang, X.; Okello, E.; Liang, J.J.; Jiang, H.; Huang, C.X. Relationship between autonomic innervation in crista terminalis and atrial arrhythmia. *J. Cardiovasc. Electrophysiol.* **2009**, *20*, 551–557. [CrossRef]

54. Yamashita, T.; Nakajima, T.; Hazama, H.; Hamada, E.; Murakawa, Y.; Sawada, K.; Omata, M. Regional differences in transient outward current density and inhomogeneities of repolarization in rabbit right atrium. *Circulation* **1995**, *92*, 3061–3069. [CrossRef] [PubMed]
55. Giles, W.R.; van Ginneken, A.C. A transient outward current in isolated cells from the crista terminalis of rabbit heart. *J. Physiol.* **1985**, *368*, 243–264. [CrossRef]
56. Wu, T.J.; Yashima, M.; Xie, F.; Athill, C.A.; Kim, Y.H.; Fishbein, M.C.; Qu, Z.; Garfinkel, A.; Weiss, J.N.; Karagueuzian, H.S.; et al. Role of pectinate muscle bundles in the generation and maintenance of intra-atrial reentry: Potential implications for the mechanism of conversion between atrial fibrillation and atrial flutter. *Circ. Res.* **1998**, *83*, 448–462. [CrossRef]
57. Baptiste, F.; Kalifa, J.; Durand, C.; Gitenay, E.; Bremondy, M.; Ayari, A.; Maillot, N.; Taormina, A.; Fofana, A.; Penaranda, G.; et al. Right atrial appendage firing in atrial fibrillation. *Front. Cardiovasc. Med.* **2022**, *9*, 997998. [CrossRef]
58. Marcus, M.B.; Shein, J.A.; Vaishnav, A.S.; Mountantonakis, S.E. Paroxysmal Atrial Fibrillation with Both Triggers and Rotational Drivers within the Right Atrial Appendage. *JACC Case Rep.* **2019**, *1*, 607–611. [CrossRef]
59. Takahashi, Y.; Sanders, P.; Rotter, M.; Haïssaguerre, M. Disconnection of the left atrial appendage for elimination of foci maintaining atrial fibrillation. *J. Cardiovasc. Electrophysiol.* **2005**, *16*, 917–919. [CrossRef]
60. Di Biase, L.; Burkhardt, J.D.; Mohanty, P.; Sanchez, J.; Mohanty, S.; Horton, R.; Gallinghouse, G.J.; Bailey, S.M.; Zagrodzky, J.D.; Santangeli, P.; et al. Left atrial appendage: An underrecognized trigger site of atrial fibrillation. *Circulation* **2010**, *122*, 109–118. [CrossRef]
61. AlTurki, A.; Huynh, T.; Dawas, A.; AlTurki, H.; Joza, J.; Healey, J.S.; Essebag, V. Left atrial appendage isolation in atrial fibrillation catheter ablation: A meta-analysis. *J. Arrhythm.* **2018**, *34*, 478–484. [CrossRef] [PubMed]
62. Di Biase, L.; Burkhardt, J.D.; Mohanty, P.; Mohanty, S.; Sanchez, J.E.; Trivedi, C.; Güneş, M.; Gökoğlan, Y.; Gianni, C.; Horton, R.P.; et al. Left Atrial Appendage Isolation in Patients with Longstanding Persistent AF Undergoing Catheter Ablation: BELIEF Trial. *J. Am. Coll. Cardiol.* **2016**, *68*, 1929–1940. [CrossRef] [PubMed]
63. Ghannam, M.; Jongnarangsin, K.; Emami, H.; Yokokawa, M.; Liang, J.J.; Saeed, M.; Oral, H.; Morady, F.; Chugh, A. Incidental left atrial appendage isolation after catheter ablation of persistent atrial fibrillation: Mechanisms and long-term risk of thromboembolism. *J. Cardiovasc. Electrophysiol.* **2023**, *34*, 1152–1161. [CrossRef] [PubMed]
64. Morillo, C.A.; Klein, G.J.; Jones, D.L.; Guiraudon, C.M. Chronic rapid atrial pacing. Structural, functional, and electrophysiological characteristics of a new model of sustained atrial fibrillation. *Circulation* **1995**, *91*, 1588–1595. [CrossRef]
65. Huang, J.L.; Tai, C.T.; Lin, Y.J.; Ting, C.T.; Chen, Y.T.; Chang, M.S.; Lin, F.Y.; Lai, W.T.; Chen, S.A. The mechanisms of an increased dominant frequency in the left atrial posterior wall during atrial fibrillation in acute atrial dilatation. *J. Cardiovasc. Electrophysiol.* **2006**, *17*, 178–188. [CrossRef]
66. Kalifa, J.; Tanaka, K.; Zaitsev, A.V.; Warren, M.; Vaidyanathan, R.; Auerbach, D.; Pandit, S.; Vikstrom, K.L.; Ploutz-Snyder, R.; Talkachou, A.; et al. Mechanisms of wave fractionation at boundaries of high-frequency excitation in the posterior left atrium of the isolated sheep heart during atrial fibrillation. *Circulation* **2006**, *113*, 626–633. [CrossRef]
67. Wu, T.J.; Doshi, R.N.; Huang, H.L.; Blanche, C.; Kass, R.M.; Trento, A.; Cheng, W.; Karagueuzian, H.S.; Peter, C.T.; Chen, P.S. Simultaneous biatrial computerized mapping during permanent atrial fibrillation in patients with organic heart disease. *J. Cardiovasc. Electrophysiol.* **2002**, *13*, 571–577. [CrossRef]
68. Bliss, D.F., 2nd; Hutchins, G.M. The dorsal mesocardium and development of the pulmonary veins in human embryos. *Am. J. Cardiovasc. Pathol.* **1995**, *5*, 55–67.
69. Douglas, Y.L.; Jongbloed, M.R.; Gittenberger-de Groot, A.C.; Evers, D.; Dion, R.A.; Voigt, P.; Bartelings, M.M.; Schalij, M.J.; Ebels, T.; DeRuiter, M.C. Histology of vascular myocardial wall of left atrial body after pulmonary venous incorporation. *Am. J. Cardiol.* **2006**, *97*, 662–670. [CrossRef]
70. Sherif, H.M. The developing pulmonary veins and left atrium: Implications for ablation strategy for atrial fibrillation. *Eur. J. Cardiothorac. Surg.* **2013**, *44*, 792–799. [CrossRef]
71. Tessari, A.; Pietrobon, M.; Notte, A.; Cifelli, G.; Gage, P.J.; Schneider, M.D.; Lembo, G.; Campione, M. Myocardial Pitx2 differentially regulates the left atrial identity and ventricular asymmetric remodeling programs. *Circ. Res.* **2008**, *102*, 813–822. [CrossRef] [PubMed]
72. Galli, D.; Domínguez, J.N.; Zaffran, S.; Munk, A.; Brown, N.A.; Buckingham, M.E. Atrial myocardium derives from the posterior region of the second heart field, which acquires left-right identity as Pitx2c is expressed. *Development* **2008**, *135*, 1157–1167. [CrossRef] [PubMed]
73. Klos, M.; Calvo, D.; Yamazaki, M.; Zlochiver, S.; Mironov, S.; Cabrera, J.A.; Sanchez-Quintana, D.; Jalife, J.; Berenfeld, O.; Kalifa, J. Atrial septopulmonary bundle of the posterior left atrium provides a substrate for atrial fibrillation initiation in a model of vagally mediated pulmonary vein tachycardia of the structurally normal heart. *Circ. Arrhythm. Electrophysiol.* **2008**, *1*, 175–183. [CrossRef] [PubMed]
74. Markides, V.; Schilling, R.J.; Ho, S.Y.; Chow, A.W.; Davies, D.W.; Peters, N.S. Characterization of left atrial activation in the intact human heart. *Circulation* **2003**, *107*, 733–739. [CrossRef] [PubMed]
75. Burstein, B.; Nattel, S. Atrial fibrosis: Mechanisms and clinical relevance in atrial fibrillation. *J. Am. Coll. Cardiol.* **2008**, *51*, 802–809. [CrossRef]
76. Xu, J.; Cui, G.; Esmailian, F.; Plunkett, M.; Marelli, D.; Ardehali, A.; Odim, J.; Laks, H.; Sen, L. Atrial extracellular matrix remodeling and the maintenance of atrial fibrillation. *Circulation* **2004**, *109*, 363–368. [CrossRef]

77. Miragoli, M.; Salvarani, N.; Rohr, S. Myofibroblasts induce ectopic activity in cardiac tissue. *Circ. Res.* **2007**, *101*, 755–758. [CrossRef]
78. Rohr, S. Myofibroblasts in diseased hearts: New players in cardiac arrhythmias? *Heart Rhythm* **2009**, *6*, 848–856. [CrossRef]
79. Yue, L.; Xie, J.; Nattel, S. Molecular determinants of cardiac fibroblast electrical function and therapeutic implications for atrial fibrillation. *Cardiovasc. Res.* **2011**, *89*, 744–753. [CrossRef]
80. Sánchez, J.; Gomez, J.F.; Martinez-Mateu, L.; Romero, L.; Saiz, J.; Trenor, B. Heterogeneous Effects of Fibroblast-Myocyte Coupling in Different Regions of the Human Atria under Conditions of Atrial Fibrillation. *Front. Physiol.* **2019**, *10*, 847. [CrossRef]
81. Oakes, R.S.; Badger, T.J.; Kholmovski, E.G.; Akoum, N.; Burgon, N.S.; Fish, E.N.; Blauer, J.J.; Rao, S.N.; DiBella, E.V.; Segerson, N.M.; et al. Detection and quantification of left atrial structural remodeling with delayed-enhancement magnetic resonance imaging in patients with atrial fibrillation. *Circulation* **2009**, *119*, 1758–1767. [CrossRef] [PubMed]
82. Marrouche, N.F.; Wilber, D.; Hindricks, G.; Jais, P.; Akoum, N.; Marchlinski, F.; Kholmovski, E.; Burgon, N.; Hu, N.; Mont, L.; et al. Association of atrial tissue fibrosis identified by delayed enhancement MRI and atrial fibrillation catheter ablation: The DECAAF study. *JAMA* **2014**, *311*, 498–506. [CrossRef] [PubMed]
83. Benito, E.M.; Cabanelas, N.; Nuñez-Garcia, M.; Alarcón, F.; Figueras, I.V.R.M.; Soto-Iglesias, D.; Guasch, E.; Prat-Gonzalez, S.; Perea, R.J.; Borràs, R.; et al. Preferential regional distribution of atrial fibrosis in posterior wall around left inferior pulmonary vein as identified by late gadolinium enhancement cardiac magnetic resonance in patients with atrial fibrillation. *Europace* **2018**, *20*, 1959–1965. [CrossRef] [PubMed]
84. Higuchi, K.; Cates, J.; Gardner, G.; Morris, A.; Burgon, N.S.; Akoum, N.; Marrouche, N.F. The Spatial Distribution of Late Gadolinium Enhancement of Left Atrial Magnetic Resonance Imaging in Patients with Atrial Fibrillation. *JACC Clin. Electrophysiol.* **2018**, *4*, 49–58. [CrossRef]
85. Chen, J.; Arentz, T.; Cochet, H.; Müller-Edenborn, B.; Kim, S.; Moreno-Weidmann, Z.; Minners, J.; Kohl, P.; Lehrmann, H.; Allgeier, J.; et al. Extent and spatial distribution of left atrial arrhythmogenic sites, late gadolinium enhancement at magnetic resonance imaging, and low-voltage areas in patients with persistent atrial fibrillation: Comparison of imaging vs. electrical parameters of fibrosis and arrhythmogenesis. *Europace* **2019**, *21*, 1484–1493. [CrossRef]
86. Peters, D.C.; Wylie, J.V.; Hauser, T.H.; Nezafat, R.; Han, Y.; Woo, J.J.; Taclas, J.; Kissinger, K.V.; Goddu, B.; Josephson, M.E.; et al. Recurrence of atrial fibrillation correlates with the extent of post-procedural late gadolinium enhancement: A pilot study. *JACC Cardiovasc. Imaging* **2009**, *2*, 308–316. [CrossRef]
87. McGann, C.; Akoum, N.; Patel, A.; Kholmovski, E.; Revelo, P.; Damal, K.; Wilson, B.; Cates, J.; Harrison, A.; Ranjan, R.; et al. Atrial fibrillation ablation outcome is predicted by left atrial remodeling on MRI. *Circ. Arrhythm. Electrophysiol.* **2014**, *7*, 23–30. [CrossRef]
88. Marrouche, N.F.; Wazni, O.; McGann, C.; Greene, T.; Dean, J.M.; Dagher, L.; Kholmovski, E.; Mansour, M.; Marchlinski, F.; Wilber, D.; et al. Effect of MRI-Guided Fibrosis Ablation vs Conventional Catheter Ablation on Atrial Arrhythmia Recurrence in Patients with Persistent Atrial Fibrillation: The DECAAF II Randomized Clinical Trial. *JAMA* **2022**, *327*, 2296–2305. [CrossRef]
89. Sanders, P.; Morton, J.B.; Davidson, N.C.; Spence, S.J.; Vohra, J.K.; Sparks, P.B.; Kalman, J.M. Electrical remodeling of the atria in congestive heart failure: Electrophysiological and electroanatomic mapping in humans. *Circulation* **2003**, *108*, 1461–1468. [CrossRef]
90. Huo, Y.; Gaspar, T.; Pohl, M.; Sitzy, J.; Richter, U.; Neudeck, S.; Mayer, J.; Kronborg, M.B.; Piorkowski, C. Prevalence and predictors of low voltage zones in the left atrium in patients with atrial fibrillation. *Europace* **2018**, *20*, 956–962. [CrossRef]
91. Liu, Z.; Xia, Y.; Guo, C.; Li, X.; Fang, P.; Yin, X.; Yang, X. Low-Voltage Zones as the Atrial Fibrillation Substrates: Relationship with Initiation, Perpetuation, and Termination. *Front. Cardiovasc. Med.* **2021**, *8*, 705510. [CrossRef] [PubMed]
92. Verma, A.; Wazni, O.M.; Marrouche, N.F.; Martin, D.O.; Kilicaslan, F.; Minor, S.; Schweikert, R.A.; Saliba, W.; Cummings, J.; Burkhardt, J.D.; et al. Pre-existent left atrial scarring in patients undergoing pulmonary vein antrum isolation: An independent predictor of procedural failure. *J. Am. Coll. Cardiol.* **2005**, *45*, 285–292. [CrossRef]
93. Vlachos, K.; Efremidis, M.; Letsas, K.P.; Bazoukis, G.; Martin, R.; Kalafateli, M.; Lioni, L.; Georgopoulos, S.; Saplaouras, A.; Efremidis, T.; et al. Low-voltage areas detected by high-density electroanatomical mapping predict recurrence after ablation for paroxysmal atrial fibrillation. *J. Cardiovasc. Electrophysiol.* **2017**, *28*, 1393–1402. [CrossRef] [PubMed]
94. Moustafa, A.; Karim, S.; Kahaly, O.; Elzanaty, A.; Meenakshisundaram, C.; Abi-Saleh, B.; Eltahawy, E.; Chacko, P. Low voltage area guided substrate modification in nonparoxysmal atrial fibrillation: A systematic review and meta-analysis. *J. Cardiovasc. Electrophysiol.* **2023**, *34*, 455–464. [CrossRef] [PubMed]
95. Zhou, Y.; Zhang, H.; Yan, P.; Zhou, P.; Wang, P.; Li, X. Efficacy of left atrial low-voltage area-guided catheter ablation of atrial fibrillation: An updated systematic review and meta-analysis. *Front. Cardiovasc. Med.* **2022**, *9*, 993790. [CrossRef]
96. Junarta, J.; Siddiqui, M.U.; Riley, J.M.; Dikdan, S.J.; Patel, A.; Frisch, D.R. Low-voltage area substrate modification for atrial fibrillation ablation: A systematic review and meta-analysis of clinical trials. *Europace* **2022**, *24*, 1585–1598. [CrossRef]
97. Miyamoto, K.; Tsuchiya, T.; Narita, S.; Yamaguchi, T.; Nagamoto, Y.; Ando, S.; Hayashida, K.; Tanioka, Y.; Takahashi, N. Bipolar electrogram amplitudes in the left atrium are related to local conduction velocity in patients with atrial fibrillation. *Europace* **2009**, *11*, 1597–1605. [CrossRef]
98. Ali, R.L.; Qureshi, N.A.; Liverani, S.; Roney, C.H.; Kim, S.; Lim, P.B.; Tweedy, J.H.; Cantwell, C.D.; Peters, N.S. Left Atrial Enhancement Correlates with Myocardial Conduction Velocity in Patients with Persistent Atrial Fibrillation. *Front. Physiol.* **2020**, *11*, 570203. [CrossRef]

99. Fukumoto, K.; Habibi, M.; Ipek, E.G.; Zahid, S.; Khurram, I.M.; Zimmerman, S.L.; Zipunnikov, V.; Spragg, D.; Ashikaga, H.; Trayanova, N.; et al. Association of Left Atrial Local Conduction Velocity with Late Gadolinium Enhancement on Cardiac Magnetic Resonance in Patients with Atrial Fibrillation. *Circ. Arrhythm. Electrophysiol.* **2016**, *9*, e002897. [CrossRef]
100. Cochet, H.; Dubois, R.; Yamashita, S.; Al Jefairi, N.; Berte, B.; Sellal, J.M.; Hooks, D.; Frontera, A.; Amraoui, S.; Zemoura, A.; et al. Relationship between Fibrosis Detected on Late Gadolinium-Enhanced Cardiac Magnetic Resonance and Re-Entrant Activity Assessed with Electrocardiographic Imaging in Human Persistent Atrial Fibrillation. *JACC Clin Electrophysiol.* **2018**, *4*, 17–29. [CrossRef]
101. Chrispin, J.; Gucuk Ipek, E.; Zahid, S.; Prakosa, A.; Habibi, M.; Spragg, D.; Marine, J.E.; Ashikaga, H.; Rickard, J.; Trayanova, N.A.; et al. Lack of regional association between atrial late gadolinium enhancement on cardiac magnetic resonance and atrial fibrillation rotors. *Heart Rhythm* **2016**, *13*, 654–660. [CrossRef] [PubMed]
102. Sohns, C.; Lemes, C.; Metzner, A.; Fink, T.; Chmelevsky, M.; Maurer, T.; Budanova, M.; Solntsev, V.; Schulze, W.H.W.; Staab, W.; et al. First-in-Man Analysis of the Relationship between Electrical Rotors from Noninvasive Panoramic Mapping and Atrial Fibrosis from Magnetic Resonance Imaging in Patients with Persistent Atrial Fibrillation. *Circ. Arrhythm. Electrophysiol.* **2017**, *10*, e004419. [CrossRef] [PubMed]
103. Nakamura, T.; Kiuchi, K.; Fukuzawa, K.; Takami, M.; Watanabe, Y.; Izawa, Y.; Suehiro, H.; Akita, T.; Takemoto, M.; Sakai, J.; et al. Late-gadolinium enhancement properties associated with atrial fibrillation rotors in patients with persistent atrial fibrillation. *J. Cardiovasc. Electrophysiol.* **2021**, *32*, 1005–1013. [CrossRef] [PubMed]
104. Schade, A.; Nentwich, K.; Costello-Boerrigter, L.C.; Halbfass, P.; Mueller, P.; Roos, M.; Barth, S.; Krug, J.; Szoelloesi, G.A.; Lapp, H.; et al. Spatial Relationship of Focal Impulses, Rotors and Low Voltage Zones in Patients with Persistent Atrial Fibrillation. *J. Cardiovasc. Electrophysiol.* **2016**, *27*, 507–514. [CrossRef]
105. King, J.H.; Huang, C.L.; Fraser, J.A. Determinants of myocardial conduction velocity: Implications for arrhythmogenesis. *Front. Physiol.* **2013**, *4*, 154. [CrossRef]
106. Guichard, J.B.; Naud, P.; Xiong, F.; Qi, X.; L'Heureux, N.; Hiram, R.; Tardif, J.C.; Cartier, R.; Da Costa, A.; Nattel, S. Comparison of Atrial Remodeling Caused by Sustained Atrial Flutter versus Atrial Fibrillation. *J. Am. Coll. Cardiol.* **2020**, *76*, 374–388. [CrossRef]
107. Nattel, S. Molecular and Cellular Mechanisms of Atrial Fibrosis in Atrial Fibrillation. *JACC Clin. Electrophysiol.* **2017**, *3*, 425–435. [CrossRef]
108. Zhang, Z.; Vlcek, J.; Pauly, V.; Hesse, N.; Bauer, J.; Chataut, K.R.; Maderspacher, F.; Volz, L.S.; Buchberger, K.; Xia, R.; et al. Atrial fibrosis heterogeneity is a risk for atrial fibrillation in pigs with ischaemic heart failure. *Eur. J. Clin. Investig.* **2024**, *54*, e14137. [CrossRef]
109. Jadidi, A.S.; Cochet, H.; Shah, A.J.; Kim, S.J.; Duncan, E.; Miyazaki, S.; Sermesant, M.; Lehrmann, H.; Lederlin, M.; Linton, N.; et al. Inverse relationship between fractionated electrograms and atrial fibrosis in persistent atrial fibrillation: Combined magnetic resonance imaging and high-density mapping. *J. Am. Coll. Cardiol.* **2013**, *62*, 802–812. [CrossRef]
110. Haissaguerre, M.; Shah, A.J.; Cochet, H.; Hocini, M.; Dubois, R.; Efimov, I.; Vigmond, E.; Bernus, O.; Trayanova, N. Intermittent drivers anchoring to structural heterogeneities as a major pathophysiological mechanism of human persistent atrial fibrillation. *J. Physiol.* **2016**, *594*, 2387–2398. [CrossRef]
111. Tanaka, K.; Zlochiver, S.; Vikstrom, K.L.; Yamazaki, M.; Moreno, J.; Klos, M.; Zaitsev, A.V.; Vaidyanathan, R.; Auerbach, D.S.; Landas, S.; et al. Spatial distribution of fibrosis governs fibrillation wave dynamics in the posterior left atrium during heart failure. *Circ. Res.* **2007**, *101*, 839–847. [CrossRef] [PubMed]
112. Kazbanov, I.V.; ten Tusscher, K.H.; Panfilov, A.V. Effects of Heterogeneous Diffuse Fibrosis on Arrhythmia Dynamics and Mechanism. *Sci. Rep.* **2016**, *6*, 20835. [CrossRef] [PubMed]
113. Xie, F.; Qu, Z.; Garfinkel, A.; Weiss, J.N. Electrophysiological heterogeneity and stability of reentry in simulated cardiac tissue. *Am. J. Physiol. Heart Circ. Physiol.* **2001**, *280*, H535–H545. [CrossRef] [PubMed]
114. Saha, M.; Roney, C.H.; Bayer, J.D.; Meo, M.; Cochet, H.; Dubois, R.; Vigmond, E.J. Wavelength and Fibrosis Affect Phase Singularity Locations during Atrial Fibrillation. *Front. Physiol.* **2018**, *9*, 1207. [CrossRef] [PubMed]
115. Zahid, S.; Cochet, H.; Boyle, P.M.; Schwarz, E.L.; Whyte, K.N.; Vigmond, E.J.; Dubois, R.; Hocini, M.; Haïssaguerre, M.; Jaïs, P.; et al. Patient-derived models link re-entrant driver localization in atrial fibrillation to fibrosis spatial pattern. *Cardiovasc. Res.* **2016**, *110*, 443–454. [CrossRef] [PubMed]
116. Morgan, R.; Colman, M.A.; Chubb, H.; Seemann, G.; Aslanidi, O.V. Slow Conduction in the Border Zones of Patchy Fibrosis Stabilizes the Drivers for Atrial Fibrillation: Insights from Multi-Scale Human Atrial Modeling. *Front. Physiol.* **2016**, *7*, 474. [CrossRef]
117. Lu, Z.; Scherlag, B.J.; Lin, J.; Niu, G.; Fung, K.M.; Zhao, L.; Ghias, M.; Jackman, W.M.; Lazzara, R.; Jiang, H.; et al. Atrial fibrillation begets atrial fibrillation: Autonomic mechanism for atrial electrical remodeling induced by short-term rapid atrial pacing. *Circ. Arrhythm. Electrophysiol.* **2008**, *1*, 184–192. [CrossRef]
118. Armour, J.A.; Murphy, D.A.; Yuan, B.X.; Macdonald, S.; Hopkins, D.A. Gross and microscopic anatomy of the human intrinsic cardiac nervous system. *Anat. Rec.* **1997**, *247*, 289–298. [CrossRef]
119. Patterson, E.; Po, S.S.; Scherlag, B.J.; Lazzara, R. Triggered firing in pulmonary veins initiated by in vitro autonomic nerve stimulation. *Heart Rhythm* **2005**, *2*, 624–631. [CrossRef]
120. Po, S.S.; Li, Y.; Tang, D.; Liu, H.; Geng, N.; Jackman, W.M.; Scherlag, B.; Lazzara, R.; Patterson, E. Rapid and stable re-entry within the pulmonary vein as a mechanism initiating paroxysmal atrial fibrillation. *J. Am. Coll. Cardiol.* **2005**, *45*, 1871–1877. [CrossRef]

121. Po, S.S.; Scherlag, B.J.; Yamanashi, W.S.; Edwards, J.; Zhou, J.; Wu, R.; Geng, N.; Lazzara, R.; Jackman, W.M. Experimental model for paroxysmal atrial fibrillation arising at the pulmonary vein-atrial junctions. *Heart Rhythm* **2006**, *3*, 201–208. [CrossRef] [PubMed]
122. Choi, E.K.; Zhao, Y.; Everett, T.H.t.; Chen, P.S. Ganglionated plexi as neuromodulation targets for atrial fibrillation. *J. Cardiovasc. Electrophysiol.* **2017**, *28*, 1485–1491. [CrossRef] [PubMed]
123. Rackley, J.; Nudy, M.; Gonzalez, M.D.; Naccarelli, G.; Maheshwari, A. Pulmonary vein isolation with adjunctive left atrial ganglionic plexus ablation for treatment of atrial fibrillation: A meta-analysis of randomized controlled trials. *J. Interv. Card. Electrophysiol.* **2023**, *66*, 333–342. [CrossRef] [PubMed]
124. Kampaktsis, P.N.; Oikonomou, E.K.; Choi, D.Y.; Cheung, J.W. Efficacy of ganglionated plexi ablation in addition to pulmonary vein isolation for paroxysmal versus persistent atrial fibrillation: A meta-analysis of randomized controlled clinical trials. *J. Interv. Card. Electrophysiol.* **2017**, *50*, 253–260. [CrossRef]
125. Doshi, R.N.; Wu, T.J.; Yashima, M.; Kim, Y.H.; Ong, J.J.; Cao, J.M.; Hwang, C.; Yashar, P.; Fishbein, M.C.; Karagueuzian, H.S.; et al. Relation between ligament of Marshall and adrenergic atrial tachyarrhythmia. *Circulation* **1999**, *100*, 876–883. [CrossRef]
126. Lam, A.; Küffer, T.; Hunziker, L.; Nozica, N.; Asatryan, B.; Franzeck, F.; Madaffari, A.; Haeberlin, A.; Mühl, A.; Servatius, H.; et al. Efficacy and safety of ethanol infusion into the vein of Marshall for mitral isthmus ablation. *J. Cardiovasc. Electrophysiol.* **2021**, *32*, 1610–1619. [CrossRef]
127. Valderrábano, M.; Peterson, L.E.; Swarup, V.; Schurmann, P.A.; Makkar, A.; Doshi, R.N.; DeLurgio, D.; Athill, C.A.; Ellenbogen, K.A.; Natale, A.; et al. Effect of Catheter Ablation with Vein of Marshall Ethanol Infusion vs Catheter Ablation Alone on Persistent Atrial Fibrillation: The VENUS Randomized Clinical Trial. *JAMA* **2020**, *324*, 1620–1628. [CrossRef]
128. Bosch, R.F.; Zeng, X.; Grammer, J.B.; Popovic, K.; Mewis, C.; Kühlkamp, V. Ionic mechanisms of electrical remodeling in human atrial fibrillation. *Cardiovasc. Res.* **1999**, *44*, 121–131. [CrossRef]
129. Workman, A.J.; Kane, K.A.; Rankin, A.C. The contribution of ionic currents to changes in refractoriness of human atrial myocytes associated with chronic atrial fibrillation. *Cardiovasc. Res.* **2001**, *52*, 226–235. [CrossRef]
130. Van Wagoner, D.R.; Pond, A.L.; Lamorgese, M.; Rossie, S.S.; McCarthy, P.M.; Nerbonne, J.M. Atrial L-type Ca^{2+} currents and human atrial fibrillation. *Circ. Res.* **1999**, *85*, 428–436. [CrossRef]
131. Skasa, M.; Jüngling, E.; Picht, E.; Schöndube, F.; Lückhoff, A. L-type calcium currents in atrial myocytes from patients with persistent and non-persistent atrial fibrillation. *Basic Res. Cardiol.* **2001**, *96*, 151–159. [CrossRef] [PubMed]
132. Yue, L.; Melnyk, P.; Gaspo, R.; Wang, Z.; Nattel, S. Molecular mechanisms underlying ionic remodeling in a dog model of atrial fibrillation. *Circ. Res.* **1999**, *84*, 776–784. [CrossRef] [PubMed]
133. Severs, N.J.; Bruce, A.F.; Dupont, E.; Rothery, S. Remodelling of gap junctions and connexin expression in diseased myocardium. *Cardiovasc. Res.* **2008**, *80*, 9–19. [CrossRef] [PubMed]
134. van der Velden, H.M.; Ausma, J.; Rook, M.B.; Hellemons, A.J.; van Veen, T.A.; Allessie, M.A.; Jongsma, H.J. Gap junctional remodeling in relation to stabilization of atrial fibrillation in the goat. *Cardiovasc. Res.* **2000**, *46*, 476–486. [CrossRef]
135. van der Velden, H.M.; van Kempen, M.J.; Wijffels, M.C.; van Zijverden, M.; Groenewegen, W.A.; Allessie, M.A.; Jongsma, H.J. Altered pattern of connexin40 distribution in persistent atrial fibrillation in the goat. *J. Cardiovasc. Electrophysiol.* **1998**, *9*, 596–607. [CrossRef]
136. Polontchouk, L.; Haefliger, J.A.; Ebelt, B.; Schaefer, T.; Stuhlmann, D.; Mehlhorn, U.; Kuhn-Regnier, F.; De Vivie, E.R.; Dhein, S. Effects of chronic atrial fibrillation on gap junction distribution in human and rat atria. *J. Am. Coll. Cardiol.* **2001**, *38*, 883–891. [CrossRef]
137. Yeh, H.I.; Lai, Y.J.; Lee, S.H.; Chen, S.T.; Ko, Y.S.; Chen, S.A.; Severs, N.J.; Tsai, C.H. Remodeling of myocardial sleeve and gap junctions in canine superior vena cava after rapid pacing. *Basic Res. Cardiol.* **2006**, *101*, 269–280. [CrossRef]
138. Kostin, S.; Klein, G.; Szalay, Z.; Hein, S.; Bauer, E.P.; Schaper, J. Structural correlate of atrial fibrillation in human patients. *Cardiovasc. Res.* **2002**, *54*, 361–379. [CrossRef]
139. Jennings, M.M.; Donahue, J.K. Connexin Remodeling Contributes to Atrial Fibrillation. *J. Atr. Fibrillation* **2013**, *6*, 839. [CrossRef]
140. Sakabe, M.; Fujiki, A.; Nishida, K.; Sugao, M.; Nagasawa, H.; Tsuneda, T.; Mizumaki, K.; Inoue, H. Enalapril prevents perpetuation of atrial fibrillation by suppressing atrial fibrosis and over-expression of connexin43 in a canine model of atrial pacing-induced left ventricular dysfunction. *J. Cardiovasc. Pharmacol.* **2004**, *43*, 851–859. [CrossRef]
141. Elvan, A.; Huang, X.D.; Pressler, M.L.; Zipes, D.P. Radiofrequency catheter ablation of the atria eliminates pacing induced sustained atrial fibrillation and reduces connexin 43 in dogs. *Circulation* **1997**, *96*, 1675–1685. [CrossRef] [PubMed]
142. Chaldoupi, S.M.; Loh, P.; Hauer, R.N.; de Bakker, J.M.; van Rijen, H.V. The role of connexin40 in atrial fibrillation. *Cardiovasc. Res.* **2009**, *84*, 15–23. [CrossRef] [PubMed]
143. Nattel, S.; Heijman, J.; Zhou, L.; Dobrev, D. Molecular Basis of Atrial Fibrillation Pathophysiology and Therapy: A Translational Perspective. *Circ. Res.* **2020**, *127*, 51–72. [CrossRef] [PubMed]
144. Imanaga, I.; Hai, L.; Ogawa, K.; Matsumura, K.; Mayama, T. Phosphorylation of connexin in functional regulation of the cardiac gap junction. *Exp. Clin. Cardiol.* **2004**, *9*, 161–164. [PubMed]
145. TenBroek, E.M.; Lampe, P.D.; Solan, J.L.; Reynhout, J.K.; Johnson, R.G. Ser364 of connexin43 and the upregulation of gap junction assembly by cAMP. *J. Cell Biol.* **2001**, *155*, 1307–1318. [CrossRef]

146. Saidi Brikci-Nigassa, A.; Clement, M.J.; Ha-Duong, T.; Adjadj, E.; Ziani, L.; Pastre, D.; Curmi, P.A.; Savarin, P. Phosphorylation controls the interaction of the connexin43 C-terminal domain with tubulin and microtubules. *Biochemistry* **2012**, *51*, 4331–4342. [CrossRef]
147. Takanari, H.; Bourgonje, V.J.; Fontes, M.S.; Raaijmakers, A.J.; Driessen, H.; Jansen, J.A.; van der Nagel, R.; Kok, B.; van Stuijvenberg, L.; Boulaksil, M.; et al. Calmodulin/CaMKII inhibition improves intercellular communication and impulse propagation in the heart and is antiarrhythmic under conditions when fibrosis is absent. *Cardiovasc. Res.* **2016**, *111*, 410–421. [CrossRef]
148. Zhong, C.; Chang, H.; Wu, Y.; Zhou, L.; Wang, Y.; Wang, M.; Wu, P.; Qi, Z.; Zou, J. Up-regulated Cx43 phosphorylation at Ser368 prolongs QRS duration in myocarditis. *J. Cell Mol. Med.* **2018**, *22*, 3537–3547. [CrossRef]
149. Forrester, S.J.; Booz, G.W.; Sigmund, C.D.; Coffman, T.M.; Kawai, T.; Rizzo, V.; Scalia, R.; Eguchi, S. Angiotensin II Signal Transduction: An Update on Mechanisms of Physiology and Pathophysiology. *Physiol. Rev.* **2018**, *98*, 1627–1738. [CrossRef]
150. Nattel, S.; Dobrev, D. Electrophysiological and molecular mechanisms of paroxysmal atrial fibrillation. *Nat. Rev. Cardiol.* **2016**, *13*, 575–590. [CrossRef]
151. Wakili, R.; Voigt, N.; Kääb, S.; Dobrev, D.; Nattel, S. Recent advances in the molecular pathophysiology of atrial fibrillation. *J. Clin. Investig.* **2011**, *121*, 2955–2968. [CrossRef] [PubMed]
152. Stacy, M.R.; Lin, B.A.; Thorn, S.L.; Lobb, D.C.; Maxfield, M.W.; Novack, C.; Zellars, K.N.; Freeburg, L.; Akar, J.G.; Sinusas, A.J.; et al. Regional heterogeneity in determinants of atrial matrix remodeling and association with atrial fibrillation vulnerability postmyocardial infarction. *Heart Rhythm* **2022**, *19*, 847–855. [CrossRef] [PubMed]
153. Ihara, K.; Sasano, T. Role of Inflammation in the Pathogenesis of Atrial Fibrillation. *Front. Physiol.* **2022**, *13*, 862164. [CrossRef] [PubMed]
154. Willar, B.; Tran, K.V.; Fitzgibbons, T.P. Epicardial adipocytes in the pathogenesis of atrial fibrillation: An update on basic and translational studies. *Front. Endocrinol.* **2023**, *14*, 1154824. [CrossRef]
155. Venteclef, N.; Guglielmi, V.; Balse, E.; Gaborit, B.; Cotillard, A.; Atassi, F.; Amour, J.; Leprince, P.; Dutour, A.; Clément, K.; et al. Human epicardial adipose tissue induces fibrosis of the atrial myocardium through the secretion of adipo-fibrokines. *Eur. Heart. J.* **2015**, *36*, 795–805a. [CrossRef]
156. Nalliah, C.J.; Bell, J.R.; Raaijmakers, A.J.A.; Waddell, H.M.; Wells, S.P.; Bernasochi, G.B.; Montgomery, M.K.; Binny, S.; Watts, T.; Joshi, S.B.; et al. Epicardial Adipose Tissue Accumulation Confers Atrial Conduction Abnormality. *J. Am. Coll. Cardiol.* **2020**, *76*, 1197–1211. [CrossRef]
157. Abe, I.; Teshima, Y.; Kondo, H.; Kaku, H.; Kira, S.; Ikebe, Y.; Saito, S.; Fukui, A.; Shinohara, T.; Yufu, K.; et al. Association of fibrotic remodeling and cytokines/chemokines content in epicardial adipose tissue with atrial myocardial fibrosis in patients with atrial fibrillation. *Heart Rhythm* **2018**, *15*, 1717–1727. [CrossRef]
158. Mazurek, T.; Zhang, L.; Zalewski, A.; Mannion, J.D.; Diehl, J.T.; Arafat, H.; Sarov-Blat, L.; O'Brien, S.; Keiper, E.A.; Johnson, A.G.; et al. Human epicardial adipose tissue is a source of inflammatory mediators. *Circulation* **2003**, *108*, 2460–2466. [CrossRef]
159. Tsao, H.M.; Hu, W.C.; Wu, M.H.; Tai, C.T.; Lin, Y.J.; Chang, S.L.; Lo, L.W.; Hu, Y.F.; Tuan, T.C.; Wu, T.J.; et al. Quantitative analysis of quantity and distribution of epicardial adipose tissue surrounding the left atrium in patients with atrial fibrillation and effect of recurrence after ablation. *Am. J. Cardiol.* **2011**, *107*, 1498–1503. [CrossRef]
160. Tsai, C.T.; Lai, L.P.; Hwang, J.J.; Lin, J.L.; Chiang, F.T. Molecular genetics of atrial fibrillation. *J. Am. Coll. Cardiol.* **2008**, *52*, 241–250. [CrossRef]
161. Chen, Y.H.; Xu, S.J.; Bendahhou, S.; Wang, X.L.; Wang, Y.; Xu, W.Y.; Jin, H.W.; Sun, H.; Su, X.Y.; Zhuang, Q.N.; et al. KCNQ1 gain-of-function mutation in familial atrial fibrillation. *Science* **2003**, *299*, 251–254. [CrossRef] [PubMed]
162. Hong, K.; Bjerregaard, P.; Gussak, I.; Brugada, R. Short QT syndrome and atrial fibrillation caused by mutation in KCNH2. *J. Cardiovasc. Electrophysiol.* **2005**, *16*, 394–396. [CrossRef] [PubMed]
163. Xia, M.; Jin, Q.; Bendahhou, S.; He, Y.; Larroque, M.M.; Chen, Y.; Zhou, Q.; Yang, Y.; Liu, Y.; Liu, B.; et al. A Kir2.1 gain-of-function mutation underlies familial atrial fibrillation. *Biochem. Biophys. Res. Commun.* **2005**, *332*, 1012–1019. [CrossRef] [PubMed]
164. Roselli, C.; Rienstra, M.; Ellinor, P.T. Genetics of Atrial Fibrillation in 2020: GWAS, Genome Sequencing, Polygenic Risk, and Beyond. *Circ. Res.* **2020**, *127*, 21–33. [CrossRef]
165. Gudbjartsson, D.F.; Arnar, D.O.; Helgadottir, A.; Gretarsdottir, S.; Holm, H.; Sigurdsson, A.; Jonasdottir, A.; Baker, A.; Thorleifsson, G.; Kristjansson, K.; et al. Variants conferring risk of atrial fibrillation on chromosome 4q25. *Nature* **2007**, *448*, 353–357. [CrossRef]
166. Ryan, A.K.; Blumberg, B.; Rodriguez-Esteban, C.; Yonei-Tamura, S.; Tamura, K.; Tsukui, T.; de la Peña, J.; Sabbagh, W.; Greenwald, J.; Choe, S.; et al. Pitx2 determines left-right asymmetry of internal organs in vertebrates. *Nature* **1998**, *394*, 545–551. [CrossRef]
167. Campione, M.; Ros, M.A.; Icardo, J.M.; Piedra, E.; Christoffels, V.M.; Schweickert, A.; Blum, M.; Franco, D.; Moorman, A.F. Pitx2 expression defines a left cardiac lineage of cells: Evidence for atrial and ventricular molecular isomerism in the iv/iv mice. *Dev. Biol.* **2001**, *231*, 252–264. [CrossRef]
168. Chinchilla, A.; Daimi, H.; Lozano-Velasco, E.; Dominguez, J.N.; Caballero, R.; Delpón, E.; Tamargo, J.; Cinca, J.; Hove-Madsen, L.; Aranega, A.E.; et al. PITX2 insufficiency leads to atrial electrical and structural remodeling linked to arrhythmogenesis. *Circ. Cardiovasc. Genet.* **2011**, *4*, 269–279. [CrossRef]
169. Kirchhof, P.; Kahr, P.C.; Kaese, S.; Piccini, I.; Vokshi, I.; Scheld, H.H.; Rotering, H.; Fortmueller, L.; Laakmann, S.; Verheule, S.; et al. PITX2c is expressed in the adult left atrium, and reducing Pitx2c expression promotes atrial fibrillation inducibility and complex changes in gene expression. *Circ. Cardiovasc. Genet.* **2011**, *4*, 123–133. [CrossRef]

170. Mommersteeg, M.T.; Brown, N.A.; Prall, O.W.; de Gier-de Vries, C.; Harvey, R.P.; Moorman, A.F.; Christoffels, V.M. Pitx2c and Nkx2-5 are required for the formation and identity of the pulmonary myocardium. *Circ. Res.* **2007**, *101*, 902–909. [CrossRef]
171. Li, Z.; Hertervig, E.; Yuan, S.; Yang, Y.; Lin, Z.; Olsson, S.B. Dispersion of atrial repolarization in patients with paroxysmal atrial fibrillation. *Europace* **2001**, *3*, 285–291. [CrossRef] [PubMed]
172. Yuan, S.; Wohlfart, B.; Olsson, S.B.; Blomström-Lundqvist, C. The dispersion of repolarization in patients with ventricular tachycardia. A study using simultaneous monophasic action potential recordings from two sites in the right ventricle. *Eur. Heart J.* **1995**, *16*, 68–76. [CrossRef] [PubMed]
173. Garfinkel, A.; Kim, Y.H.; Voroshilovsky, O.; Qu, Z.; Kil, J.R.; Lee, M.H.; Karagueuzian, H.S.; Weiss, J.N.; Chen, P.S. Preventing ventricular fibrillation by flattening cardiac restitution. *Proc. Natl. Acad. Sci. USA* **2000**, *97*, 6061–6066. [CrossRef] [PubMed]
174. Pop, T.; Fleischmann, D. Alternans in human atrial monophasic action potenial. *Br. Heart. J.* **1977**, *39*, 1273–1275. [CrossRef] [PubMed]
175. Ihara, K.; Sugiyama, K.; Takahashi, K.; Yamazoe, M.; Sasano, T.; Furukawa, T. Electrophysiological Assessment of Murine Atria with High-Resolution Optical Mapping. *J. Vis. Exp.* **2018**, *32*, e56478. [CrossRef]
176. Fareh, S.; Villemaire, C.; Nattel, S. Importance of refractoriness heterogeneity in the enhanced vulnerability to atrial fibrillation induction caused by tachycardia-induced atrial electrical remodeling. *Circulation* **1998**, *98*, 2202–2209. [CrossRef]
177. Wood, M.A.; Mangano, R.A.; Schieken, R.M.; Baumgarten, C.M.; Simpson, P.M.; Ellenbogen, K.A. Modulation of atrial repolarization by site of pacing in the isolated rabbit heart. *Circulation* **1996**, *94*, 1465–1470. [CrossRef]
178. Arora, R.; Verheule, S.; Scott, L.; Navarrete, A.; Katari, V.; Wilson, E.; Vaz, D.; Olgin, J.E. Arrhythmogenic substrate of the pulmonary veins assessed by high-resolution optical mapping. *Circulation* **2003**, *107*, 1816–1821. [CrossRef]
179. Anyukhovsky, E.P.; Sosunov, E.A.; Chandra, P.; Rosen, T.S.; Boyden, P.A.; Danilo, P., Jr.; Rosen, M.R. Age-associated changes in electrophysiologic remodeling: A potential contributor to initiation of atrial fibrillation. *Cardiovasc. Res.* **2005**, *66*, 353–363. [CrossRef]
180. Ehrlich, J.R.; Cha, T.J.; Zhang, L.; Chartier, D.; Melnyk, P.; Hohnloser, S.H.; Nattel, S. Cellular electrophysiology of canine pulmonary vein cardiomyocytes: Action potential and ionic current properties. *J. Physiol.* **2003**, *551*, 801–813. [CrossRef]
181. Diker, E.; Ozdemir, M.; Aydoğdu, S.; Tezcan, U.K.; Korkmaz, S.; Kütük, E.; Göksel, S. Dispersion of repolarization in paroxysmal atrial fibrillation. *Int. J. Cardiol.* **1998**, *63*, 281–286. [CrossRef] [PubMed]
182. Namba, T.; Ashihara, T.; Nakazawa, K.; Ohe, T. Spatial heterogeneity in refractoriness as a proarrhythmic substrate: Theoretical evaluation by numerical simulation. *Jpn Circ. J.* **2000**, *64*, 121–129. [CrossRef] [PubMed]
183. Watanabe, M.; Yokoshiki, H.; Mitsuyama, H.; Mizukami, K.; Ono, T.; Tsutsui, H. Conduction and refractory disorders in the diabetic atrium. *Am. J. Physiol. Heart Circ. Physiol.* **2012**, *303*, H86–H95. [CrossRef] [PubMed]
184. Qu, Z.; Xie, F.; Garfinkel, A.; Weiss, J.N. Origins of spiral wave meander and breakup in a two-dimensional cardiac tissue model. *Ann. Biomed. Eng.* **2000**, *28*, 755–771. [CrossRef] [PubMed]
185. Ciaccio, E.J.; Peters, N.S.; Garan, H. Effects of refractory gradients and ablation on fibrillatory activity. *Comput. Biol. Med.* **2018**, *95*, 175–187. [CrossRef]
186. Seitz, J.; Bars, C.; Théodore, G.; Beurtheret, S.; Lellouche, N.; Bremondy, M.; Ferracci, A.; Faure, J.; Penaranda, G.; Yamazaki, M.; et al. AF Ablation Guided by Spatiotemporal Electrogram Dispersion without Pulmonary Vein Isolation: A Wholly Patient-Tailored Approach. *J. Am. Coll. Cardiol.* **2017**, *69*, 303–321. [CrossRef]
187. Qu, Z.; Garfinkel, A.; Chen, P.S.; Weiss, J.N. Mechanisms of discordant alternans and induction of reentry in simulated cardiac tissue. *Circulation* **2000**, *102*, 1664–1670. [CrossRef]
188. Colman, M.A.; Aslanidi, O.V.; Kharche, S.; Boyett, M.R.; Garratt, C.; Hancox, J.C.; Zhang, H. Pro-arrhythmogenic effects of atrial fibrillation-induced electrical remodelling: Insights from the three-dimensional virtual human atria. *J. Physiol.* **2013**, *591*, 4249–4272. [CrossRef]
189. Gaborit, N.; Le Bouter, S.; Szuts, V.; Varro, A.; Escande, D.; Nattel, S.; Demolombe, S. Regional and tissue specific transcript signatures of ion channel genes in the non-diseased human heart. *J. Physiol.* **2007**, *582*, 675–693. [CrossRef]
190. Voigt, N.; Trausch, A.; Knaut, M.; Matschke, K.; Varró, A.; Van Wagoner, D.R.; Nattel, S.; Ravens, U.; Dobrev, D. Left-to-right atrial inward rectifier potassium current gradients in patients with paroxysmal versus chronic atrial fibrillation. *Circ. Arrhythm. Electrophysiol.* **2010**, *3*, 472–480. [CrossRef]
191. Zhang, H.; Garratt, C.J.; Zhu, J.; Holden, A.V. Role of up-regulation of IK1 in action potential shortening associated with atrial fibrillation in humans. *Cardiovasc. Res.* **2005**, *66*, 493–502. [CrossRef] [PubMed]
192. Berenfeld, O. The Major Role of I(K1) in Mechanisms of Rotor Drift in the Atria: A Computational Study. *Clin. Med. Insights Cardiol.* **2016**, *10*, 71–79. [CrossRef] [PubMed]
193. Calvo, C.J.; Deo, M.; Zlochiver, S.; Millet, J.; Berenfeld, O. Attraction of rotors to the pulmonary veins in paroxysmal atrial fibrillation: A modeling study. *Biophys. J.* **2014**, *106*, 1811–1821. [CrossRef] [PubMed]
194. Honarbakhsh, S.; Schilling, R.J.; Orini, M.; Providencia, R.; Keating, E.; Finlay, M.; Sporton, S.; Chow, A.; Earley, M.J.; Lambiase, P.D.; et al. Structural remodeling and conduction velocity dynamics in the human left atrium: Relationship with reentrant mechanisms sustaining atrial fibrillation. *Heart Rhythm* **2019**, *16*, 18–25. [CrossRef]
195. Qu, Z.; Weiss, J.N. Cardiac Alternans: From Bedside to Bench and Back. *Circ. Res.* **2023**, *132*, 127–149. [CrossRef]
196. Al Abed, A.; Lovell, N.H.; Dokos, S. Local Heterogeneous Electrical Restitution Properties of Rabbit Atria. *J. Cardiovasc. Electrophysiol.* **2016**, *27*, 743–753. [CrossRef]

197. Kim, B.S.; Kim, Y.H.; Hwang, G.S.; Pak, H.N.; Lee, S.C.; Shim, W.J.; Oh, D.J.; Ro, Y.M. Action potential duration restitution kinetics in human atrial fibrillation. *J. Am. Coll. Cardiol.* **2002**, *39*, 1329–1336. [CrossRef]
198. Narayan, S.M.; Franz, M.R.; Clopton, P.; Pruvot, E.J.; Krummen, D.E. Repolarization alternans reveals vulnerability to human atrial fibrillation. *Circulation* **2011**, *123*, 2922–2930. [CrossRef]
199. Hiromoto, K.; Shimizu, H.; Furukawa, Y.; Kanemori, T.; Mine, T.; Masuyama, T.; Ohyanagi, M. Discordant repolarization alternans-induced atrial fibrillation is suppressed by verapamil. *Circ. J.* **2005**, *69*, 1368–1373. [CrossRef]
200. Liu, T.; Xiong, F.; Qi, X.Y.; Xiao, J.; Villeneuve, L.; Abu-Taha, I.; Dobrev, D.; Huang, C.; Nattel, S. Altered calcium handling produces reentry-promoting action potential alternans in atrial fibrillation-remodeled hearts. *JCI Insight* **2020**, *5*, e133754. [CrossRef]
201. Maesen, B.; Zeemering, S.; Afonso, C.; Eckstein, J.; Burton, R.A.; van Hunnik, A.; Stuckey, D.J.; Tyler, D.; Maessen, J.; Grau, V.; et al. Rearrangement of atrial bundle architecture and consequent changes in anisotropy of conduction constitute the 3-dimensional substrate for atrial fibrillation. *Circ. Arrhythm. Electrophysiol.* **2013**, *6*, 967–975. [CrossRef] [PubMed]
202. Frontera, A.; Pagani, S.; Limite, L.R.; Peirone, A.; Fioravanti, F.; Enache, B.; Cuellar Silva, J.; Vlachos, K.; Meyer, C.; Montesano, G.; et al. Slow Conduction Corridors and Pivot Sites Characterize the Electrical Remodeling in Atrial Fibrillation. *JACC Clin. Electrophysiol.* **2022**, *8*, 561–577. [CrossRef] [PubMed]
203. Ueda, N.; Yamamoto, M.; Honjo, H.; Kodama, I.; Kamiya, K. The role of gap junctions in stretch-induced atrial fibrillation. *Cardiovasc. Res.* **2014**, *104*, 364–370. [CrossRef] [PubMed]
204. Koura, T.; Hara, M.; Takeuchi, S.; Ota, K.; Okada, Y.; Miyoshi, S.; Watanabe, A.; Shiraiwa, K.; Mitamura, H.; Kodama, I.; et al. Anisotropic conduction properties in canine atria analyzed by high-resolution optical mapping: Preferential direction of conduction block changes from longitudinal to transverse with increasing age. *Circulation* **2002**, *105*, 2092–2098. [CrossRef] [PubMed]
205. Konings, K.T.; Smeets, J.L.; Penn, O.C.; Wellens, H.J.; Allessie, M.A. Configuration of unipolar atrial electrograms during electrically induced atrial fibrillation in humans. *Circulation* **1997**, *95*, 1231–1241. [CrossRef]
206. Nademanee, K.; McKenzie, J.; Kosar, E.; Schwab, M.; Sunsaneewitayakul, B.; Vasavakul, T.; Khunnawat, C.; Ngarmukos, T. A new approach for catheter ablation of atrial fibrillation: Mapping of the electrophysiologic substrate. *J. Am. Coll. Cardiol.* **2004**, *43*, 2044–2053. [CrossRef]
207. Wu, S.H.; Jiang, W.F.; Gu, J.; Zhao, L.; Wang, Y.L.; Liu, Y.G.; Zhou, L.; Gu, J.N.; Xu, K.; Liu, X. Benefits and risks of additional ablation of complex fractionated atrial electrograms for patients with atrial fibrillation: A systematic review and meta-analysis. *Int. J. Cardiol.* **2013**, *169*, 35–43. [CrossRef]
208. de Groot, N.; van der Does, L.; Yaksh, A.; Lanters, E.; Teuwen, C.; Knops, P.; van de Woestijne, P.; Bekkers, J.; Kik, C.; Bogers, A.; et al. Direct Proof of Endo-Epicardial Asynchrony of the Atrial Wall During Atrial Fibrillation in Humans. *Circ. Arrhythm. Electrophysiol.* **2016**, *9*, e003648. [CrossRef]
209. Kharbanda, R.K.; Kik, C.; Knops, P.; Bogers, A.; de Groot, N.M.S. First Evidence of Endo-Epicardial Asynchrony of the Left Atrial Wall in Humans. *JACC Case Rep.* **2020**, *2*, 745–749. [CrossRef]
210. Iwamiya, S.; Ihara, K.; Furukawa, T.; Sasano, T. Sacubitril/valsartan attenuates atrial conduction disturbance and electrophysiological heterogeneity with ameliorating fibrosis in mice. *Front. Cardiovasc. Med.* **2024**, *11*, 1341601. [CrossRef]
211. Eijsbouts, S.C.; Majidi, M.; van Zandvoort, M.; Allessie, M.A. Effects of acute atrial dilation on heterogeneity in conduction in the isolated rabbit heart. *J. Cardiovasc. Electrophysiol.* **2003**, *14*, 269–278. [CrossRef] [PubMed]
212. Bian, W.; Tung, L. Structure-related initiation of reentry by rapid pacing in monolayers of cardiac cells. *Circ. Res.* **2006**, *98*, e29–e38. [CrossRef] [PubMed]
213. Lalani, G.G.; Schricker, A.; Gibson, M.; Rostamian, A.; Krummen, D.E.; Narayan, S.M. Atrial conduction slows immediately before the onset of human atrial fibrillation: A bi-atrial contact mapping study of transitions to atrial fibrillation. *J. Am. Coll. Cardiol.* **2012**, *59*, 595–606. [CrossRef] [PubMed]
214. Schricker, A.A.; Lalani, G.G.; Krummen, D.E.; Rappel, W.J.; Narayan, S.M. Human atrial fibrillation initiates via organized rather than disorganized mechanisms. *Circ. Arrhythm. Electrophysiol.* **2014**, *7*, 816–824. [CrossRef] [PubMed]
215. Duytschaever, M.; Danse, P.; Eysbouts, S.; Allessie, M. Is there an optimal pacing site to prevent atrial fibrillation?: An experimental study in the chronically instrumented goat. *J. Cardiovasc. Electrophysiol.* **2002**, *13*, 1264–1271. [CrossRef] [PubMed]
216. van Schie, M.S.; Ramdat Misier, N.L.; Razavi Ebrahimi, P.; Heida, A.; Kharbanda, R.K.; Taverne, Y.; de Groot, N.M.S. Premature atrial contractions promote local directional heterogeneities in conduction velocity vectors. *Europace* **2023**, *25*, 1162–1171. [CrossRef]
217. Kumagai, K.; Tojo, H.; Noguchi, H.; Yasuda, T.; Ogawa, M.; Nakashima, H.; Zhang, B.; Saku, K. Effects of the NA+ channel blocker pilsicainide on the electrophysiologic properties of pulmonary veins in patients with atrial fibrillation. *J. Cardiovasc. Electrophysiol.* **2004**, *15*, 1396–1401. [CrossRef]
218. Spach, M.S.; Dolber, P.C.; Heidlage, J.F. Influence of the passive anisotropic properties on directional differences in propagation following modification of the sodium conductance in human atrial muscle. A model of reentry based on anisotropic discontinuous propagation. *Circ. Res.* **1988**, *62*, 811–832. [CrossRef]
219. Spach, M.S.; Dolber, P.C.; Heidlage, J.F. Interaction of inhomogeneities of repolarization with anisotropic propagation in dog atria. A mechanism for both preventing and initiating reentry. *Circ. Res.* **1989**, *65*, 1612–1631. [CrossRef]
220. Spach, M.S.; Boineau, J.P. Microfibrosis produces electrical load variations due to loss of side-to-side cell connections: A major mechanism of structural heart disease arrhythmias. *Pacing Clin. Electrophysiol.* **1997**, *20*, 397–413. [CrossRef]

221. Jaïs, P.; Hsu, L.F.; Rotter, M.; Sanders, P.; Takahashi, Y.; Rostock, T.; Sacher, F.; Hocini, M.; Clémenty, J.; Haïssaguerre, M. Mitral isthmus ablation for atrial fibrillation. *J. Cardiovasc. Electrophysiol* **2005**, *16*, 1157–1159. [CrossRef] [PubMed]
222. Hocini, M.; Jaïs, P.; Sanders, P.; Takahashi, Y.; Rotter, M.; Rostock, T.; Hsu, L.F.; Sacher, F.; Reuter, S.; Clémenty, J.; et al. Techniques, evaluation, and consequences of linear block at the left atrial roof in paroxysmal atrial fibrillation: A prospective randomized study. *Circulation* **2005**, *112*, 3688–3696. [CrossRef] [PubMed]
223. Narayan, S.M.; Baykaner, T.; Clopton, P.; Schricker, A.; Lalani, G.G.; Krummen, D.E.; Shivkumar, K.; Miller, J.M. Ablation of rotor and focal sources reduces late recurrence of atrial fibrillation compared with trigger ablation alone: Extended follow-up of the CONFIRM trial (Conventional Ablation for Atrial Fibrillation with or without Focal Impulse and Rotor Modulation). *J. Am. Coll. Cardiol.* **2014**, *63*, 1761–1768. [CrossRef] [PubMed]
224. Fareh, S.; Bénardeau, A.; Thibault, B.; Nattel, S. The T-type Ca^{2+} channel blocker mibefradil prevents the development of a substrate for atrial fibrillation by tachycardia-induced atrial remodeling in dogs. *Circulation* **1999**, *100*, 2191–2197. [CrossRef]
225. Li, D.; Shinagawa, K.; Pang, L.; Leung, T.K.; Cardin, S.; Wang, Z.; Nattel, S. Effects of angiotensin-converting enzyme inhibition on the development of the atrial fibrillation substrate in dogs with ventricular tachypacing-induced congestive heart failure. *Circulation* **2001**, *104*, 2608–2614. [CrossRef]
226. Liao, J.; Ebrahimi, R.; Ling, Z.; Meyer, C.; Martinek, M.; Sommer, P.; Futyma, P.; Di Vece, D.; Schratter, A.; Acou, W.J.; et al. Effect of SGLT-2 inhibitors on arrhythmia events: Insight from an updated secondary analysis of >80,000 patients (the SGLT2i-Arrhythmias and Sudden Cardiac Death). *Cardiovasc. Diabetol.* **2024**, *23*, 78. [CrossRef]

International Journal of *Molecular Sciences*

MDPI

Article

Enhanced Late I_{Na} Induces Intracellular Ion Disturbances and Automatic Activity in the Guinea Pig Pulmonary Vein Cardiomyocytes

Taro Saito, Mahiru Suzuki, Aiko Ohba, Shogo Hamaguchi, Iyuki Namekata * and Hikaru Tanaka

Department of Pharmacology, Faculty of Pharmaceutical Sciences, Toho University, 2-2-1 Miyama Funabashi, Chiba 274-8510, Japan; 3022005s@st.toho-u.jp (T.S.); mahiru-s@iuhw.ac.jp (M.S.); obaaiko2@gmail.com (A.O.); shogo.hamaguchi@phar.toho-u.ac.jp (S.H.); htanaka@phar.toho-u.ac.jp (H.T.)

* Correspondence: iyuki@phar.toho-u.ac.jp; Tel.: +81-474722088

Abstract: The effects of enhanced late I_{Na}, a persistent component of the Na^+ channel current, on the intracellular ion dynamics and the automaticity of the pulmonary vein cardiomyocytes were studied with fluorescent microscopy. Anemonia viridis toxin II (ATX- II), an enhancer of late I_{Na}, caused increases in the basal Na^+ and Ca^{2+} concentrations, increases in the number of Ca^{2+} sparks and Ca^{2+} waves, and the generation of repetitive Ca^{2+} transients. These phenomena were inhibited by eleclazine, a blocker of the late I_{Na}; SEA0400, an inhibitor of the Na^+/Ca^{2+} exchanger (NCX); H89, a protein kinase A (PKA) inhibitor; and KN-93, a Ca^{2+}/calmodulin-dependent protein kinase II (CaMKII) inhibitor. These results suggest that enhancement of late I_{Na} in the pulmonary vein cardiomyocytes causes disturbance of the intracellular ion environment through activation of the NCX and Ca^{2+}-dependent enzymes. Such mechanisms are probably involved in the ectopic electrical activity of the pulmonary vein myocardium.

Keywords: pulmonary vein cardiomyocytes; automaticity; intracellular Na^+; Ca^{2+} overload; late I_{Na}; Na^+/Ca^{2+} exchanger

check for updates

Citation: Saito, T.; Suzuki, M.; Ohba, A.; Hamaguchi, S.; Namekata, I.; Tanaka, H. Enhanced Late I_{Na} Induces Intracellular Ion Disturbances and Automatic Activity in the Guinea Pig Pulmonary Vein Cardiomyocytes. *Int. J. Mol. Sci.* **2024**, *25*, 8688. https://doi.org/10.3390/ijms25168688

Academic Editors: Yosuke Okamoto and Kunichika Tsumoto

Received: 30 June 2024
Revised: 30 July 2024
Accepted: 7 August 2024
Published: 9 August 2024

1. Introduction

The present study was undertaken to clarify the intracellular mechanisms for the induction of automaticity in the pulmonary vein myocardium by the persistent component of the voltage-dependent Na^+ channel current. The pulmonary vein is the blood vessel for the blood flow from the lung to the heart. Spontaneous pulsation of the pulmonary vein independent of the cardiac pacemaker has been repeatedly observed in various animal species [1–3]. This is due to a myocardial layer in the pulmonary vein wall connected to the left atrial myocardium capable of generating spontaneous or triggered action potentials. Since the clinical report that the automatic activity of the pulmonary vein myocardium plays a central role in the generation and maintenance of atrial fibrillation [4,5], the automaticity of the pulmonary vein myocardium has received attention as a therapeutic target and has been studied in various animal species [1,2,6,7]. The cardiomyocytes of the pulmonary vein myocardium have a smaller inward-rectifying K^+ channel current density and less repolarizing power than the atrial myocardium [8–10], which allows otherwise masked depolarizing currents to induce a gradual depolarization of the membrane. The depolarizing membrane currents contributing to the diastolic depolarization of the pulmonary vein myocardium include the Na^+/Ca^{2+} exchanger current [11,12], hyperpolarization-activated current [7,10], and the voltage-dependent Na^+ channel current [13,14], which is the main subject of the present study.

The voltage-dependent Na^+ channel current (I_{Na}) is composed of two components. The transient component, which is referred to as peak I_{Na}, is activated by rapid depolarizations from the resting membrane potential range and rapidly inactivates within a few milliseconds. Peak I_{Na} has a large current density and is responsible for the rapid upstroke and conduction of the action potential in the working myocardium [15]. After the inactivation of the peak I_{Na}, a small fraction of the I_{Na} flows persistently [16]; such current is referred to as late I_{Na}, persistent I_{Na}, or sustained I_{Na} [17]. Such a persistent Na^+ current is also observed during a slow depolarization in the negative voltage range and has been suggested to contribute to abnormal automaticity underlying cardiac arrhythmia including atrial fibrillation [18,19]. Although the amplitude of late I_{Na} is small in the normal myocardium, it may be enhanced under various acquired and congenital conditions related to myocardial automaticity such as heart failure, ischemia, and arrhythmia including atrial fibrillation [20–22]. In the pulmonary vein myocardium, whose resting or maximum diastolic potential is about −70 mV, less negative than the −80 to −90 mV of the working myocardium [8,9], the late I_{Na} can flow depending on the situation. It was previously reported in the pulmonary vein myocardium that anemonia viridis toxin II (ATX-II), an enhancer of late I_{Na}, increased the pacemaker depolarization slope, as well as the automatic firing frequency [13,23]. Agents with inhibitory effects on late I_{Na}, such as GS-458967, reduced the pacemaker depolarization slope and automatic activity in the pulmonary vein myocardium [13]. Thus, blockade of late I_{Na} appears to be an effective pharmacological strategy for the treatment of atrial fibrillation.

While late I_{Na} can directly contribute to depolarization as a depolarizing current, it may also cause disturbances in intracellular ion homeostasis and consequently lead to arrhythmic activity. An increase in intracellular Na^+ concentration through inhibition of the Na^+/K^+-ATPase by ouabain induced automatic electrical activity in the pulmonary vein myocardium, which was accompanied by an increase in the number of Ca^{2+} sparks and Ca^{2+} waves; these were both inhibited by pretreatment with a selective NCX inhibitor, SEA0400 [11]. Thus, a probable explanation for the induction of electrical activity by an increase in intracellular Na^+ concentration in the pulmonary vein myocardium is the induction of Ca^{2+} overload through activation of Ca^{2+} influx through NCX [11]. Ca^{2+} overload is reported to cause hyperactivation of Ca^{2+}-dependent enzymes such as Ca^{2+}/calmodulin-dependent protein kinase II (CaMKII) and protein kinase A (PKA), and leads to arrhythmias in ventricular muscle [17,24–26]. If an enhancement of the late I_{Na} can induce automatic activity in the pulmonary vein myocardium through similar mechanisms, their inhibitors would be promising as drugs for the treatment of atrial fibrillation caused by pulmonary vein automaticity.

In the present study, fluorescent ion measurements and pharmacological studies were conducted with isolated pulmonary vein cardiomyocytes to clarify the following points. Firstly, does enhancement of the late I_{Na} in isolated pulmonary vein myocytes induce automatic electrical activity? Secondly, does the enhancement of the late I_{Na} cause intracellular Ca^{2+} overload through NCX function? Finally, are Ca^{2+}-dependent enzymes involved in the induction of automatic activity? Given the positive results obtained, their implication for drug selection and development was discussed.

2. Results

2.1. The Spontaneous and Evoked Ca^{2+} Transients of Isolated Pulmonary Vein Cardiomyocytes

Pulmonary vein cardiomyocytes isolated from guinea pigs were spindle shaped and had transverse striations. (Figure 1A). They were loaded with the Ca^{2+}-sensitive fluorescent probe, Fluo-4, and observed with a confocal microscope. Among the 787 isolated guinea pig pulmonary vein cardiomyocytes examined, 220 cells showed spontaneous Ca^{2+} transients, which is a rapid rise in the Ca^{2+} concentration in the entire subsarcolemmal cytoplasm (Figure 1B); the average firing frequency of the Ca^{2+} transient was 1.85 ± 0.13 times in 5 s. During the early phase of the Ca^{2+} transient, fluorescence intensity increased from the subsarcolemmal region and then spread to the cell interior (Figure 1B). In the quiescent pulmonary vein cardiomyocytes, action potentials were elicited by field electrical stimulation, and the resulting Ca^{2+} transients were observed; the rise in Ca^{2+} concentration started from the subsarcolemmal region and spread to the central region of the cell (Figure 1C).

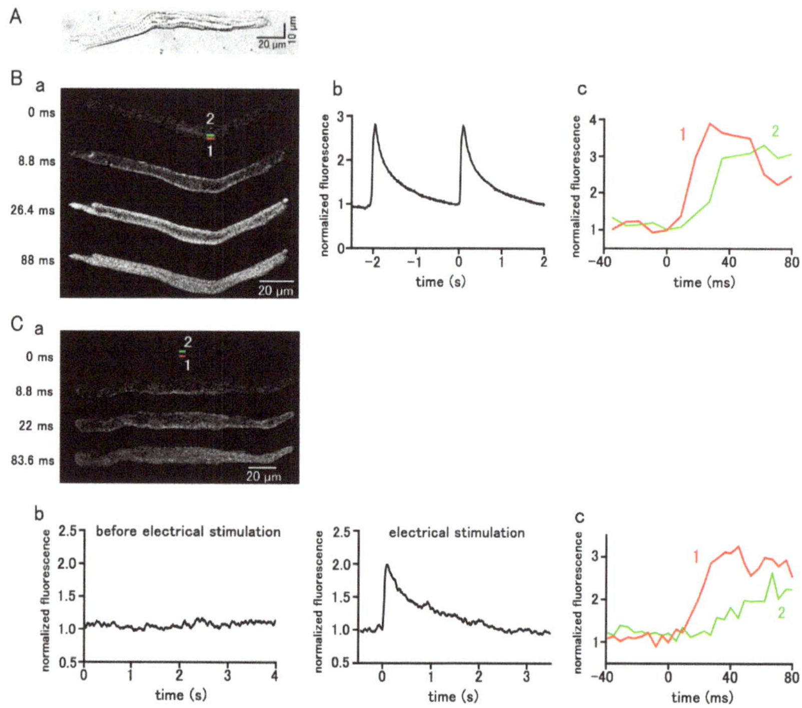

Figure 1. Spontaneous and stimulation-evoked Ca^{2+} transients in pulmonary vein myocytes. (**A**) A typical differential interference contrast (DIC) image. (**B**,**C**) Spontaneous (**B**) and electrical-stimulation-evoked (**C**) Ca^{2+} transients. Typical x−y images of a cell loaded with Fluo−4/AM (**B**(**a**),**C**(**a**)). Time course of fluorescence intensity of the whole cell (**B**(**b**),**C**(**b**)) and quantified in rectangular (1 × 4 μm) regions located at 0–1 μm (1 red) and 4–5 μm (2 green) from the sarcolemma (**B**(**c**),**C**(**c**)), as shown in the top panel in (**B**(**a**),**C**(**a**)).

2.2. *Effect of Enhanced Late I_{Na} on Automatic Activity and Intracellular Ca^{2+} Dynamics*

Treatment of quiescent pulmonary vein myocytes with 100 nM ATX-II induced automatic Ca^{2+} transients. Among the 40 isolated pulmonary vein cardiomyocytes examined, automatic Ca^{2+} transients were induced in 18 cells. The firing frequency of Ca^{2+} transients increased in a time-dependent manner (Figure 2); the average frequency was 1.25 ± 0.25 times in 5 s at 3 min and 2.25 ± 0.25 times in 5 s at 5 min. Regarding the automatic Ca^{2+} transient induced by ATX-II, the rise in the Ca^{2+} concentration first occurred at the subsarcolemmal region and spread to the central region of the cell, similarly to electrical-stimulation-evoked Ca^{2+} transients. Inhibition of late I_{Na} by pretreatment with 10 μM eleclazine significantly reduced the incidence of Ca^{2+} transients induced by ATX-II (Figure 2). Such induction of automatic activity by ATX-II was less frequently observed in atrial cardiomyocytes. Among the 56 isolated atrial cardiomyocytes examined, automatic Ca^{2+} transients were induced only in seven cells.

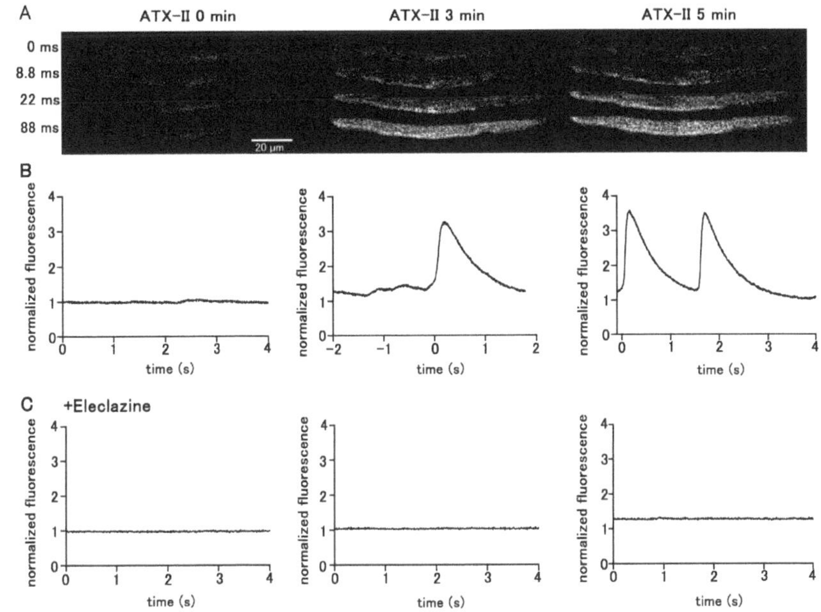

Figure 2. Induction of Ca^{2+} transients by ATX-II in quiescent pulmonary vein cardiomyocytes. Typical x−y images (**A**) and time course (**B**) of a cell loaded with Fluo−4 before ATX-II treatment and at 3 and 5 min after treatment with 100 nM ATX-II. Pretreatment of eleclazine (10 μM) inhibited the induction of automatic activity (**C**).

The basal Ca^{2+} fluorescence intensity of the cells was increased by ATX-II in a time-dependent manner, regardless of the induction of automatic Ca^{2+} transients (Figure 3). Pretreatment of the cells with 10 μM eleclazine for 5 min significantly suppressed the ATX-II-induced increase in basal Ca^{2+} concentration (Figure 3).

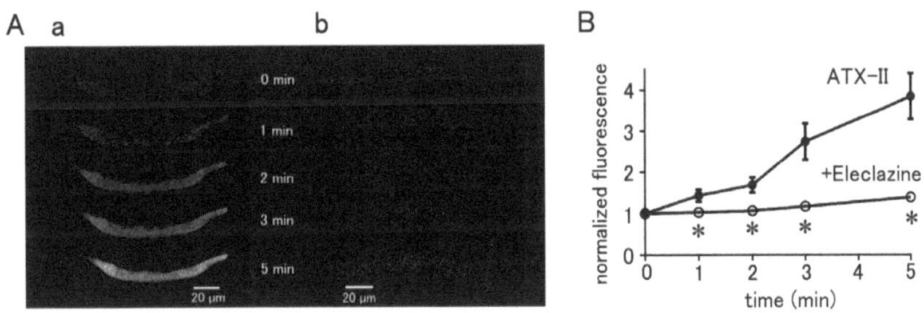

Figure 3. ATX II induced changes in intracellular Ca^{2+} concentration. (**A**) Typical x−y images of pulmonary vein cardiomyocytes loaded with Fluo−4 after treatment with 100 nM ATX-II alone (**a**) and in the presence of 10 μM eleclazine (**b**). (**B**) Time course of the changes in basal Ca^{2+} concentration induced by 100 nM ATX-II alone (closed circles) and in the presence of 10 μM eleclazine (open circles). Each point with vertical bars represents the mean ± S.E.M of 30–40 cells. Asterisks indicate significant differences ($p < 0.05$) from the corresponding values in ATX-II-treated cells.

2.3. Effect of Enhanced Late I_{Na} on Intracellular Na^+ Concentration

To measure the intracellular Na^+ concentration, the pulmonary vein cardiomyocytes were loaded with SBFI/AM and observed by epifluorescence microscopy. SBFI was excited at 340 nm when bound to Na^+ and at 380 nm when unbound, and the ratio of their fluorescence (F_{340}/F_{380}) can be taken to quantitatively measure Na^+ concentration. Fluorescence excited at either wavelength was found to be homogeneously distributed in the cell (Figure 4A). The intracellular Na^+ concentration could be quantified using a calibration curve based on a staircase graph of F_{340}/F_{380} obtained by adjusting the Na^+ concentrations in the extracellular fluid to 0, 5, 10, 15, and 20 mM (Figure 4B). The intracellular Na^+ concentration in quiescent pulmonary vein cardiomyocytes was 11.19 ± 0.16 mM (n = 23). Intracellular Na^+ concentration was increased by 3.91 ± 0.7 mM in 5 min by treatment with 100 nM ATX-II (n = 7; Figure 4C,D). This was significantly higher than the 0.73 ± 0.1 mM increase in Na^+ concentration seen in the time control (n = 7). Pretreatment with 10 μM eleclazine significantly suppressed the ATX-II-induced increase in Na^+ concentration; the ATX-II-induced increase in Na^+ concentration was 0.06 ± 0.2 mM (n = 7).

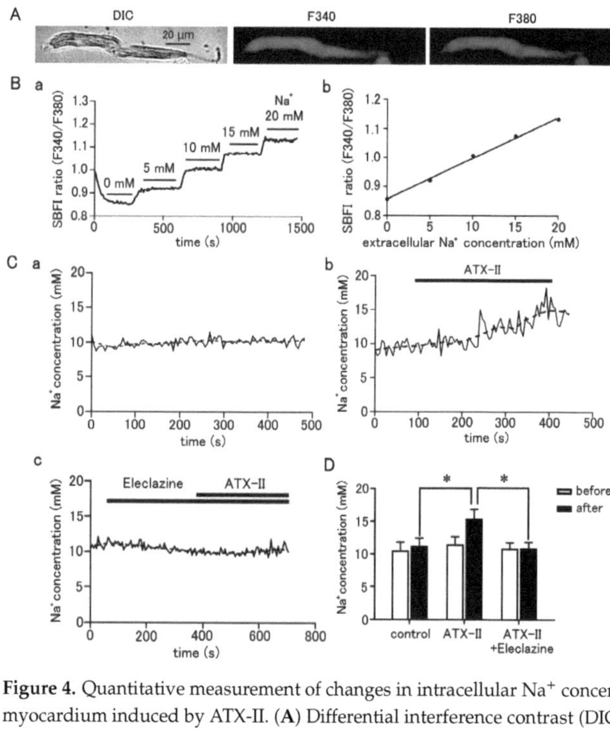

Figure 4. Quantitative measurement of changes in intracellular Na^+ concentration in pulmonary vein myocardium induced by ATX-II. (**A**) Differential interference contrast (DIC) images and fluorescence images excited at 340 nm and 380 nm in pulmonary vein cardiomyocytes loaded with SBFI. (**B**) Determination of intracellular Na^+ concentration in pulmonary vein cardiomyocytes. After measuring the Na^+ fluorescence ratio, calibration of the fluorescence ratio to Na^+ concentration was performed as described in the methods section (**B**(**a**)); the calibration curve was prepared by plotting the fluorescence intensity ratio of SBFI obtained in (**B**(**a**)) for each extracellular Na^+ concentration (**B**(**b**)). (**C**) Effect of 100 nM ATX-II on intracellular Na^+ concentration. Time course of Na^+ concentration for time control (**C**(**a**)), ATX-II (**C**(**b**)), and ATX-II in the presence of 10 μM eleclazine (**C**(**c**)). The solid line and the dashed line represent the Na^+ concentration obtained every 5 s and its 15-point moving average, respectively. (**D**) Intracellular Na^+ concentration at 5 min after ATX-II treatment. Each column with vertical bars represents the mean ± S.E.M. of 7 cells. Asterisks indicate significant differences ($p < 0.05$) from the corresponding values.

2.4. Contribution of NCX to Automatic Activity

Isolated pulmonary vein cardiomyocytes were loaded with SBFI and Fluo-4, and intracellular Na^+ and Ca^{2+} concentrations were simultaneously measured by epifluorescence microscopy. ATX-II (100 nM) caused increases in both Na^+ and Ca^{2+} concentrations, but it took 1.23 ± 0.2 min for the Na^+ concentration to reach half of its peak value and 3.46 ± 0.2 min for the Ca^{2+} concentration ($n = 5$; Figure 5A). Thus, ATX-II induced increases in intracellular Na^+ and Ca^{2+} concentrations sequentially in this order. Furthermore, pretreatment with 1 μM of SEA0400, a selective NCX inhibitor, significantly reduced the ATX-II-induced increase in basal Ca^{2+} concentration (Figure 5B).

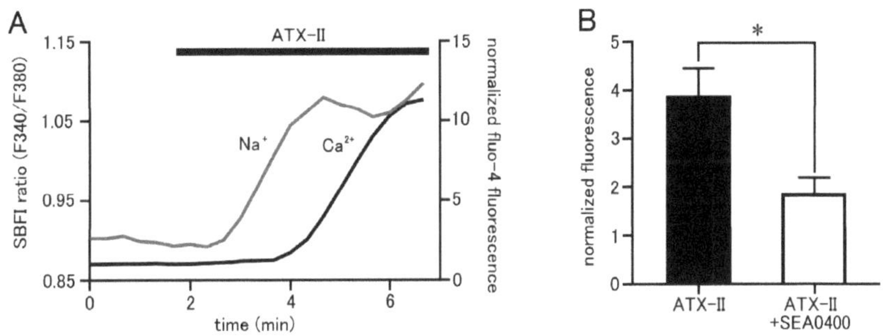

Figure 5. ATX-II-induced changes in intracellular ion concentrations and the involvement of NCX. (**A**) Time course of respective fluorescence intensity in cells loaded with SBFI and Fluo−4. The gray line shows the fluorescence intensity ratio of SBFI, and the black line shows the fluorescence intensity of Fluo-4. Note that ATX-II (100 nM) induced an increase in the fluorescence intensity ratio of SBFI, followed by an increase in Fluo-4 fluorescence intensity. (**B**) Effect of SEA0400 (1 μM) on the elevation of basal Ca^{2+} concentration induced by ATX-II. Each column with vertical bars represents the mean ± S.E.M. of 30–40 cells. Asterisks indicate significant differences ($p < 0.05$) from the corresponding values in ATX-II-treated cells.

2.5. ATX-II-Induced Ca^{2+} Oscillation and Involvement of Ca^{2+}-Dependent Enzymes

The effect of ATX-II on intracellular Ca^{2+} dynamics appeared in the form of an increase in basal Ca^{2+} concentration, Ca^{2+} sparks, and Ca^{2+} waves as well as Ca^{2+} transients (Figure 6A). These Ca^{2+} oscillations often occurred after 3 min of ATX-II treatment. The increase in basal Ca^{2+} concentration by 100 nM ATX-II occurred in all cells examined; it was suppressed by pretreatment with 5 μM H89 ($n = 30$), a PKA inhibitor, and 3 μM KN-93 ($n = 30$), a CaMKII inhibitor (Figure 6B). Such suppression was not observed with 3 μM KN-92 ($n = 20$), an inert analog of KN-93. The percentage of cells in which these Ca^{2+} oscillations were induced by ATX-II was significantly higher than the time control, and Ca^{2+} transients were observed in approximately half of the cells (Figure 6C). Pretreatment with eleclazine, SEA0400, H89, and KN-93 reduced the incidence of Ca^{2+} oscillations induced by ATX-II (Figure 6C).

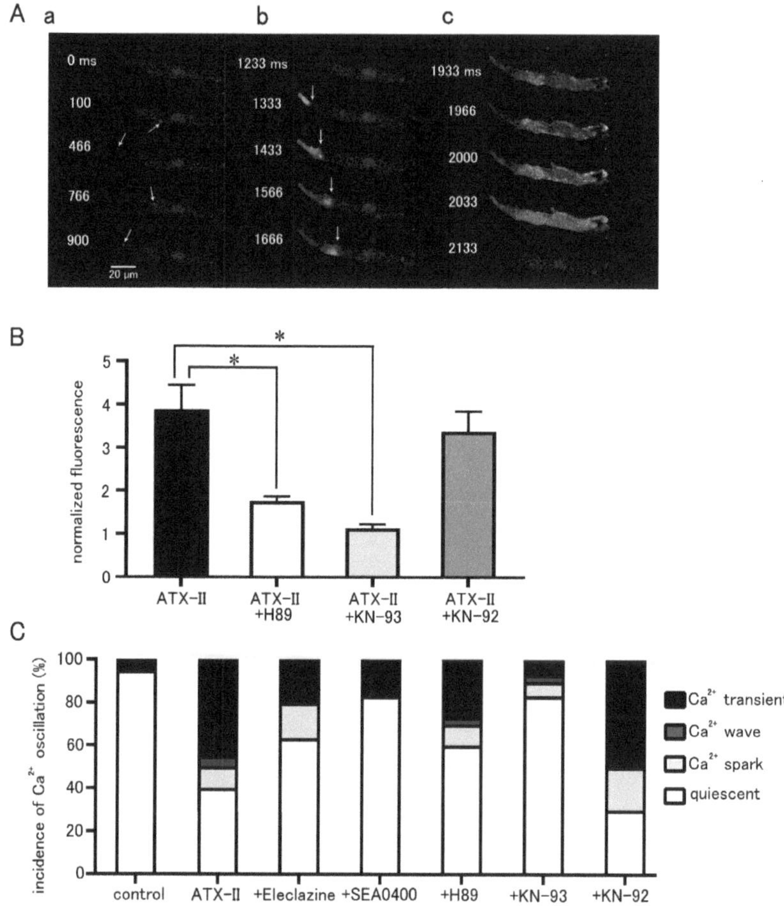

Figure 6. Induction of intracellular Ca^{2+} oscillation by ATX-II and the effects of pharmacological agents. (**A**) Typical x−y images of a Fluo-4-loaded pulmonary vein cardiomyocyte after ATX-II treatment (100 nM) showing Ca^{2+} sparks (arrows; (**a**)), a Ca^{2+} wave (arrows indicate wavefront of Ca^{2+} wave; (**b**)), and a Ca^{2+} transient (**c**). The images in ((**a**–**c**)) were obtained in this order in the same cell. (**B**) The effects of H89 (5 µM), KN-93 (3 µM), and KN-92 (3 µM) on the ATX-II-induced increase in basal Ca^{2+} concentration. Asterisks indicate significant differences ($p < 0.05$) from the corresponding control values. (**C**) The effects of eleclazine (10 µM), H89 (5 µM), KN-93 (3 µM), and KN-92 (3 µM) on the incidence of ATX-II-induced Ca^{2+} oscillations. Each column with vertical bars in (**B**,**C**) represents the mean ± S.E.M of 20–40 cells.

3. Discussion

3.1. ATX-II-Induced Changes in Intracellular Ca^{2+} Concentration

The present study was undertaken to examine whether a pharmacological enhancement of the late I_{Na} can induce automatic activity in the pulmonary vein cardiomyocytes and to clarify the intracellular mechanisms involved. We first confirmed the relationship between action potential generation and the Ca^{2+} transient, a temporal increase in Ca^{2+} concentration throughout the cell. Both in the spontaneous Ca^{2+} transients and action-potential evoked Ca^{2+} transients, the rise in Ca^{2+} concentration occurred initially in the subsarcolemmal region and then spread to the cell center (Figure 1). This is a characteristic feature of Ca^{2+} transients in cardiomyocytes lacking T-tubules, including the atrial and pulmonary vein cardiomyocytes [27]. In cells lacking T-tubules, the trans-sarcolemmal Ca^{2+} influx elicited by an action potential first triggers Ca^{2+} release from the Ca^{2+} release channel (ryanodine receptor) in the sarcoplasmic reticulum (SR) in the subsarcolemmal region [27–29]. Next, in the central region lacking T-tubules, a propagated Ca^{2+}-induced Ca^{2+} release mechanism in the SR activates the Ca^{2+} release from SR [28,29]. This pattern of rise in Ca^{2+} indicates the firing of action potentials. The same pattern was observed in ATX-II-induced Ca^{2+} transients (Figure 2), indicating that they were accompanied by action potentials. A characteristic feature of ATX-II-induced Ca^{2+} transients is that they were preceded by a gradual rise in basal Ca^{2+} concentration (Figure 3). This suggests that the rise in Ca^{2+} concentration is a cause of action potential generation. The rise in basal Ca^{2+} concentration, as well as the Ca^{2+} transient, was inhibited by eleclazine, confirming that both phenomena were caused by enhancement of late I_{Na} (Figures 2 and 3).

3.2. NCX Is Involved in ATX-II-Induced Increases in Intracellular Na^+ and Ca^{2+} Concentrations

ATX-II caused an increase in intracellular Na^+ concentration in quiescent pulmonary vein cardiomyocytes, which was inhibited by eleclazine (Figure 4). These results also indicated that late I_{Na} could be activated at negative membrane potentials near the resting potential. This is consistent with our previous results with voltage-clamped pulmonary vein cardiomyocytes that the late I_{Na} can flow at membrane potentials near the resting potential of about −70 mV [13,30]. On the other hand, in the ventricular cardiomyocytes, ATX-II increased the Na^+ concentration in electrically stimulated cardiomyocytes but had no effects on Na^+ concentration in quiescent cardiomyocytes [31]. Our results also showed that ATX-II tended not to induce automatic activity in atrial cardiomyocytes. This is probably because the resting membrane potential of the atrial and ventricular myocardium of about −80 to −90 mV is more negative than that of the pulmonary vein myocardium [8,9]; the late I_{Na} is unlikely to flow near the resting membrane potential in the working myocardium. The pulmonary vein myocardium has a smaller inward rectifying K^+ channel current density and less repolarizing power than the working myocardium [8–10]; this allows the membrane potential to stay in a less negative voltage range in which the late I_{Na} can flow.

Simultaneous measurement of intracellular Na^+ and Ca^{2+} concentrations revealed that the ATX-II-induced increase in intracellular Na^+ concentration was followed by an increase in Ca^{2+} concentration (Figure 5A); this order suggested the involvement of the reverse-mode NCX. The NCX has two modes of action: the forward mode, in which Na^+ is taken into the cell and Ca^{2+} is excreted, and the reverse mode, in which Na^+ is excreted from the cell and Ca^{2+} is taken into the cell. The forward mode is dominant in normal cardiomyocytes, but under conditions with elevated intracellular Na^+ concentration such as ischemia–reperfusion or cardiac glycoside treatment, the NCX can function in the reverse mode and cause an increase in intracellular Ca^{2+} concentration [32]. In the present study, the treatment of quiescent cardiomyocytes with ATX-II increased the Na^+ concentration from 11 to 15 mM, and, as mentioned above, the resting membrane potential of the pulmonary vein myocardium is less negative than that of the working myocardium [8,9]; these are conditions which favor Ca^{2+} entry through the reverse-mode NCX. Some reports suggested that the reverse-mode NCX may be more dominant at intracellular Na^+ concentrations above 12 mM [33,34]. The result that SEA0400, a selective inhibitor of NCX, inhibited the increase in basal Ca^{2+} concentration by ATX-II (Figure 5B) also confirmed the involvement of NCX. Among the three known types of NCX, the expression of the SEA0400-sensitive NCX1 was reported in the pulmonary vein myocardium of rats [12].

3.3. ATX-II Ca^{2+} Overload Triggers Further Spontaneous Activity

ATX-II induced not only an increase in basal Ca^{2+} concentration but also Ca^{2+} oscillations such as Ca^{2+} sparks and Ca^{2+} waves, which are local non-propagating increases in Ca^{2+} concentration that occur independently of action potentials [28,29]. Ca^{2+} sparks, which are the unit of Ca^{2+} release from the SR Ca^{2+} release channel, occur in a small area of 1 to 2 μm in diameter, while on Ca^{2+} waves, the Ca^{2+} release propagates within the cell at a velocity of about 100 μm/s [27]. Confocal imaging in pulmonary vein tissue showed that Ca^{2+} waves, as well as Ca^{2+} sparks, do not propagate between connected cells, which confirms that they occur independently of action potential firing [35]. Ca^{2+} waves and Ca^{2+} sparks are considered to reflect spontaneous Ca^{2+} leaks from the sarcoplasmic reticulum which are enhanced in Ca^{2+} overload myocytes [36–38]. The increase in cytoplasmic Ca^{2+} concentration not only stimulates the SR Ca^{2+} release channel from the cytoplasmic side through the Ca^{2+}-induced Ca^{2+} release mechanism but also causes an increase in Ca^{2+} concentration in the lumen of the SR, which affects the Ca^{2+} release channel and increases the Ca^{2+} sparks and Ca^{2+} waves. The abnormal increase in cytoplasmic Ca^{2+} causes hyperactivation of voltage-dependent ion channels, Ca^{2+}-activated channels, and Ca^{2+}-driven transporters, which leads to spontaneous electrical activity and arrhythmias [39]. The induction of Ca^{2+} sparks, Ca^{2+} waves, and Ca^{2+} transients in the pulmonary vein myocardium was also observed with ouabain, which increases intracellular Na^+ concentration through inhibition of the Na^+/K^+ ATPase [11]. We also observed that the Ca^{2+} waves and Ca^{2+} transients were suppressed by carbachol, which hyperpolarized the resting membrane potential of the pulmonary vein myocardium to the same level as the working myocardium [8]. Thus, less repolarizing power of the pulmonary vein cardiomyocytes appears to play a permissive role in the generation of Ca^{2+} overload and the subsequent induction of automatic activity.

Intracellular Ca^{2+} overload in cardiomyocytes not only causes alterations in ion channels and transporters but also causes activation of various Ca^{2+}-dependent enzymes [40]. The ATX-II-induced increase in basal Ca^{2+} concentration and generation of Ca^{2+} oscillation was suppressed in the presence of H89, a PKA inhibitor, and KN-93, a CaMKII inhibitor (Figure 6). A similar suppression of automatic activity after inhibition of CaMKII and PKA has been reported in rabbit pulmonary vein myocardium [41]. The activation of adenylate cyclase (AC) and PKA was reported in cardiomyocytes from the rat pulmonary vein [42] and guinea pig atria [43]. CaMKII, a multifunctional serine/threonine protein kinase is activated by the Ca^{2+}/calmodulin complex when intracellular Ca^{2+} levels are increased [44,45]. CaMKII could phosphorylate the Na^+ channels and Ca^{2+} channels [46], which could cause membrane depolarization. The hyperphosphorylation of the respective target sites on ryanodine receptors and phospholamban by CaMKII and PKA triggered a significant increase in Ca^{2+} leaks from SR [43]. Thus, enhancement of late I_{Na} and the resulting intracellular Ca^{2+} overload causes hyperactivation of various Ca^{2+}-dependent ion channels, transporters, and enzymes, which in turn causes further accumulation of intracellular Ca^{2+} (Figure 7).

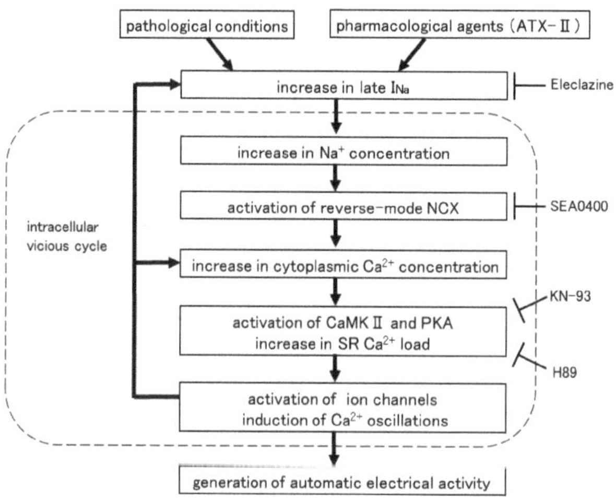

Figure 7. Intracellular mechanism for the induction of automatic activity by enhanced late I_{Na}. Arrows denote stimulation and T-shaped lines indicate inhibition.

3.4. Clinical Implications of This Study

The present study showed that induction of pulmonary vein automaticity by enhancement of the late I_{Na} is partly mediated by mechanisms including the NCX and enzymes such as CaMKII and PKA. This implies that inhibitors of these mechanisms may be promising as drugs for the treatment of atrial fibrillation caused by pulmonary vein automaticity. Ranolazine, which selectively blocks the late I_{Na}, was shown to be anti-arrhythmic in human patients in some situations such as post-operative atrial fibrillation [47]; the mechanisms of action still remain to be clarified because the drug also has effects on many other ion channels and the α-adrenoceptor. Agents with higher selectivity towards late I_{Na} such as GS-458967 [13] and NCC-3902 [30] have been shown to inhibit pulmonary vein automaticity and to be effective against animal models of atrial fibrillation but have not been applied to human patients at present. Some of the class I antiarrhythmic drugs were shown to have inhibitory effects on the late I_{Na} and reduce the diastolic depolarization slope of the isolated guinea pig pulmonary vein myocardium [48]; whether such effects contribute to the antiarrhythmic effect of these drugs in human patients is unknown at present. A novel inhibitor of the NCX, SAR296968, was reported to reduce the frequency of Ca^{2+} sparks in isolated human atrial cardiomyocytes [49]; whether such effects can lead to antiarrhythmic effects in vivo awaits further investigation because the effect of NCX inhibitors appears to largely vary depending on the experimental condition [50,51]. Some of the class I antiarrhythmic drugs were reported to have inhibitory effects on the NCX, but only at concentrations higher than the therapeutic blood concentration [52]. A CaMKII inhibitor, AS105, which is more potent and selective than KN-93, was reported to inhibit SR Ca^{2+} leaks in human atrial cardiomyocytes [53,54]. Whether the mechanisms clarified in this study are involved in the pulmonary vein automaticity of human patients and whether these inhibitors are effective in clinical practice await further investigation.

3.5. Conclusions

The present results suggested that enhancement of late I_{Na} in pulmonary vein cardiomyocytes causes Ca^{2+} overload through the reverse-mode NCX and induces automatic activity. Such mechanisms are particularly likely to occur in the pulmonary vein myocardium, whose repolarizing membrane current density is lower than the working myocardium.

4. Materials and Methods

4.1. General

All experiments were performed in compliance with the Guiding Principles for the Care and Use of Laboratory Animals of the Japanese Pharmacological Society and were approved by the Toho University Animal Care and User Committee (21-41-507, 7 May 2022).

4.2. Isolation of Pulmonary Vein Cardiomyocytes

Pulmonary vein cardiomyocytes were obtained according to the methods described in our previous study [13,27,30]. Male Hartley guinea pigs (weight, 300–450 g) were anesthetized with isoflurane, and the hearts with lungs were isolated and perfused via the aorta with a Tyrode's solution of the following composition (mM concentration): NaCl 143, KCl 4, $MgCl_2$ 0.5, $CaCl_2$ 1.8, NaH_2PO_4 0.33, glucose 5.5, and HEPES 5 (pH 7.4, gassed with 100% O_2, and warmed to 36 °C). The heart was further perfused successively with nominally Ca^{2+}-free Tyrode's solution and the same solution containing 0.1 mg/mL protease (type XIV; Sigma-Aldrich, St. Louis, MO, USA) and 0.5 mg/mL collagenase (YK-102; Yakult, Tokyo, Japan) for about 20 min. After the washout of the enzymes, the cardiomyocytes were isolated by gentle disruption of the pulmonary vein.

4.3. Measurement of Intracellular Ca^{2+} Dynamics

For the measurement of intracellular Ca^{2+} dynamics, the isolated pulmonary vein cardiomyocytes were loaded with 5 μM Fluo-4/AM (Dojindo, Kumamoto, Japan) for about 30 min and superfused with the Tyrode's solution described above at room temperature. The cardiomyocytes were observed with a rapid-scanning laser confocal microscope (A1R; Nikon Corporation, Tokyo, Japan) so that the Ca^{2+} sparks and Ca^{2+} waves could be detected. The objective lens was Apochromat ×40, 1.15 numerical aperture (water immersion). The excitation wavelength was 488 nm, and the emission in the wavelength range of 500 to 550 nm was detected. The data were analyzed with computer software, NIS elements (Nikon Corporation). The fluorescence intensity of Fluo-4 at each time point was normalized against the basal intensity. The scanning was performed every 4.4 ms at 1024 × 64 pixels or every 8.8 ms at 1024 × 128 pixels. For the triggering of action potentials, field electrical stimulation by a rectangle voltage pulse of 3 ms duration was applied to cells through a platinum wire electrode paired with an electric stimulator (SEN-3303; Nihon Kohden Corporation, Tokyo, Japan).

4.4. Measurement of Intracellular Na^+ Concentration

Isolated cardiomyocytes were loaded with SBFI (5 μM SBFI/AM) and 0.05% pluronic F-127 at 36 °C (Invitrogen, Carlsbad, CA, USA), and the total Na^+ fluorescence from single cardiomyocytes was measured with epifluorescence microscopy. The cells were excited at 340 and 380 nm from a xenon lamp, and the emission (>500 nm) was separated with a dichroic mirror, detected by a cooled CCD camera (C6790, Hamamatsu Photonics, Shizuoka, Japan) at a time resolution of 5 s, and ratioed after correction of background fluorescence (Aquacosmos software, version 2.52, Hamamatsu Photonics, Shizuoka, Japan). A cooled CCD camera was used because a rapid-scanning laser confocal microscope is not suitable for excitation using two wavelengths. In situ calibration of SBFI was performed by exposing the myocytes to various extracellular Na^+ solutions in the presence of 10 μM ouabain, 2 μM gramicidin, and 40 μM monensin. The Na^+ concentration was increased from 0 mM to 20 mM in 5 mM steps.

4.5. Simultaneous Measurement of Intracellular Ca^{2+} and Na^+ Concentrations

For the simultaneous measurement of intracellular Ca^{2+} and Na^+ concentration, the isolated pulmonary vein cardiomyocytes were loaded with Fluo-4/AM and SBFI/AM (see above). The cells were excited at 480 nm for Fluo-4 from an LED lamp (COLIBRI, Carl Zeiss, Oberkochen, Germany), and 340 and 380 nm for SBFI from a mercury lamp (HXP120V, Zeiss, Oberkochen, Germany), and the emission (>500 nm) was separated with a dichroic mirror, detected by a digital camera (AxioCam MRm, Carl Zeiss, Oberkochen, Germany) at a time resolution of 10 s, and ratioed after correction of background fluorescence (Zen Pro, Carl Zeiss).

4.6. Chemicals

ATX-II and H89 were purchased from Alomone Labs (Jerusalem, Israel); gramicidin, monensin, ouabain, and SEA0400 were purchased from Sigma-Aldrich (St. Louis, USA); eleclazine was purchased from Cosmo Bio (Tokyo, Japan); and KN-93 and KN-92 were purchased from Wako Pure Chemical Industries Ltd. (Osaka, Japan). ATX-II was dissolved in distilled water, and monensin was dissolved in ethanol. All other chemicals were dissolved in dimethyl sulfoxide.

4.7. Data Analytics and Statistics

All data were expressed as the mean ± standard error of the mean (S.E.M.). Statistical significance between means was evaluated by Welch's *t*-test, or one-way repeated-measures ANOVA followed by Tukey's or Dunnett's multiple comparisons. A *p-value* less than 0.05 was considered statistically significant.

Author Contributions: Conceptualization, I.N. and H.T.; methodology, I.N. and H.T.; software, S.H.; validation, T.S., M.S. and A.O.; formal analysis, T.S. and M.S.; investigation, T.S., M.S. and A.O.; resources, I.N. and H.T.; data curation, T.S.; writing—original draft preparation, T.S.; writing—review and editing, I.N. and H.T.; visualization, T.S.; supervision, H.T.; project administration, I.N. and H.T.; funding acquisition, I.N., S.H. and H.T. All authors have read and agreed to the published version of the manuscript.

Funding: This study was supported in part by JSPS KAKENHI, grant numbers JP20K16013, JP20K07299, and JP20K07091. T.S. received the Nagai Memorial Research Scholarship from the Pharmaceutical Society of Japan (N-231501).

Institutional Review Board Statement: The study was approved by the Ethics Committee of Toho University (21-55-362; 29 March 2021) and was conducted in accordance with the "Guiding Principles for the Care and Use of Laboratory Animals" approved by the Japanese Pharmacological Society.

Informed Consent Statement: Not applicable.

Data Availability Statement: Data are contained within the article.

Conflicts of Interest: The authors declare no conflicts of interest.

References

1. Nattel, S. Basic electrophysiology of the pulmonary veins and their role in atrial fibrillation: Precipitators, perpetuators, and perplexers. *J. Cardiovasc. Electrophysiol.* **2003**, *14*, 1372–1375. [CrossRef] [PubMed]
2. Namekata, I.; Tsuneoka, Y.; Tanaka, H. Electrophysiological and pharmacological properties of the pulmonary vein myocardium. *Biol. Pharm. Bull.* **2013**, *36*, 2–7. [CrossRef] [PubMed]
3. Chen, Y.; Chen, S.; Chen, Y.; Yeh, H.; Chang, M.; Lin, C. Electrophysiology of single cardiomyocytes isolated from rabbit pulmonary veins: Implication in initiation of focal atrial fibrillation. *Basic. Res. Cardiol.* **2002**, *97*, 26–34. [CrossRef] [PubMed]
4. Haissaguerre, M.; Jais, P.; Shah, D.C.; Takahashi, A.; Hocini, M.; Quiniou, G.; Garrigue, S.; Le Mouroux, A.; Le Metayer, P.; Clementy, J. Spontaneous initiation of atrial fibrillation by ectopic beats originating in the pulmonary veins. *N. Engl. J. Med.* **1998**, *339*, 659–666. [CrossRef] [PubMed]
5. Chen, S.A.; Hsieh, M.H.; Tai, C.T.; Tsai, C.F.; Prakash, V.S.; Yu, W.C.; Hsu, T.L.; Ding, Y.A.; Chang, M.S. Initiation of atrial fibrillation by ectopic beats originating from the pulmonary veins: Electrophysiological characteristics, pharmacological responses, and effects of radiofrequency ablation. *Circulation* **1999**, *100*, 1879–1886. [CrossRef] [PubMed]
6. de Bakker, J.M.T.; Ho, S.Y.; Hocini, M. Basic and clinical electrophysiology of pulmonary vein ectopy. *Cardiovasc. Res.* **2002**, *54*, 287–294. [CrossRef] [PubMed]
7. Takagi, D.; Okamoto, Y.; Ohba, T.; Yamamoto, H.; Ono, K. Comparative study of hyperpolarization-activated currents in pulmonary vein cardiomyocytes isolated from rat, guinea pig, and rabbit. *J. Physiol. Sci.* **2020**, *70*, 6. [CrossRef] [PubMed]
8. Tsuneoka, Y.; Irie, M.; Tanaka, Y.; Sugimoto, T.; Kobayashi, Y.; Kusakabe, T.; Kato, K.; Hamaguchi, S.; Namekata, I.; Tanaka, H. Permissive role of reduced inwardly-rectifying potassium current density in the automaticity of the guinea pig pulmonary vein myocardium. *J. Pharmacol. Sci.* **2017**, *133*, 195–202. [CrossRef] [PubMed]
9. Ehrlich, J.R.; Cha, T.; Zhang, L.; Chartier, D.; Melnyk, P.; Hohnloser, S.H.; Nattel, S. Cellular electrophysiology of canine pulmonary vein cardiomyocytes: Action potential and ionic current properties. *J. Physiol.* **2003**, *551*, 801–813. [CrossRef]
10. Okamoto, Y.; Kawamura, K.; Nakamura, Y.; Ono, K. Pathological impact of hyperpolarization-activated chloride current peculiar to rat pulmonary vein cardiomyocytes. *J. Mol. Cell Cardiol.* **2014**, *66*, 53–62. [CrossRef]
11. Namekata, I.; Tsuneoka, Y.; Takahara, A.; Shimada, H.; Sugimoto, T.; Takeda, K.; Nagaharu, M.; Shigenobu, K.; Kawanishi, T.; Tanaka, H. Involvement of the Na^+/Ca^{2+} exchanger in the automaticity of guinea-pig pulmonary vein myocardium as revealed by SEA0400. *J. Pharmacol. Sci.* **2009**, *110*, 111–116. [CrossRef]
12. Okamoto, Y.; Takano, M.; Ohba, T.; Ono, K. Arrhythmogenic coupling between the Na^+-Ca^{2+} exchanger and inositol 1,4,5-triphosphate receptor in rat pulmonary vein cardiomyocytes. *J. Mol. Cell. Cardiol.* **2012**, *52*, 988–997. [CrossRef] [PubMed]
13. Irie, M.; Hiiro, H.; Hamaguchi, S.; Namekata, I.; Tanaka, H. Involvement of the persistent Na^+ current in the diastolic depolarization and automaticity of the guinea pig pulmonary vein myocardium. *J. Pharmacol. Sci.* **2019**, *141*, 9–16. [CrossRef]
14. Malecot, C.O.; Bredeloux, P.; Findlay, I.; Maupoil, V. A TTX-sensitive resting Na^+ permeability contributes to the catecholaminergic automatic activity in rat pulmonary vein. *J. Cardiovasc. Electrophysiol.* **2015**, *26*, 311–319. [CrossRef]
15. Cohen, C.J.; Bean, B.P.; Tsien, R.W. Maximal upstroke velocity as an index of available sodium conductance. Comparison of maximal upstroke velocity and voltage clamp measurements of sodium current in rabbit Purkinje fibers. *Circ. Res.* **1984**, *54*, 636–651. [CrossRef]
16. Crill, W.E. Persistent sodium current in mammalian central neurons. *Annu. Rev. Physiol.* **1996**, *58*, 349–362. [CrossRef]
17. Antzelevitch, C.; Nesterenko, V.; Shryock, J.C.; Rajamani, S.; Song, Y.; Belardinelli, L. The role of late I_{Na} in development of cardiac arrhythmias. *Handb. Exp. Pharmacol.* **2014**, *221*, 137–168. [CrossRef] [PubMed]

18. Song, Y.; Shryock, J.C.; Belardinelli, L. A slowly inactivating sodium current contributes to spontaneous diastolic depolarization of atrial myocytes. *Am. J. Physiol. Heart Circ. Physiol.* **2009**, *297*, H1254–H1262. [CrossRef]
19. Rota, M.; Vassalle, M. Patch-clamp analysis in canine cardiac Purkinje cells of a novel sodium component in the pacemaker range. *J. Physiol.* **2003**, *548*, 147–165. [CrossRef]
20. Sossalla, S.; Kallmeyer, B.; Wagner, S.; Mazur, M.; Maurer, U.; Toischer, K.; Schmitto, J.D.; Seipelt, R.; Schondube, F.A.; Hasenfuss, G.; et al. Altered Na^+ currents in atrial fibrillation effects of ranolazine on arrhythmias and contractility in human atrial myocardium. *J. Am. Coll. Cardiol.* **2010**, *55*, 2330–2342. [CrossRef]
21. Wimmer, N.J.; Stone, P.H. Anti-anginal and anti-ischemic effects of late sodium current inhibition. *Cardiovasc. Drugs Ther.* **2013**, *27*, 69–77. [CrossRef] [PubMed]
22. Horvath, B.; Bers, D.M. The late sodium current in heart failure: Pathophysiology and clinical relevance. *ESC Heart Fail.* **2014**, *1*, 26–40. [CrossRef] [PubMed]
23. Lu, Y.; Cheng, C.; Chen, Y.; Chen, S.; Chen, Y. ATX-II-induced pulmonary vein arrhythmogenesis related to atrial fibrillation and long QT syndrome. *Eur. J. Clin. Investig.* **2012**, *42*, 823–831. [CrossRef]
24. Hegyi, B.; Polonen, R.; Hellgren, K.T.; Ko, C.Y.; Ginsburg, K.S.; Bossuyt, J.; Mercola, M.; Bers, D.M. Cardiomyocyte Na^+ and Ca^{2+} mishandling drives vicious cycle involving CaMKII, ROS, and ryanodine receptors. *Basic Res. Cardiol.* **2021**, *116*, 58–59. [CrossRef] [PubMed]
25. Belardinelli, L.; Giles, W.R.; Rajamani, S.; Karagueuzian, H.S.; Shryock, J.C. Cardiac late Na^+ current: Proarrhythmic effects, roles in long QT syndromes, and pathological relationship to CaMKII and oxidative stress. *Heart Rhythm.* **2015**, *12*, 440–448. [CrossRef] [PubMed]
26. Sag, C.M.; Mallwitz, A.; Wagner, S.; Hartmann, N.; Schotola, H.; Fischer, T.H.; Ungeheuer, N.; Herting, J.; Shah, A.M.; Maier, L.S.; et al. Enhanced late INa induces proarrhythmogenic SR Ca leak in a CaMKII-dependent manner. *J. Mol. Cell. Cardiol.* **2014**, *76*, 94–105. [CrossRef]
27. Namekata, I.; Tanaka, Y.; Ohmori, T.; Tsuneoka, Y.; Hamaguchi, S.; Tanaka, H. Cell Morphology and Early-phase Ca^{2+} Transients of Guinea-Pig Pulmonary Vein Cardiomyocytes Compared with Atrial and Ventricular Cardiomyocytes. *Bioimages* **2019**, *27*, 1–12.
28. Tanaka, H.; Nishimaru, K.; Sekine, T.; Kawanishi, T.; Nakamura, R.; Yamagaki, K.; Shigenobu, K. Two-dimensional millisecond analysis of intracellular Ca^{2+} sparks in cardiac myocytes by rapid scanning confocal microscopy: Increase in amplitude by isoproterenol. *Biochem. Biophys. Res. Commun.* **1997**, *233*, 413–418. [CrossRef]
29. Tanaka, H.; Kawanishi, T.; Matsuda, T.; Takahashi, M.; Shigenobu, K. Intracellular free Ca^{2+} movements in cultured cardiac myocytes as shown by rapid scanning confocal microscopy. *J. Cardiovasc. Pharmacol.* **1996**, *27*, 761–769. [CrossRef]
30. Namekata, I.; Hiiro, H.; Odaka, R.; Saito, T.; Hamaguchi, S.; Tsukamoto, T.; Ishikawa, R.; Katayama, Y.; Kondo, Y.; Tanaka, H. Inhibitory Effect of a Late Sodium Current Blocker, NCC-3902, on the Automaticity of the Guinea Pig Pulmonary Vein Myocardium. *Biol. Pharm. Bull.* **2022**, *45*, 1644–1652. [CrossRef]
31. Kornyeyev, D.; El-Bizri, N.; Hirakawa, R.; Nguyen, S.; Viatchenko-Karpinski, S.; Yao, L.; Rajamani, S.; Belardinelli, L. Contribution of the late sodium current to intracellular sodium and calcium overload in rabbit ventricular myocytes treated by anemone toxin. *Am. J. Physiol. Heart Circ. Physiol.* **2016**, *310*, H426–H435. [CrossRef] [PubMed]
32. Bers, D.M.; Weber, C.R. Na/Ca exchange function in intact ventricular myocytes. *Ann. N. Y. Acad. Sci.* **2002**, *976*, 500–512. [CrossRef] [PubMed]
33. Noble, D.; Noble, P.J. Late sodium current in the pathophysiology of cardiovascular disease: Consequences of sodium-calcium overload. *Heart* **2006**, *92* (Suppl. 4), iv1–iv5. [CrossRef] [PubMed]
34. Despa, S.; Islam, M.A.; Weber, C.R.; Pogwizd, S.M.; Bers, D.M. Intracellular Na^+ concentration is elevated in heart failure but Na/K pump function is unchanged. *Circulation* **2002**, *105*, 2543–2548. [CrossRef] [PubMed]
35. Namekata, I.; Tsuneoka, Y.; Akiba, A.; Nakamura, H.; Shimada, H.; Takahara, A.; Tanaka, H. Intracellular Calcium and Membrane Potential Oscillations in the Guinea Pig and Rat Pulmonary Vein Myocardium. *Bioimages* **2010**, *18*, 11–22.
36. Fujiwara, K.; Tanaka, H.; Mani, H.; Nakagami, T.; Takamatsu, T. Burst emergence of intracellular Ca^{2+} waves evokes arrhythmogenic oscillatory depolarization via the Na^+-Ca^{2+} exchanger: Simultaneous confocal recording of membrane potential and intracellular Ca^{2+} in the heart. *Circ. Res.* **2008**, *103*, 509–518. [CrossRef] [PubMed]
37. Zima, A.V.; Bovo, E.; Bers, D.M.; Blatter, L.A. Ca^{2+} spark-dependent and -independent sarcoplasmic reticulum Ca^{2+} leak in normal and failing rabbit ventricular myocytes. *J. Physiol.* **2010**, *588*, 4743–4757. [CrossRef] [PubMed]
38. Mattiazzi, A.; Argenziano, M.; Aguilar-Sanchez, Y.; Mazzocchi, G.; Escobar, A.L. Ca^{2+} Sparks and Ca^{2+} waves are the subcellular events underlying Ca^{2+} overload during ischemia and reperfusion in perfused intact hearts. *J. Mol. Cell. Cardiol.* **2015**, *79*, 69–78. [CrossRef] [PubMed]
39. Lin, D.; Lee, W.; Chien, Y.; Chen, T.; Yang, K. The link between abnormalities of calcium handling proteins and catecholaminergic polymorphic ventricular tachycardia. *Tzu Chi Med. J.* **2021**, *33*, 323–331. [CrossRef]
40. Bers, D.M.; Guo, T. Calcium signaling in cardiac ventricular myocytes. *Ann. N. Y. Acad. Sci.* **2005**, *1047*, 86–98. [CrossRef]
41. Chan, C.; Lin, F.; Chen, Y.; Lin, Y.; Higa, S.; Chen, S.; Chen, Y. Glucagon-like Peptide-1 Receptor Activation Reduces Pulmonary Vein Arrhythmogenesis and Regulates Calcium Homeostasis. *Int. J. Mol. Sci.* **2023**, *24*, 13100. [CrossRef] [PubMed]
42. Okamoto, Y.; Aung, N.Y.; Tanaka, M.; Takeda, Y.; Takagi, D.; Igarashi, W.; Ishii, K.; Yamakawa, M.; Ono, K. Preferential Expression of Ca^{2+}-Stimulable Adenylyl Cyclase III in the Supraventricular Area, including Arrhythmogenic Pulmonary Vein of the Rat Heart. *Biomolecules* **2022**, *12*, 724. [CrossRef] [PubMed]

43. Fischer, T.H.; Herting, J.; Mason, F.E.; Hartmann, N.; Watanabe, S.; Nikolaev, V.O.; Sprenger, J.U.; Fan, P.; Yao, L.; Popov, A.; et al. Late INa increases diastolic SR-Ca2+-leak in atrial myocardium by activating PKA and CaMKII. *Cardiovasc. Res.* **2015**, *107*, 184–196. [CrossRef] [PubMed]
44. Couchonnal, L.F.; Anderson, M.E. The role of calmodulin kinase II in myocardial physiology and disease. *Physiology* **2008**, *23*, 151–159. [CrossRef] [PubMed]
45. Hudmon, A.; Schulman, H. Structure-function of the multifunctional Ca^{2+}/calmodulin-dependent protein kinase II. *Biochem. J.* **2002**, *364*, 593–611. [CrossRef] [PubMed]
46. Maier, L.S.; Bers, D.M. Role of Ca^{2+}/calmodulin-dependent protein kinase (CaMK) in excitation-contraction coupling in the heart. *Cardiovasc. Res.* **2007**, *73*, 631–640. [CrossRef]
47. Kaplan, A.D.; Joca, H.C.; Boyman, L.; Greiser, M. Calcium Signaling Silencing in Atrial Fibrillation: Implications for Atrial Sodium Homeostasis. *Int. J. Mol. Sci.* **2021**, *22*, 10513. [CrossRef] [PubMed]
48. Irie, M.; Hiiro, H.; Kato, S.; Kuramochi, M.; Hamaguchi, S.; Namekata, I.; Tanaka, H. Differential effects of class I antiarrhythmic drugs on the guinea pig pulmonary vein myocardium: Inhibition of automatic activity correlates with blockade of a diastolic sodium current component. *J. Pharmacol. Sci.* **2020**, *143*, 325–329. [CrossRef] [PubMed]
49. Hegner, P.; Drzymalski, M.; Biedermann, A.; Memmel, B.; Durczok, M.; Wester, M.; Floerchinger, B.; Provaznik, Z.; Schmid, C.; Zausig, Y.; et al. SAR296968, a Novel Selective Na^+/Ca^{2+} Exchanger Inhibitor, Improves Ca^{2+} Handling and Contractile Function in Human Atrial Cardiomyocytes. *Biomedicines* **2022**, *10*, 1932. [CrossRef] [PubMed]
50. Christ, T.; Kovacs, P.P.; Acsai, K.; Knaut, M.; Eschenhagen, T.; Jost, N.; Varro, A.; Wettwer, E.; Ravens, U. Block of Na^+/Ca^{2+} exchanger by SEA0400 in human right atrial preparations from patients in sinus rhythm and in atrial fibrillation. *Eur. J. Pharmacol.* **2016**, *788*, 286–293. [CrossRef]
51. Nagy, N.; Toth, N.; Nanasi, P.P. Antiarrhythmic and Inotropic Effects of Selective Na^+/Ca^{2+} Exchanger Inhibition: What Can We Learn from the Pharmacological Studies? *Int. J. Mol. Sci.* **2022**, *23*, 14651. [CrossRef] [PubMed]
52. Watanabe, Y.; Koide, Y.; Kimura, J. Topics on the Na^+/Ca^{2+} Exchanger: Pharmacological Characterization of Na+/Ca2+ Exchanger Inhibitors. *J. Pharmacol. Sci.* **2006**, *102*, 7–16. [CrossRef] [PubMed]
53. Grandi, E.; Dobrev, D. Non-ion channel therapeutics for heart failure and atrial fibrillation: Are CaMKII inhibitors ready for clinical use? *J. Mol. Cell Cardiol.* **2018**, *121*, 300–303. [CrossRef]
54. Neef, S.; Steffens, A.; Pellicena, P.; Mustroph, J.; Lebek, S.; Ort, K.R.; Schulman, H.; Maier, L.S. Improvement of cardiomyocyte function by a novel pyrimidine-based CaMKII-inhibitor. *J. Mol. Cell. Cardiol.* **2018**, *115*, 73–81. [CrossRef]

International Journal of
Molecular Sciences

MDPI

Article

Eicosapentaenoic Acid Rescues Cav1.2-L-Type Ca^{2+} Channel Decline Caused by Saturated Fatty Acids via Both Free Fatty Acid Receptor 4-Dependent and -Independent Pathways in Cardiomyocytes

Masaki Morishima [1,2,*], Pu Wang [3], Kosuke Horii [2], Kazuki Horikawa [4] and Katsushige Ono [3,5,*]

1 Department of Food Science and Nutrition, Faculty of Agriculture, Kindai University, Nara 6318505, Japan
2 Department of Applied Biological Chemistry, Graduate School of Agriculture, Kindai University, Nara 6318505, Japan; 2444670003v@nara.kindai.ac.jp
3 Department of Pathophysiology, Oita University School of Medicine, Yufu 8795593, Japan; wangpucardio@163.com
4 Department of Optical Imaging, Advanced Research Promotion Center, Tokushima University, Tokushima 7708503, Japan; horikawa.kazuki@tokushima-u.ac.jp
5 Oita Shimogori Hospital, Oita 8700926, Japan
* Correspondence: mmoris@nara.kindai.ac.jp (M.M.); ono@oita-u.ac.jp (K.O.); Tel.: +81-742-43-2751 (M.M.); +81-97-569-1021 (K.O.)

Abstract: Dietary intake of omega-3 polyunsaturated fatty acids (eicosapentaenoic acid, EPA) exerts antiarrhythmic effects, although the mechanisms are poorly understood. Here, we investigated the possible beneficial actions of EPA on saturated fatty acid-induced changes in the L-type Ca^{2+} channel in cardiomyocytes. Cardiomyocytes were cultured with an oleic acid/palmitic acid mixture (OAPA) in the presence or absence of EPA. Beating rate reduction in cardiomyocytes caused by OAPA were reversed by EPA. EPA also retrieved a reduction in Cav1.2 L-type Ca^{2+} current, mRNA, and protein caused by OAPA. Immunocytochemical analysis revealed a distinct downregulation of the Cav1.2 channel caused by OAPA with a concomitant decrease in the phosphorylated component of a transcription factor adenosine-3′,5′-cyclic monophosphate (cAMP) response element binding protein (CREB) in the nucleus, which were rescued by EPA. A free fatty acid receptor 4 (FFAR4) agonist TUG-891 reversed expression of *Cav1.2* and *CREB* mRNA caused by OAPA, whereas an FFAR4 antagonist AH-7614 abolished the effects of EPA. Excessive reactive oxygen species (ROS) accumulation caused by OAPA decreased *Cav1.2* and *CREB* mRNA expressions, which was reversed by an ROS scavenger. Our data suggest that EPA rescues cellular Cav1.2-Ca^{2+} channel decline caused by OAPA lipotoxicity and oxidative stresses via both free fatty acid receptor 4-dependent and -independent pathways.

Keywords: eicosapentaenoic acid; oleic acid; palmitic acid; omega-3 polyunsaturated fatty acid; PUFA; FFAR4; CREB; Cav1.2; L-type Ca^{2+} channel

Citation: Morishima, M.; Wang, P.; Horii, K.; Horikawa, K.; Ono, K. Eicosapentaenoic Acid Rescues Cav1.2-L-Type Ca^{2+} Channel Decline Caused by Saturated Fatty Acids via Both Free Fatty Acid Receptor 4-Dependent and -Independent Pathways in Cardiomyocytes. *Int. J. Mol. Sci.* **2024**, *25*, 7570. https://doi.org/10.3390/ijms25147570

Academic Editor: Demetrios A. Arvanitis

Received: 29 April 2024
Revised: 3 July 2024
Accepted: 8 July 2024
Published: 10 July 2024

1. Introduction

A large number of epidemiologic studies demonstrate that higher saturated fat intake is associated with an increased risk of sudden cardiac death, suggesting that the effects of dietary saturated fat may be sufficient to cause heart diseases [1–4]. High levels of circulating saturated fatty acids are also associated with diabetes, obesity, and hyperlipidemia [5,6]. In the heart, the accumulation of saturated fatty acids has been proposed to play a role in the development of heart failure and arrhythmias [6–8]. The two major circulating fatty acids are saturated palmitic acid (C16:0) and monounsaturated oleic acid (C18:1) [9,10]. Palmitic acid is one of the most abundant fatty acids in human/animals and its overload induces lipotoxicity, resulting in apoptosis, endoplasmic reticulum stress, and reactive oxygen species production [11,12]. In addition, several recent studies have reported that plasma concentrations of palmitic acid were increased in a high-fat diet-induced obesity mouse model which has been used in a large number of publications to demonstrate lifestyle-related diseases in animals [7–10,13]. Exposure of high levels of palmitic acid to isolated cardiomyocytes is known to result in contractile dysfunction and apoptosis [7,14]. It is also recognized that palmitic acid-induced ROS production impairs cellular Ca^{2+} handling possibly through the decrease in L-type Ca^{2+} currents, increase in the open probability of SR Ca^{2+} release channels, slowing SR Ca^{2+} reuptake, and the activation of sarcolemmal Na^{+}/Ca^{2+} exchange activity, ultimately leading to reduced SR Ca^{2+} content [7,15–17]. In contrast, numerous animal experiments and human epidemiological studies [18–21] have shown that omega-3 polyunsaturated fatty acids (PUFAs) exert beneficial effects on physical health. The strongest evidence for a valuable action of PUFAs has to do with cardiovascular diseases, causing a number of physiological changes such as decreasing heart rate and lowering blood pressure [22]. Interestingly, accumulating evidence from in vitro experiments has demonstrated that omega-3 PUFAs exert antiarrhythmic effects [23–26]. Eicosapentaenoic acid (EPA) has been shown to affect sodium channels [23] to protect cardiomyocytes against arrhythmias induced by high extracellular calcium, ouabain, isoproterenol, or lysophosphatidylcholine [24]. EPA and docosahexaenoic acid (DHA) are also known to regulate the activity of L-type Ca^{2+} channels, which plays an important role in reducing excessive excitability and increasing refractoriness of cardiac myocytes [25,26]. Thus, inhibition of Na^{+} and/or Ca^{2+} currents may account for the acute antiarrhythmic effects of omega-3 PUFA [23–26]. However, the underlying mechanisms associated with saturated fatty acid-induced long-term electrical changes of cardiomyocyte remain unclear [27–29].

Recently, the molecular targets for PUFAs were elucidated [30,31]. The free fatty acid receptor 4 (FFAR4) is a G protein-coupled receptor for endogenous medium- or long-chain fatty acids that attenuates metabolic diseases and inflammation. PUFAs are generally full agonists for FFAR4 [30,31] that are expressed in various cell types including cardiomyocytes [28,29]. It has been reported that high-fat diet-induced obesity and liver steatosis were more severe in FFAR4-deficient mouse than in wild-type mouse [32], suggesting the functional importance of this receptor in lipid pathology. Of note, Murphy et al. showed that FFAR4 in cardiac myocytes responds to endogenous fatty acids, reduces oxidative damage, and protects the heart from pathological stress. This could provide important translational implications for targeting FFAR4 in cardiovascular disease [33]. Furthermore, EPA was more effective than DHA at preventing lethal arrhythmias by inhibiting inflammasome and sympathetic innervation through activation of peroxisome proliferator-activated (PPAR) γ-mediated FFAR4-dependent and -independent signaling pathways after injury [34]. These results suggest that FFAR4 becomes an important site of action for EPA in the regulation of cardiac electrical activity. However, the role of EPA on cardiac electrical changes and transcriptional regulation of ion channels in the modification of these FFAR4-mediated signaling pathways has not been elucidated. Thus, the aim of this study was to investigate the possible beneficial effects of EPA and FFAR4 on saturated fatty acid-induced electrical changes in cardiac myocytes, focusing on the voltage-gated L-type Ca^{2+} channel.

2. Results

2.1. Actions of Saturated Fatty Acid on the L-Type Ca^{2+} Channel and the Cellular Excitability

To determine whether cardiomyocyte excitability and automaticity can be affected by a high level of saturated fatty acid, we examined the effects of OAPA (100 to 500 μM) on neonatal mouse cardiomyocytes for 24 h. Since cardiomyocytes' beating or automaticity was strongly influenced by the function of pacemaker ion channels, we first measured the mRNAs of these pacemaker ion channels in cardiomyocytes with OAPA by use of real-time PCR. In preliminary studies, OAPA significantly reduced the expression of *HCN4* and *KCNJ3* mRNAs. Interestingly, at the same time, the expression of *Cav1.2* L-type Ca^{2+} channel mRNAs were drastically reduced by OAPA. The expression of *Cav1.3* L-type Ca^{2+} channel mRNAs was also reduced by OAPA. Because the *Cav1.3* mRNA was expressed at much lower levels than that of *Cav1.2*, approximately ~3% of *Cav1.2* in neonatal cardiomyocytes, we only focused on the Cav1.2 L-type Ca^{2+} channel in this study. To investigate the possible effect of saturated fatty acid on cardiac automaticity in neonatal mouse cardiomyocytes, the spontaneous beating rate of cardiomyocyte was measured by live cell imaging system. As shown in Figure 1B, OAPA treatment significantly decreased the spontaneous beating rate in a dose-dependent manner. Interestingly, polyunsaturated fatty acid EPA also affected the expression of *Cav1.2* mRNA (Figure 1C). Although a 50 μM or higher concentration of EPA significantly decreased *Cav1.2* mRNA expression, lower concentration of EPA (1 μM or 10 μM) had a negligible effect on *Cav1.2* expression (Figure 1C). Given these results, we conclude that OAPA, and EPA with very high concentration, reduce the beating rate of cardiomyocytes in a concentration-dependent manner concomitant with the reduction in *Cav1.2* L-type Ca^{2+} channel expression in cardiomyocytes.

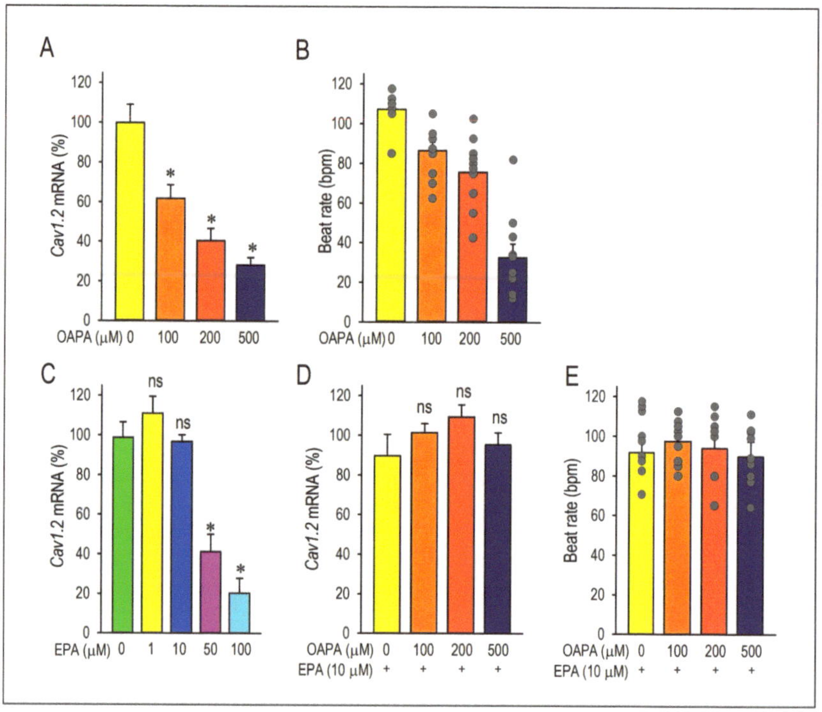

Figure 1. Long-term effects of fatty acid (OAPA) and EPA on the Ca^{2+} channel expression and cardiac automaticity. (**A**,**B**) Effects of OAPA on the expression of L-type Ca^{2+} channel isoforms (*Cav1.2*) mRNA and mean spontaneous beating rate (not normalized) of cardiomyocytes after application for 24 h. Cardiomyocytes were cultured with 100–500 μM palmitic/oleic acid (OAPA: 2:1) for 24 h. (**C**). Effects of EPA on the expression of *Cav1.2* mRNA. (**D**,**E**) Effects of 10 μM EPA on the expression of *Cav1.2* mRNA and mean spontaneous beating rate in combined with 100–500 μM OAPA. Data were normalized to *Cav1.2* mRNA expression in non-treated cardiomyocytes, which was designated as 100. Data are expressed as the means ± SE (n = 8). * $p < 0.05$ compared with the control group (EPA 0 μM or OAPA 0 μM). "ns" indicates statistical non-significance ($p > 0.05$) compared to EPA (-) or OAPA (-).

2.2. EPA Rescued OAPA-Induced Reduction in L-Type Ca^{2+} Channel

The combined actions of OAPA and EPA on Cav1.2 expression and beating rate were then examined based on the results in Figure 1B,E. Interestingly, the reduction in *Cav1.2* mRNA expression caused by OAPA was all rescued by the presence of EPA (Figure 1A,D).

Of note, even the highest concentration of OAPA (500 μM) was without effect on *Cav1.2* mRNA expression when 10 μM EPA was applied together (Figure 1D). These results suggest the rescue effect of EPA on the OAPA-mediated decline of the L-type Ca^{2+} channel when applied for 24 h. Consistent with these results, a decrease in the beating rate caused by OAPA was canceled in the presence of EPA (Figure 1E). In addition, a significant reduction in the protein expression levels of Cav1.2 caused by OAPA was completely rescued by EPA verified by Western blotting (Figure 2A), while EPA had an unchanged Cav1.2 protein level when OAPA was absent (p = 0.514). Furthermore, immunocytochemical analysis revealed that OAPA suppressed the expression of the Cav1.2 channel protein, which was completely rescued by EPA (Figure 2B,C). These results were all consistent with the changes in *Cav1.2* mRNA as shown in Figure 1. Functional modification of Cav1.2 was then confirmed by the electrophysiological method to explore the changes in the L-type Ca^{2+} channel current ($I_{Ca.L}$). Since the saturated fatty acid OAPA strongly decreased the expression level of Cav1.2, which was blocked by EPA, a rescue of L-type Ca^{2+} channel current by EPA was expected. In conventional whole-cell patch clamp experiments using neonatal rat cardiomyocytes, $I_{Ca.L}$ was recorded. Results in Figure 3 demonstrate that OAPA significantly reduced $I_{Ca.L}$ when applied for 24 h, which was completely rescued by 10 μM EPA. Importantly, the current (I)–voltage (V) relationship was unchanged in the presence of OAPA and EPA, suggesting that the gating properties of the Cav1.2 channel was not modified by OAPA and/or EPA (Figure 3B).

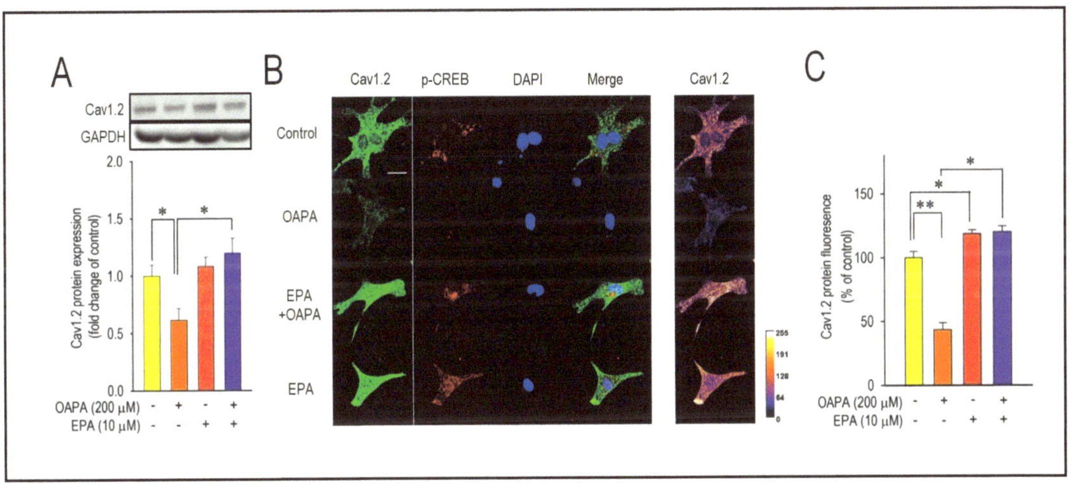

Figure 2. Effect of OAPA and EPA on the expression of Cav1.2 protein in neonatal mouse cardiomyocytes. (**A**) Representative Western blot and summary graph of Cav1.2 level after application with OAPA (200 μM) in the presence or absence of EPA (10 μM) for 24 h. (**B**) Expression and distribution of Cav1.2 and p-CREB as assessed by immunocytochemistry procedure. Cardiomyocytes were exposed to OAPA (200 μM) with or without EPA (10 μM) for 24 h. Cav1.2 for stained in green, p-CREB in red, and DAPI staining to visualize nuclei in blue. Scale bar = 20 μm. (**C**) The percentage of Cav1.2 fluorescence intensity was calculated. The signal intensity of Cav1.2 in non-treated myocytes (OAPA (-), EPA (-)) was set as 100%. Data are expressed as mean ± SE (n = 4). Asterisks indicate significant differences (* $p < 0.05$, ** $p < 0.01$).

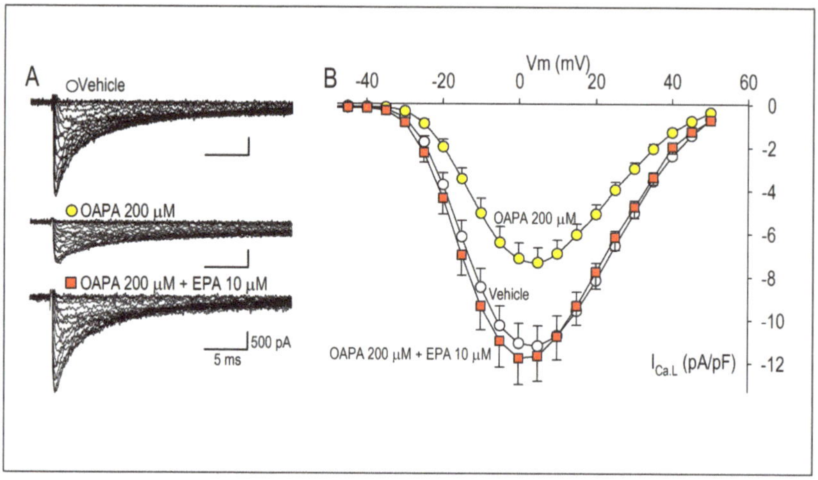

Figure 3. Effects of OAPA and EPA on $I_{Ca.L}$ in rat neonatal cardiomyocytes. Cardiomyocytes were cultured with OAPA (200 μM) in the presence or absence of EPA (10 μM) for 24 h. Representative $I_{Ca.L}$ traces in the vehicle and OAPA with or without EPA applied for 24 h (**A**), and their group data of current (I)–voltage (V) relationship (**B**). Current traces were obtained from a holding potential of −40 mV to test potentials up to 50 mV with 10 mV increments. Data are expressed as mean ± SD (n = 7).

Activation of the transcription factor adenosine-3′,5′-cyclic monophosphate (cAMP) response element binding protein (CREB) as detected by phosphorylation of CREB has been recognized as an index of the initiation of Cav1.2 transcription [35]. Accordingly, the expression and the changes in *CREB* mRNA by OAPA were explored in the presence or absence of EPA. Importantly, *CREB* mRNA was reduced by OAPA in a dose-dependent manner, which was highly consistent with the changes in *Cav1.2* mRNA (Figure 1A) and Cav1.2 protein (Figure 2A). In accordance with the regulation of *Cav1.2* mRNA, EPA rescued the changes in *CREB* mRNA. A reduction in the phosphorylated component of CREB in the nucleus was also rescued by EPA, strongly suggesting that EPA rescues Cav1.2 protein through a transcriptional pathway involving CREB (Figure 4B).

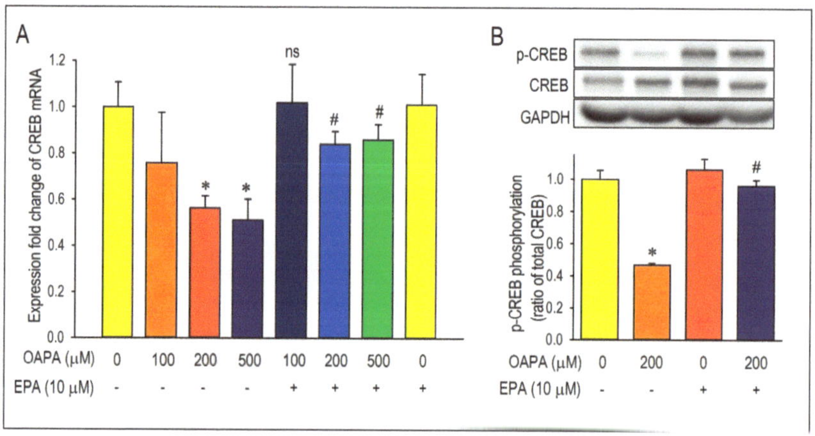

Figure 4. Long-term effect of OAPA and EPA on *CREB* mRNA expression and CREB phosphorylation in neonatal mouse cardiomyocyte. (**A**) Cardiomyocytes were cultured with 100–500 μM OAPA in the

presence or absence of EPA (10 μM) for 24 h to assess the expression of *CREB* mRNA. (**B**) Changes in phosphorylated CREB (p-CREB) protein level in the nucleus of cardiomyocytes. Relative levels of proteins were determined by densitometry of the immunoblots. Data were normalized by taking the value of the control groups as 1.0. Data are expressed as mean ± SE (n = 5). * $p < 0.05$, vs. non-treated cardiomyocytes (OAPA (-), EPA (-)). # $p < 0.05$, vs. cardiomyocytes (OAPA (200 μM), EPA (-)) or (OAPA (500 μM), EPA (-)). "ns" indicates statistical non-significance vs. (EPA (-), OAPA (100 μM).

2.3. Actions of FFAR4 in Cardiomyocytes

Although the downregulation of *Cav1.2* and *CREB* mRNA expression caused by OAPA and the rescue by EPA were confirmed, their signal pathways are largely unknown. To clarify the intracellular signal pathway responsible for the action of EPA, the functional role of the EPA receptor FFAR4 was examined. To our knowledge, the function of FFAR4 is poorly known in cardiac tissue and/or cardiomyocytes. We investigated whether FFAR4 could be detected in the adult mouse heart and neonatal mouse ventricular cardiomyocytes by using real-time PCR analysis. Figure 5A clearly demonstrates that FFAR4 is detectably expressed in adult and neonatal hearts. Notably, there was an approximately 25-fold difference in *FFAR4* mRNA expression between the atrium and the ventricle in the adult mouse heart; surprisingly, *FFAR4* was highly expressed in the atrium. Based on the result that *FFAR4* was appreciably expressed in neonatal ventricular cardiomyocytes, FFAR4-mediated signal pathways to regulate Cav1.2 and CREB were examined by use of neonatal mouse ventricular cardiomyocytes; an FFAR4 antagonist AH7614 and an FFAR4 agonist TUG-891 were applied [36]. It is worthy of note that the actions of EPA to rescue *Cav1.2* and *CREB* were blocked by an FFRA4 antagonist AH-7614 (Figure 5B,C). Even more, the reduction in *Cav1.2* and *CREB* mRNA caused by OAPA was rescued by an FFAR4 agonist TUG-891 mimicking the effect of EPA (Figure 6B,C). These results firmly suggest the mechanism of FFAR4-mediated L-type Ca^{2+} channel regulation pathway led by EPA in cardiomyocytes.

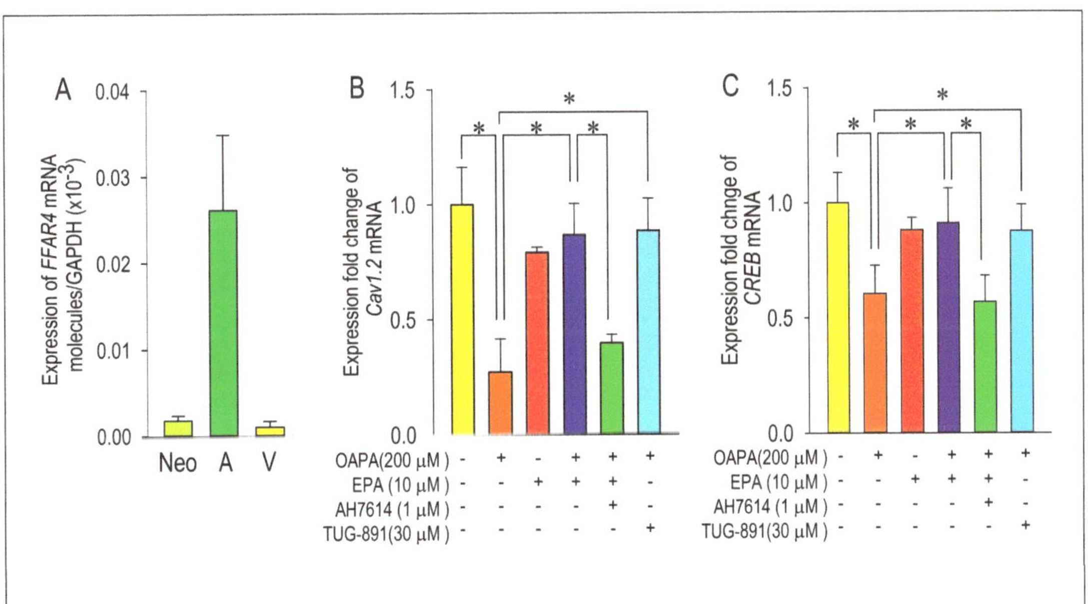

Figure 5. FFAR4 actions on Cav1.2 and CREB expression. (**A**) *FFAR4* mRNA expression levels in cardiomyocytes from neonatal mouse ventricle, adult mouse atrium, and adult mouse ventricle.

(**B**,**C**) Cardiomyocytes were cultured with OAPA, EPA, a selective FFAR4 antagonist AH7614, and an FFAR4 agonist TUG-891 for 24 h. Data were normalized to *Cav1.2* mRNA expression in non-treated cardiomyocytes, which was designated as 100. Data are expressed as mean ± SE (n = 6). Asterisks indicate significant differences (* $p < 0.05$).

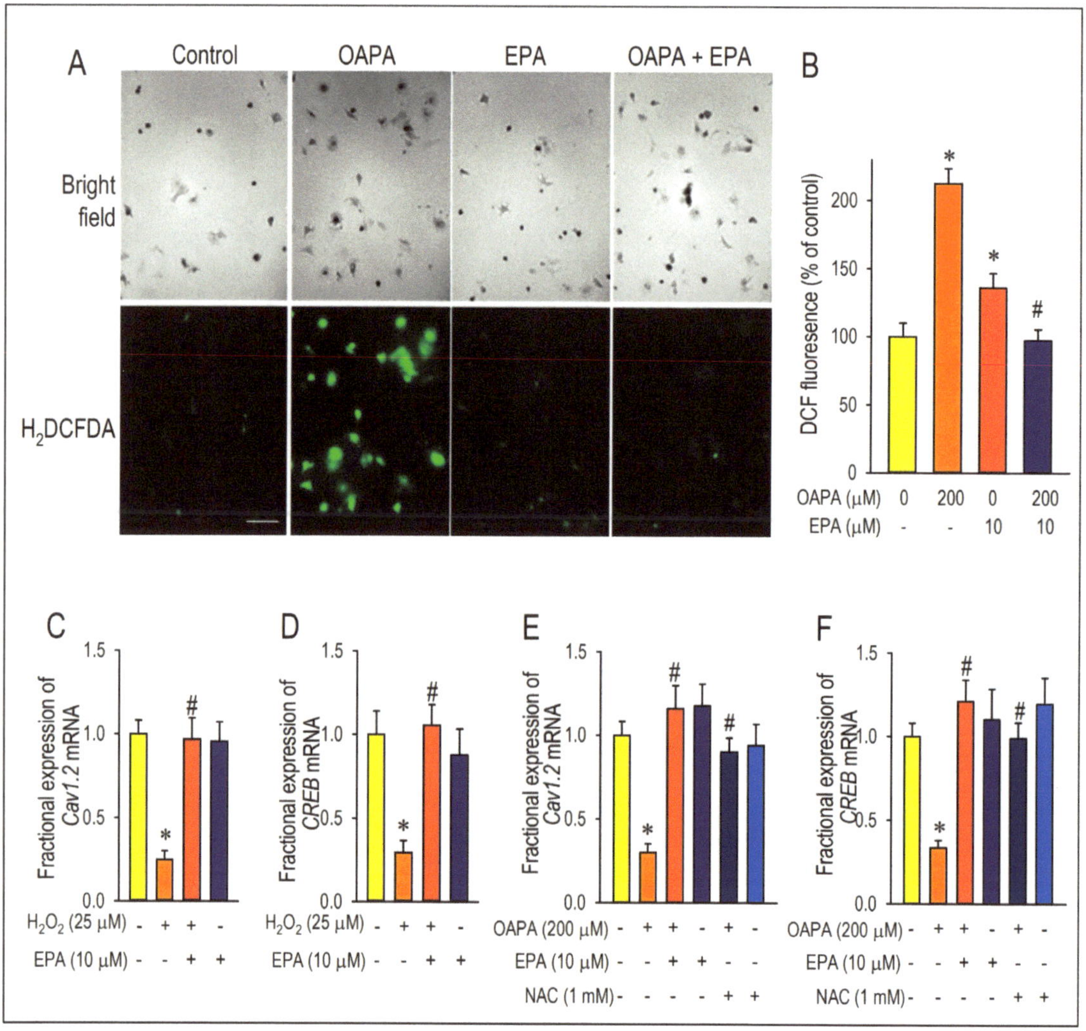

Figure 6. Demonstration of actions of ROS for the expression of Cav1.2 and CREB. Cardiomyocytes were incubated with OAPA (200 μM) in the presence or absence of EPA (10 μM) for 24 h and stained with CMH_2DCFDA. Representative images (**A**) and quantitative results are shown (**B**). Fluorescence intensity of non-treated myocytes (OAPA (-), EPA (-)) was set as 100%., and data are expressed as mean ± SE. Scale bar = 50 μm. (**C**,**D**) Regulation of *Cav1.2* and *CREB* mRNA expression by oxidative stresses using H_2O_2 (25 μM) applied for 24 h. (**E**,**F**) Effects of ROS scavenger on *Cav1.2* and *CREB* mRNA expressions. Cardiomyocytes were exposed to OAPA (200 μM) for 24 h in the presence or absence of EPA (10 μM) with/without an ROS scavenger NAC (1 mM). Data are expressed as mean ± SE (n = 6). * $p < 0.05$, vs. non-treated cardiomyocytes (OAPA (-), EPA (-)) (**B**), (H_2O_2 (-), EPA (-)) (**C**,**D**) or (OAPA (-), EPA (-), NAC (-)) (**E**,**F**). # $p < 0.05$, vs. cardiomyocytes (OAPA (+), EPA (-)) (**B**), (H_2O_2 (+), EPA (-)) (**C**,**D**) or (OAPA (+), EPA (-), NAC (-)) (**E**,**F**).

2.4. Actions of ROS for Cav1.2 Expression

To investigate the effect of OAPA treatment on the level of oxidative stress, the amount of reactive oxygen species (ROS) accumulation was measured in cardiomyocytes. As shown in Figure 6A,B, the level of ROS was relatively low in cardiomyocytes exposed to the normal culture medium. On the other hand, OAPA treatment caused ROS accumulation in cardiomyocytes, which was rescued by EPA co-treatment. Intriguingly, EPA alone increased the ROS accumulation to a certain degree, although we did not examine the action any further in this study. Based on these results, it is suggested that EPA may act as a scavenger to remove ROS in cardiomyocytes. Roles of ROS for the modulation of Cav1.2 and CREB expression were further examined by use of hydrogen peroxide (H_2O_2) which induces oxidative stress in cardiomyocytes (Figure 6C,D). When cardiomyocytes were loaded with H_2O_2 for 24 h, the expression of *Cav1.2* and *CREB* mRNA was markedly suppressed. However, the simultaneous administration of EPA nearly completely rescued the reduction in *Cav1.2* and *CREB* mRNA. The ROS scavenging effects on OAPA insults for expression of *Cav1.2* and *CREB* mRNA were examined by use of an ROS scavenger N-acetylcysteine (NAC) in comparison with EPA. NAC distinctly rescued expression of *Cav1.2* and *CREB* mRNA, suggesting that OAPA downregulates Cav1.2 and CREB via the mechanism associated with ROS accumulation. Taken together, it is concluded that OAPA-induced ROS accumulation modulates Cav1.2-L-type Ca^{2+} channel transcription through an FFAR4-independent pathway.

3. Discussion

The major findings of this study were that (1) the long-term application of OAPA decreased *Cav1.2* mRNA/protein, $I_{Ca.L}$ and the spontaneous beating of cardiomyocyte, (2) EPA application reversed the decline of Cav1.2 channel caused by OAPA, (3) an FFAR4 agonist TUG-891 reversed expression of *Cav1.2* and *CREB* mRNA caused by OAPA, (4) an FFAR4 antagonist AH-7614 abolished the effects of EPA on *Cav1.2* and *CREB* mRNA caused by OAPA, (5) OAPA increased ROS production, while the action was eliminated by EPA, (6) an ROS generator H_2O_2 decreased the expression of *Cav1.2* and *CREB*, which was prevented EPA, and (7) suppressions of *Cav1.2* and *CREB* mRNA by OAPA was prevented by an ROS scavenger NAC.

Intracellular Ca^{2+} homeostasis is a critical determinant of cardiac function, and the levels of intracellular Ca^{2+} are cooperatively regulated by the sarcolemmal Ca^{2+}-ATPase, several types of the Ca^{2+} channels, the sodium–calcium exchanger, and other regulatory proteins. At the same time, intracellular Ca^{2+} signaling plays an essential role in cardiac gene expression and cardiogenesis [37,38]. In this context, it is not surprising that EPA modifies the expression of Cav1.2 through the intracellular Ca^{2+} regulatory pathway (Figure 7). We believe that the present study reveals the novel mechanism via which EPA administration rescues the downregulation of voltage-gated L-type Ca^{2+} channels caused by excessive amounts of saturated fatty acids and/or ROS in cardiomyocytes. In recent years, although little is known about the intracellular molecular mechanisms by which EPA exerts cardioprotective effects in cardiomyocytes, there has been increasing interest in FFAR4, a selective receptor for EPA [33]. Because EPA-induced FFAR4 activation is known to signal through modulation of Gq/11 proteins to stimulate CaMK activity in various cell types [33,36], and because CREB-dependent regulation of Cav1.2 channel transcription has been reported in cardiomyocytes [35], it is proposed that EPA regulates Cav1.2 channel expression possibly through the FFAR4-CaMK-CREB signaling pathway (Figure 7). Of note, FFAR4-independent signal pathway of EPA for the regulation of Cav1.2 expression is also proposed in this study. Since EPA has many double bonds and long-chain carbons, the incorporation of EPA into the plasma lipids within the plasma membrane can alter its properties and influence the function of various membrane proteins including ion channels and receptors [27]. Furthermore, the intracellular concentration of EPA can easily be upregulated following extracellular EPA treatment [39]. Taken together, the ROS-scavenging effects of EPA to upregulate CREB and Cav1.2 could be attributed to its FFAR4-independent pathway at the same time. Interestingly, ROS generation by OAPA and ROS scavenging effect of PUFA are well recognized in various types of cells including cardiomyocytes [40,41]; PUFAs are known to activate antioxidant enzymes such as catalase and superoxide dismutase [41]. Although the cardioprotection effect of PUFAs are documented by their antioxidant mechanism [41], the mechanism of their regulatory actions on ion channel transcription is largely unknown, which obviously needs to be further elucidated.

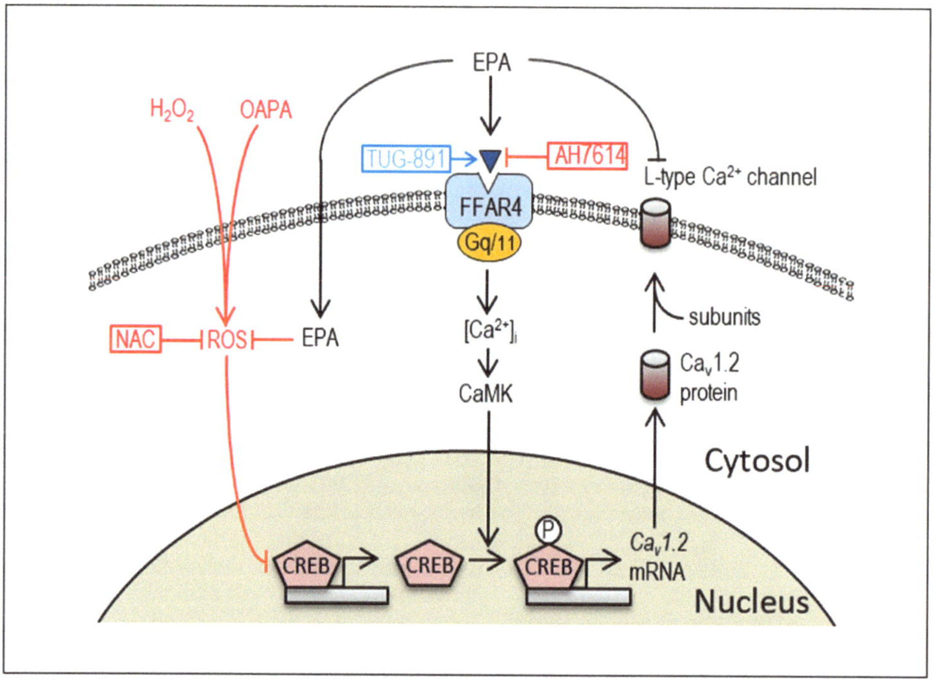

Figure 7. The proposed molecular mechanism of EPA and OAPA on the Cav1.2-L-type Ca^{2+} channel expression. EPA modulates Cav1.2 expression via FFAR4-depenent and -independent signal pathways apart from the acute inhibitory action on the L-type Ca^{2+} channel [26]. Postulated actions of an ROS scavenger NAC, a selective FFAR4 antagonist AH7614, and an FFAR4 agonist TUG-891 on CREB-associated pathways are also shown.

Several studies have revealed an association between omega-3 PUFA intake and a lower risk of cardiovascular events including arrhythmias. More specifically, omega-3 PUFA is known to modify cardiac excitability by regulating ion channel functions in cardiomyocytes [42,43]. EPA is believed to prevent atrial and ventricular arrhythmias in animal experiments by inhibiting voltage-gated ion channels such as the Na^+ channel [43,44], the Ca^{2+} channels [45], and some types of the K^+ channels [46]. These findings suggest that the antiarrhythmic actions of EPA are mediated by direct interaction with membrane ion channels. On the contrary, long-term beneficial actions of PUFAs were to limit atrial remodeling that predispose patients to develop atrial fibrillation [47]. These observations suggest that PUFAs may act as antiarrhythmic nutrients to modify cardiac excitability when applied for long-term periods as well. Several independent studies have reported the effects of PUFAs to modulate expressions of ion channels in the heart; EPA suppressed expression levels of K^+ channels and their related genes, *Kir6.2*, *Kcna5*, *Kcnd2*, *KChIP2* [48,49], and Na^+ channel mRNA [50]. Interestingly, in partially agreement with our findings, Xu at al. recently demonstrated that long-term application of fish oil upregulated the expression of Cav1.2 channel protein in the rabbit heart [46], which is in the sharp contrast to the previous studies reporting an antagonistic action of EPA to the voltage-gated Ca^{2+} channel [51]. A potentiation of the Ca^{2+} channel by long-term application of EPA could be accordingly considered as a compensatory action that maintains intracellular Ca^{2+} concentration reduction caused by the short-term effect. However, the long-term effect of EPA on the Ca^{2+} channel may be more complicated than the above postulation. It is known that ingested EPA is present in the blood in the form of phospholipids. PUFAs bind to and are incorporated into phospholipids in cell membranes because they are structurally unsaturated. EPA may affect membrane fluidity, lipid microdomain formation, and transmembrane signaling. Consistently, an animal study demonstrated a significant increase in EPA and DHA concentrations in the ventricular myocardium of mouse supporting the notion that EPA acts, at least in part, directly on cardiomyocytes to maintain the electrical properties of ion channels present on the cell membrane [46]. Although this study did not assess the mechanism further, a high concentration of EPA can reduce the expression of the Ca^{2+} channel by itself (Figure 1C). Considering that the human serum concentration of EPA approximately ranges from 14 μM to 100 μM [52], it is assumed that the long-term action of EPA on the Ca^{2+} channel ex pression could be affected by various factors including concentrations of EPA and co-existence of other saturated/unsaturated fatty acids, FFAR4 density on the plasma membrane, endogenous intensity of CaMK signals, and the magnitude of CREB-dependent transcription in cells. Although several publications describe that omega-3 fatty acids prevent ventricular arrhythmias [53], a much larger number of studies document that EPA prevents atrial fibrillation in animal studies and clinical observation [53,54]. It is crucial to note that the expression level of FFAR4 in the atrium was 25 times larger than that in the ventricle (Figure 5A). Although electrophysiological actions of EPA on cardiomyocytes appear not only via FFAR4, a dense distribution of FFAR4 in the atrium suggests that actions of EPA on the Ca^{2+} channel are highly expected in the atrium. A negative chronotropic effect of EPA [55] may also be associated with the FFAR4 expression distinction in the heart. Relatively high concentrations of EPA by itself reduced expression of *Cav1.2 mRNA* (Figure 1C), in spite of the fact that EPA rescued the reduction in them caused by OAPA. Taken together, it is suggested that EPA is able to decrease automaticity or responsiveness of the sinus node. Because the sinus node is located within the atrial tissue and is functionally connected to the atrial cardiomyocytes, the impact of FFAR4 activation in the atrium could influence pacemaker rhythm accordingly. In humans, the expression of FFAR4 decreases in cardiomyocytes with heart failure [33]. Thus, it is likely that expression of FFAR4 in the atrium may change in the pathological conditions such as in atrial fibrillation, suggesting the importance of FFAR4 density in the evaluation of EPA action on the pathological condition of the heart.

Limitations of this study include uncertainty about whether these results are directly applicable to humans with dyslipidemia. Results in this study were obtained from animal experiments with isolated mouse/rat cardiomyocytes and in vitro analysis that used real-time PCR and patch clamp analysis but was not applied to human body. Although the impact of an FFR4 agonist TUG-891 on the expression of Cav1.2 and CREB was robust, it is uncertain whether the FFAR4-dependent effect of EPA is more dominant than the FFAR4-independent effect to rescue the L-type Ca2+ channel, which needs further studies. Also, the underlying mechanisms that connect FFAR4 and Cav1.2 remain largely unclear and require further investigation.

4. Materials and Methods

4.1. Chemicals

Reagents were obtained from Sigma Aldrich (St. Louis, MO, USA) or WAKO (Osaka, Japan) unless otherwise indicated. Triton X-100 was purchased from MP Biomedicals (Aurora, OH, USA). Fetal bovine serum was obtained from Biosera (Biosera, Nuaillé, Chile). Collagenase type IV was purchased from Worthington (Lakewood, NJ, USA). H_2DCF-DA (2′,7′-dichlorodihydrofluorescein diacetate), ProLong Diamond Antifade Mountant with 4′,6-diamidino-2-phenylindole dihydrochloride (DAPI) were from Molecular Probes (Eugene, OR, USA). Anti-CREB antibody (1:1000, Cell Signaling, Beverly, MA, USA), phosphospecific antibody against anti-pCREB (Ser 133), and Alexa Fluor 488 and 594-conjugated secondary antibodies were from Cell Signaling (Danvers, MA, USA). All reagents from commercial sources were of analytical grade.

4.2. Isolation of Neonatal Mouse Cardiomyocytes

C57BL/6 mouse and Wistar rats (Japan SLC, Inc., Shizuoka, Japan) were provided with food and water ad libitum and the room temperature was maintained at 25 ± 1 °C in a 12 h light/12 h dark cycle. Neonatal mouse or rat cardiomyocytes were enzymatically isolated and cultured as previously described [56,57]. The cardiomyocytes were maintained at 37 °C under 5% CO_2 in Dulbecco's modified Eagle's medium (DMEM) supplemented with 10% fetal bovine serum for 24 h. After 24 h of culture, >70% of the cells adhered to the substrates and started to exhibit spontaneous beating. For electrophysiological experiments, we used isolated neonatal rat cardiomyocytes.

4.3. Preparation of OAPA and Cell Culture

Oleic acid (OA) was solubilized in anhydrous methanol, and then 100 mM stock solution was prepared. Palmitic acid (PA) bovine serum albumin (BSA) conjugate was prepared through soaping PA with 0.5 N sodium hydroxide (NaOH) and mixing with BSA. Briefly, 100 mM stock solution of PA in 0.5 N NaOH was incubated at 70 °C for 30 min. Then, we prepared a mixture in OA at a concentration of 500 μM and PA at a concentration of 250 μM. Both fatty acids were complexed to BSA in a ratio of 2:1 (OA:PA) and total concentration of 100–500 μM. Because in vitro studies have shown that the PA at physiological concentrations exhibits a dose-dependent cytotoxic effect associated with ROS production and apoptosis or necrosis in neonatal cardiomyocytes [9], we tried to examine a combination of OA with PA, which is more suitable for simulation of simple electrical changes than PA alone. Medium supplementation with OA (500 μM)/PA (250 μM) mixture (OAPA) in the presence or absence of EPA (10 μM) for 24 h. To investigate the effect of removing accumulated ROS, cardiomyocytes were exposed to an OAPA for 24 h in the presence or absence of 1 mM NAC, an ROS scavenger.

4.4. Measurement of Intracellular ROS Accumulation

Intracellular ROS accumulation was detected using H_2DCFDA, whose green fluorescence signal is increased by its oxidation by ROS. Cells were incubated with 2 μM H_2DCFDA for 50 min at 37 °C and rinsed with HEPES buffer twice before observation. Images were acquired and digitized on a BIOLEVO BZ-9000 epifluorescence microscope (Keyence, Osaka, Japan), and analyzed at 200× magnification using the associated software (Keyence).

4.5. Electrophysiological Measurements

Whole-cell voltage clamp experiments were performed as described previously [58]. L-type Ca^{2+} channel current ($I_{Ca.L}$) was recorded from a holding potential (V_H) of −50 mV followed by various test potentials. $I_{Ca.L}$ density was obtained by normalizing $I_{Ca.L}$ to the cell capacitance. All experiments were conducted at 37 °C. For measuring $I_{Ca.L}$, the bath solution was composed by Na^+- and K^+-free solution contained (mM): Tetraethylammonium chloride (TEA-Cl) 120, CsCl 6, 4-aminopyridine (4-AP) 5, $MgCl_2$ 0.5, 4,4P-diisothiocyanostilbene-2,2P-disulfonic acid (DIDS) 0.1, HEPES 10, $CaCl_2$ 1.8, and glucose 10 (pH of 7.4 adjusted with TEA-OH). The pipette solution contained (mM): CsCl 130, Mg-ATP 2, EGTA 5, and HEPES 10 (pH of 7.2 adjusted with 1 M CsOH).

4.6. Quantitative Real-Time PCR

Total RNA was extracted from rat neonatal ventricular cardiomyocytes using TRIzol (Invitrogen, Carlsbad, CA, USA) 24 h after the treatment with agents described above. The single-stranded cDNA was synthesized from 1 µg of total RNA using Transcriptor First Strand cDNA Synthesis Kit (Roche Molecular System Inc., Alameda, CA, USA). Real-time PCR was performed on Light Cycler (Roche) using the FastStart DNA Master SYBR Green I (Roche) as a detection reagent. Forward and reverse primer sequences, respectively, for mouse L-type Ca^{2+} channel isoforms and transcription factor were designed from their sequence in the GeneBank database as follows (accession numbers are indicated in parentheses): CACNA1C (NM_001255999), forward 5′-ACATCTTCGTGGGTTTCGTC-3′, reverse 5′-TGTTGAGCAGGATGAGAACG-3′; CACNA1D (NM_028981), 5′-CTTTTGGAGCCTTCTT GCAC-3′, reverse 5′-CTGGACTGAATCCCAAAGGA-3′; FFAR4 (NM_181748), forward 5′-GCCCAACCGCATAGGAGAAA-3′, reverse 5′-GTCTTGTTGGGACACTCGGA-3′; CREB (NM_133828), forward 5′-TGGAGTTGTTATGGCGTCCT-3′, and reverse 5′-CGACATTCTC TTGCTGCCTC-3′. Glyceraldehydes-3-phosphate dehydrogenase (GAPDH; GU214026) mRNA was used as an internal control. Data were calculated by $2^{-\Delta\Delta CT}$ and presented as fold change in transcripts for Cav1.2 genes in myocytes and normalized to GAPDH (defined as 1.0 fold).

4.7. Western Blot Analysis

Cultured neonatal mouse cardiomyocytes were treated with OAPA in the presence or absence of EPA for 24 h in DMEM. After the treatments, cardiomyocytes were washed twice with ice cold PBS and harvested using cell scraper, and then lysed in RIPA buffer containing protease inhibitor mixture and phosphatase inhibitors on ice for 1 h. The extracted protein concentration was determined by the BCA protein assay kit (Pierce). Samples containing 40 µg were denatured at 95 °C for 5 min in loading buffer [Tris-HCl (pH 6.8) 250 mM, 4% SDS, 1% β-mercaptoethanol, 1% bromophenol blue, and 20% glycerol], separated by SDS polyacrylamide gel electrophoresis using 10% polyacrylamide gel, and then transferred from the gel to a PVDF membrane (Hybond-P; GE Healthcare Bio-Sciences, Piscataway, NJ, USA). To prevent nonspecific binding, the blotted membranes were blocked with 5% skim milk in tris-buffered saline (TBS) with 0.1% Tween 20 (TBST) for 1 h at room temperature and then probed overnight at 4 °C with an anti-CREB antibody (1:1000, Cell Signaling, Beverly, MA, USA), phosphospecific antibody against anti-pCREB (Ser 133) (1:1000, Cell Signaling), and Cav1.2 (1:200, Alomone Labs, Jerusalem, Israel). The blot was visualized with anti-rabbit IgG horseradish peroxidaseconjugated secondary antibodies (1:2000, American Qualex, San Clemente, CA, USA) and an ECL prime Western Blotting Detection System (GE Healthcare Bio-Sciences). GAPDH antibody (1:1000, Proteintech, Wuhan, China) was used as loading control. The relative target protein levels were quantified by densitometry normalized to GAPDH on the same membrane. Band density was measured using Image J 1.51 software (National Institute of Health, Bethesda, MD, USA).

4.8. Immunocytochemistry

The details of the experiments protocol were performed as described previously [56]. Primary antibodies against Phospho-CREB (1:200, Cell Signaling, Beverly, MA, USA) and Cav1.2 (1:200, Alomone Labs Ltd., Jerusalem, Israel) were applied following incubation with the appropriate fluorescence-labeled secondary antibodies (Cell Signaling) for 1 h at room temperature. After several washes, the samples were air-dried, mounted with a drop of ProLong Diamond Antifade Mountant with DAPI (Molecular Probes) and subjected to microscopy. Images were obtained with a confocal microscope system (A1R, Nikon, Tokyo, Japan) equipped with a PlanFluor 60× objective lens and excitation lasers (488 and 561 nm, Melles Griot, New York, NY, USA). Images were saved in TIFF format and analyzed by ImageJ software (Wayne Rasband, National Institutes of Health).

4.9. Statistical Analysis

Statistical analysis was conducted using SigmaPlot 14.0 (SigmaPlot version 14.0-Systat Software, Inc., London, UK). All data are expressed as mean ± SE. The significance of differences was determined by one-way ANOVA followed by Tukey's test. Values of $p < 0.05$ were considered statistically significant.

Author Contributions: Conceptualization, M.M. and K.O.; methodology, M.M. and K.H. (Kazuki Horikawa); validation, M.M. and K.O.; formal analysis. M.M.; investigation, M.M., K.H. (Kosuke Horii) and P.W.; data curation, M.M.; writing—original draft preparation, M.M.; writing—review and editing, M.M. and K.O.; project administration, M.M.; funding acquisition, M.M. All authors have read and agreed to the published version of the manuscript.

Funding: This work was supported by KAKENHI grants #20K11636, #23K10815 to M.M., and the 2023 Kindai University Research Enhancement Grant (IP006) to M.M.

Institutional Review Board Statement: The study was conducted according to the Kindai University Animal Experimentation Regulations (Approval number: KAAG-2020-015, and KAAG-2023-003) and were carried out according to the guidelines for animal research of the Physiological Society of Japan. Also, this study was conducted according to the guidelines of the Declaration of Helsinki, and approved by the Ethics Review Committee for Animal Experimentation of Oita University School of Medicine (No. G004006), and were carried out according to the guidelines for animal research of the Physiological Society of Japan to minimize the number of animals used, as well as their suffering.

Informed Consent Statement: Not applicable.

Data Availability Statement: The data in this study are available from the corresponding authors on reasonable request.

Acknowledgments: The authors would like to thank M. Matsuda and H. Murakami for their technical support.

Conflicts of Interest: The authors declare no conflicts of interest.

References

1. Jouven, X.; Charles, M.A.; Desnos, M.; Ducimetiere, P. Circulating nonesterified fatty acid level as a predictive risk factor for sudden death in the population. *Circulation* **2001**, *104*, 756–761. [CrossRef]
2. Oliver, M.F.; Kurien, V.A.; Greenwood, T.W. Relation between serum-free-fatty acids and arrhythmias and death after acute myocardial infarction. *Lancet* **1968**, *291*, 710–714. [CrossRef]
3. Chiuve, S.E.; Rimm, E.B.; Sandhu, R.K.; Bernstein, A.M.; Rexrode, K.M.; Manson, J.E.; Willett, W.C.; Albert, C.M. Dietary fat quality and risk of sudden cardiac death in women. *Am. J. Clin. Nutr.* **2012**, *96*, 498–507. [CrossRef]
4. Lemaitre, R.N.; King, I.B.; Sotoodehnia, N.; Knopp, R.H.; Mozaffarian, D.; McKnight, B.; Rea, T.D.; Rice, K.; Friedlander, Y.; Lumley, T.S.; et al. Endogenous red blood cell membrane fatty acids and sudden cardiac arrest. *Metabolism* **2010**, *59*, 1029–1034. [CrossRef]
5. Sharma, S.; Adrogue, J.V.; Golfman, L.; Uray, I.; Lemm, J.; Youker, K.; Noon, G.P.; Frazier, O.H.; Taegtmeyer, H. Intramyocardial lipid accumulation in the failing human heart resembles the lipotoxic rat heart. *FASEB J.* **2004**, *18*, 1692–1700. [CrossRef]
6. Joseph, L.C.; Avula, U.M.R.; Wan, E.Y.; Reyes, M.V.; Lakkadi, K.R.; Subramanyam, P.; Nakanishi, K.; Homma, S.; Pajvani, U.B.; Thorp, E.B.; et al. Dietary saturated fat promotes arrhythmia by activating NOX2 (NADPH oxidase 2). *Circ. Arrhythmia Electrophysiol.* **2019**, *12*, e007573. [CrossRef]
7. Fauconnier, J.; Andersson, D.C.; Zhang, S.J.; Lanner, J.T.; Wibom, R.; Katz, A.; Bruton, J.D.; Westerblad, H. Effects of palmitate on Ca^{2+} handling in adult control and ob/ob cardiomyocytes. *Diabetes* **2007**, *56*, 1136–1142. [CrossRef] [PubMed]
8. Aromolaran, A.S. Mechanisms of electrical remodeling in lipotoxic guinea pig heart. *Biochem. Biophys. Res. Commun.* **2019**, *519*, 639–644. [CrossRef] [PubMed]
9. de Vries, J.E.; Vork, M.M.; Roemen, T.H.; de Jong, Y.F.; Cleutjens, J.P.; van der Vusse, G.J.; van Bilsen, M. Saturated but not monounsaturated fatty acids induce apoptotic cell death in neonatal rat ventricular myocytes. *J. Lipid Res.* **1997**, *38*, 1384–1394. [CrossRef] [PubMed]
10. Abdelmagid, S.A.; Clarke, S.E.; Nielsen, D.E.; Badawi, A.; El-Sohemy, A.; Mutch, D.M.; Ma, D.W.L. Comprehensive profiling of plasma fatty acid concentrations in young healthy canadean adults. *PLoS ONE* **2015**, *10*, e0116195.
11. Wende, A.R.; Abel, E.D. Lipotoxicity in the heart. *Biochem. Biophys. Acta* **2010**, *1801*, 311–319. [CrossRef] [PubMed]
12. Cetrullo, S.; D'Adamo, S.; Panichi, V.; Borzi, R.M.; Pignatti, C.; Flamigni, F. Modulation of fatty acid-related genes in the response of H9c2 cardiac cells to palmitate and n-3 polyunsaturated fatty acids. *Cells* **2020**, *9*, 537. [CrossRef]
13. Goldberg, I.J.; Trent, C.M.; Schulze, P.C. Lipid metabolism and toxicity in the heart. *Cell Metab.* **2012**, *15*, 805–812. [CrossRef]
14. Miller, T.A.; LeBrasseur, N.K.; Cote, G.M.; Trucillo, M.P.; Pimentel, D.R.; Ido, Y.; Ruderman, N.B.; Sawyer, D.B. Oleate prevents palmitate-induced cytotoxic stress in cardiac myocytes. *Biochim. Biophys. Res. Commun.* **2005**, *336*, 309–315. [CrossRef]

15. Mazumder, P.K.; O'Neill, B.T.; Roberts, M.W.; Buchanan, J.; Yun, U.J.; Cooksey, R.C.; Boudina, S.; Abel, E.D. Impaired cardiac efficiency and increased fatty acid oxidation in insulin-resistant ob/ob mouse hearts. *Diabetes* **2004**, *53*, 2366–2374. [CrossRef]
16. Kourie, J.I. Interaction of reactive oxygen species with ion transport mechanisms. *Am. J. Physiol.* **1998**, *275*, C1–C24. [CrossRef]
17. Kawakami, M.; Okabe, E. Superoxide anion radical-triggered Ca^{2+} release from cardiac sarcoplasmic reticulum through ryanodine receptor Ca^{2+} channel. *Mol. Pharmacol.* **1998**, *53*, 497–503. [CrossRef]
18. He, K.; Song, Y.; Daviglus, M.L.; Liu, K.; Van Horn, L.; Dyer, A.R.; Greenland, P. Accumulated evidence on fish consumption and coronary heart disease mortality: A meta-analysis of cohort studies. *Circulation* **2004**, *109*, 2705–2711. [CrossRef] [PubMed]
19. Kris-Etherton, P.M.; Harris, W.S.; Appel, L.J. Fish consumption, fish oil, omega-3 fatty acids, and cardiovascular disease. *Circulation* **2002**, *106*, 2747–2757. [CrossRef] [PubMed]
20. Gissi-HF, I. Effect of n-3 polyunsaturated fatty acids in patients with chronic heart failure (the GISSI-HF trial): A randomized, double-blind, placebo-controlled trial. *Lancet* **2008**, *372*, 1223–1230. [CrossRef]
21. O'Keefe, J.H.; Abuissa, H.; Sastre, A.; Steinhaus, D.M.; Harris, W.S. Effects of omega-3 fatty acids on resting heart rate, heart rate variability in men with healed myocardial infractions and depressed ejection fractions. *Am. J. Cardiol.* **2006**, *97*, 1127–1130. [CrossRef] [PubMed]
22. Macleod, J.C.; Macknight, A.D.C.; Rodrigo, G.C. The electrical and mechanical response of adult guinea pig and rat ventricular myocytes to ω3 polyunsaturated fatty acids. *Eur. J. Pharmacol.* **1998**, *356*, 261–270. [CrossRef]
23. Xiao, Y.F.; Ma, L.; Wang, S.Y.; Josephson, M.E.; Wang, G.K.; Morgan, J.P.; Leaf, A. Potent block of inactivation-deficient Na^+ channels by n-3 polyunsaturated fatty acids. *Am. J. Physiol. Cell Physiol.* **2006**, *290*, C362–C370. [CrossRef] [PubMed]
24. Calder, P.C. N-3 fatty acids and cardiovascular disease: Evidence explained and mechanisms explored. *Clin. Sci.* **2004**, *107*, 1–11. [CrossRef] [PubMed]
25. Ferrier, G.R.; Redondo, I.; Zhu, J.; Murphy, M.G. Differential effects of docosahezaenoic acid on contractions and L-type Ca^{2+} current in adult cardiac myocytes. *Cardovasc. Res.* **2002**, *54*, 601–610. [CrossRef] [PubMed]
26. Hallaq, H.; Smith, T.W.; Leaf, A. Modulation of dihydropyridine-sensitive calcium channels in heart cells by fish oil fatty acids. *Proc. Natl. Acad. Sci. USA* **1992**, *89*, 1760–1764. [CrossRef] [PubMed]
27. Endo, J.; Arita, M. Cardioprotective mechanism of omega-3 polyunsaturated fatty acids. *J. Cardiol.* **2016**, *67*, 22–27. [CrossRef]
28. Mozaffarian, D.; Wu, J.H. Omega-3 fatty acids and cardiovascular disease: Effects on risk factors, molecular pathways, and clinical events. *J. Am. Coll. Cardiol.* **2011**, *58*, 2047–2067. [CrossRef] [PubMed]
29. Denys, A.; Hichami, A.; Khan, N.A. Eicosapentaenoic acid and docosahexaenoic acid modulate MAP kinase (ERK1/ERK2) signaling in human T cells. *J. Lipid Res.* **2001**, *42*, 2015–2020. [CrossRef]
30. Oh, D.Y.; Talukdar, S.; Bae, E.J.; Imamura, T.; Morinaga, H.; Fan, W.Q.; Li, P.; Lu, W.J.; Watkins, S.M.; Olefsky, J.M. GPR120 is an omega-3 fatty acid receptor mediating potent anti-inflammatory and insulin sensitizing effects. *Cell* **2010**, *142*, 687–698. [CrossRef]
31. Christiansen, E.; Watterson, K.R.; Stocker, C.J.; Sokol, E.; Jenkins, L.; Simon, K.; Grundmann, M.; Petersen, R.K.; Wargent, E.T.; Hudson, B.D.; et al. Activity of dietary fatty acids on FFA1 and FFA4 and characterization of pinolenic acid as a dual FFA1/FFA4 agonist with potential effect against metabolic diseases. *Br. J. Nutr.* **2015**, *113*, 1677–1688. [CrossRef]
32. Ichimura, A.; Hirasawa, A.; Poulain-Godefroy, O.; Bonnefond, A.; Hara, T.; Yengo, L.; Kimura, I.; Leloire, A.; Liu, N.; Iida, K.; et al. Dysfunction of lipid sensor GPR120 leads to obesity in both mouse and human. *Nature* **2012**, *483*, 350–354. [CrossRef]
33. Murphy, K.A.; Harsch, B.A.; Healy, C.L.; Joshi, S.S.; Huang, S.; Walker, R.E.; Wagner, B.M.; Ernste, K.M.; Huang, W.; Block, R.C.; et al. Free fatty acid receptor 4 responds to endogenous fatty acids to protect the heart from pressure overload. *Cardiovasc. Res.* **2022**, *118*, 1061–1073. [CrossRef]
34. Wang, C.P.; Lee, C.C.; Wu, D.Y.; Chen, S.; Lee, T.M. Differential effects of EPA and DHA on PPARγ-mediated sympathetic innervation in infarcted rat hearts by GPR120-dependent and -independent mechanisms. *J. Nutr. Biochem.* **2022**, *103*, 108950. [CrossRef]
35. Morishima, M.; Tahara, S.; Wang, Y.; Ono, K. Oxytocin downregulates the Cav1.2 L-type Ca^{2+} channel via Gi/cAMP/PKA/CREB signaling pathway in cardiomyocytes. *Membranes* **2021**, *11*, 234. [CrossRef]
36. Kang, S.; Huang, J.; Lee, B.-K.; Jung, Y.-S.; Im, E.; Koh, J.-M.; Im, D.-S. Omega-3 polyunsaturated fatty acids protect human hepatoma cells from developing steatosis through FFA4 (GPR120). *BBA Mol. Cell Biol. Lipids* **2018**, *1863*, 105–116. [CrossRef]
37. Porter, G.A., Jr.; Makuck, R.F.; Rivkees, S.A. Intracellular calcium plays an essential role in cardiac development. *Dev. Dyn.* **2003**, *227*, 280–290. [CrossRef]
38. Wakili, R.; Voigt, N.; Kääb, S.; Dobrev, D.; Nattel, S. Recent advances in the molecular pathophysiology of atrial fibrillation. *J. Clin. Investig.* **2011**, *121*, 2955–2968. [CrossRef]
39. Tachtsis, B.; Whitfield, J.; Hawley, J.A.; Hoffman, N.J. Omega-3 polyunsaturated fatty acids mitigate palmitate-induced impairments in skeletal muscle cell viability and differentiation. *Front. Physiol.* **2020**, *11*, 563. [CrossRef]
40. Rocca, C.; Bartolo, A.D.; Guzzi, R.; Crocco, M.C.; Rago, V.; Romeo, N.; Perrotta, I.; Francesco, E.M.D.; Muoio, M.G.; Granieri, M.C.; et al. Palmitate-induced cardiac lipotoxicity is relieved by the redox-active motif of SELENOT through improving mitochondrial function and regulating metabolic state. *Cells* **2023**, *12*, 1042. [CrossRef]
41. Farías, J.G.; Carrasco-Pozo, C.; Carrasco Loza, R.; Sepúlveda, N.; Álvarez, P.; Quezada, M.; Quiñones, J.; Molina, V.; Castillo, R.L. Polyunsaturated fatty acid induces cardioprotection against ischemia-reperfusion through the inhibition of NF-kappaB and induction of Nrf2. *Exp. Biol. Med.* **2017**, *242*, 1104–1114. [CrossRef]

42. Moreno, C.; Macias, A.; Prieto, A.; de la Cruz, A.; Gonzalez, T.; Valenzuela, C. Effects of n-3 polyunsaturated fatty acids on cardiac ion channels. *Front. Physiol.* **2012**, *3*, 245. [CrossRef]
43. Bohannon, B.M.; de la Cruz, A.; Wu, X.; Jowais, J.J.; Perez, M.E.; Dykxhoorn, D.M.; Liin, S.I.; Larsson, H.P. Polyunsaturated fatty acid analogues differentially affect cardiac Nav, Cav, and Kv channels through unique mechanisms. *eLife* **2020**, *9*, e51453. [CrossRef]
44. Xiao, Y.F.; Wright, S.N.; Wang, G.K.; Morgan, J.P.; Leaf, A. Fatty acids suppress voltage-gated Na^+ currents in HEK293t cells transfected with the alpha-subunit of the cardiac Na^+ channel. *Proc. Natl. Acad. Sci. USA* **1998**, *95*, 2680–2685. [CrossRef]
45. Xiao, Y.F.; Gomez, A.M.; Morgan, J.P.; Lederer, W.J.; Leaf, A. Suppression of voltage-gated L-type Ca^{2+} currents by polyunsaturated fatty acids in adult and neonatal rat ventricular myocytes. *Proc. Natl. Acad. Sci. USA* **1997**, *94*, 4182–4187. [CrossRef]
46. Xu, X.; Jiang, M.; Wang, Y.; Smith, T.; Baumgarten, C.M.; Wood, M.A.; Tseng, G.N. Long-term fish oil supplementation induces cardiac electrical remodeling by changing channel protein expression in the rabbit model. *PLoS ONE* **2010**, *5*, e10140. [CrossRef]
47. Sheikh, O.; Hei, A.G.V.; Battisha, A.; Hammad, T.; Pham, S.; Chilton, R. Cardiovascular, electrophysiologic, and hematologic effects of omega-3 fatty acids beyond reducing hypertriglyceridemia: As it pertains to the recently published REDUCE-IT trial. *Cardiovasc. Diabetol.* **2019**, *18*, 84. [CrossRef]
48. Tsuburaya, R.; Yasuda, S.; Ito, Y.; Shiroto, T.; Gao, J.Y.; Ito, K.; Shimokawa, H. Eicosapentaenoic acid reduces ischemic ventricular fibrillation via altering monophasic action potential in pigs. *J. Mol. Cell. Cardiol.* **2011**, *51*, 329–336. [CrossRef]
49. Tong, M.; Wang, J.; Ji, Y.; Chen, X.; Wang, J.; Wang, S.; Ruan, L.; Cui, H.; Zhou, Y.; Zhang, Q.; et al. Effect of eicosapentaenoic acid and pitavastatin on electrophysiology and anticoagulant gene expression in mouse with rapid atrial pacing. *Exp. Ther. Med.* **2017**, *14*, 2310–2316. [CrossRef]
50. Kang, J.X.; Li, Y.; Leaf, A. Regulation of sodium channel gene expression by class I antiarrhythmic drugs and n-3 polyunsaturated fatty acids in cultured neonatal rat cardiac myocytes. *Proc. Natl. Acad. Sci. USA* **1997**, *94*, 2724–2728. [CrossRef]
51. Leaf, A. The electrophysiological basis for the antiarrhythmic actions of polyunsaturated fatty acids. *Eur. Heart J. Suppl.* **2001**, *3*, D98–D105. [CrossRef]
52. Dumancas, G.G.; Kimani, M.; Purdie, N.; Reilly, L. Partial least squares (PLS1) algorithm for quantitating cholesterol and polyunsaturated fatty acids in human serum. *J. Biotech Res.* **2010**, *2*, 121–130.
53. von Schacky, C. Omega-3 fatty acids: Anti-arrhythmic, pro-arrhythmic, or both? *Fornt. Physiol.* **2012**, *3*, 88. [CrossRef]
54. Kitamura, K.; Shibata, R.; Tsuji, Y.; Shimano, M.; Inden, Y.; Murohara, T. Eicosapentaenoic acid prevents atrial fibrillation associated with heart failure in a rabbit model. *Am. J. Physiol. Heart Circ. Physiol.* **2011**, *300*, H1814–H1821. [CrossRef] [PubMed]
55. Mozaffarian, D.; Geelen, A.; Brouwer, I.A.; Geleijnse, J.M.; Zock, P.L.; Katan, M.B. Effect of fish oil on heart rate in humans: A meta-analysis of randomized controlled trials. *Circulation* **2005**, *112*, 1945–1952. [CrossRef]
56. Morishima, M.; Horikawa, K.; Funaki, M. Cardiomyocytes cultured on mechanically compliant substrates, but not on conventional culture devices, exhibit prominent mitochondrial dysfunction due to reactive oxygen species and insulin resistance under high glucose. *PLoS ONE* **2018**, *13*, e0201891. [CrossRef]
57. Wang, Y.; Morishima, M.; Dan, L.; Takahashi, N.; Saikawa, T.; Nattel, S.; Ono, K. Binge alcohol exposure triggers atrial fibrillation through T-type Ca^{2+} channel upregulation via protein kinase C (PKC)/glycogen synthesis kinase 3β (GSK3β)/nuclear factor of activated T-cells (NFAT) signaling-an experimental account of holiday heart syndrome. *Circ. J.* **2020**, *84*, 1931–1940.
58. Wang, Y.; Morishima, M.; Zheng, M.; Uchino, T.; Mannen, K.; Takahashi, A.; Nakaya, Y.; Komuro, I.; Ono, K. Transcription factors Csx/Nkx2.5 and GATA4 distinctly regulate expression of Ca^{2+} channels in neonatal rat cardiomyocytes. *J. Mol. Cell. Cardiol.* **2007**, *42*, 1045–1053. [CrossRef]

International Journal of
Molecular Sciences

MDPI

Article

Arrhythmogenic Ventricular Remodeling by Next-Generation Bruton's Tyrosine Kinase Inhibitor Acalabrutinib

Yanan Zhao [1,†], Praloy Chakraborty [1,†], Julianna Tomassetti [1], Tasnia Subha [1], Stéphane Massé [1], Paaladinesh Thavendiranathan [1,2], Filio Billia [1,2], Patrick F. H. Lai [1], Husam Abdel-Qadir [1,2,3] and Kumaraswamy Nanthakumar [1,*]

[1] Toronto General Hospital Research Institute, University Health Network, Toronto, ON M5G 2M1, Canada; ynzhao17@mails.jlu.edu.cn (Y.Z.); praloy.chakraborty@uhn.ca (P.C.); juliannatomassetti101@gmail.com (J.T.); tasniasubha@outlook.com (T.S.); stephane.masse@uhn.ca (S.M.); dinesh.thavendiranathan@uhn.ca (P.T.); phyllis.billia@uhn.ca (F.B.); patrick.lai@uhn.ca (P.F.H.L.); husam.abdel-qadir@uhn.ca (H.A.-Q.)

[2] Ted Rogers Centre for Heart Research, Toronto, ON M5G 1M1, Canada

[3] Women's College Hospital, Toronto, ON M5S 1B2, Canada

* Correspondence: kumar.nanthakumar@uhn.ca; Tel.: +1-(416)340-4442

† These authors contributed equally to this work.

Abstract: Cardiac arrhythmias remain a significant concern with Ibrutinib (IBR), a first-generation Bruton's tyrosine kinase inhibitor (BTKi). Acalabrutinib (ABR), a next-generation BTKi, is associated with reduced atrial arrhythmia events. However, the role of ABR in ventricular arrhythmia (VA) has not been adequately evaluated. Our study aimed to investigate VA vulnerability and ventricular electrophysiology following chronic ABR therapy in male Sprague–Dawley rats utilizing epicardial optical mapping for ventricular voltage and Ca^{2+} dynamics and VA induction by electrical stimulation in ex-vivo perfused hearts. Ventricular tissues were snap-frozen for protein analysis for sarcoplasmic Ca^{2+} and metabolic regulatory proteins. The results show that both ABR and IBR treatments increased VA vulnerability, with ABR showing higher VA regularity index (RI). IBR, but not ABR, is associated with the abbreviation of action potential duration (APD) and APD alternans. Both IBR and ABR increased diastolic Ca^{2+} leak and Ca^{2+} alternans, reduced conduction velocity (CV), and increased CV dispersion. Decreased SERCA2a expression and AMPK phosphorylation were observed with both treatments. Our results suggest that ABR treatment also increases the risk of VA by inducing proarrhythmic changes in Ca^{2+} signaling and membrane electrophysiology, as seen with IBR. However, the different impacts of these two BTKi on ventricular electrophysiology may contribute to differences in VA vulnerability and distinct VA characteristics.

Keywords: ibrutinib; acalabrutinib; action potential; calcium cycling; ventricular arrhythmia; electrical remodeling

check for updates

Citation: Zhao, Y.; Chakraborty, P.; Tomassetti, J.; Subha, T.; Massé, S.; Thavendiranathan, P.; Billia, F.; Lai, P.F.H.; Abdel-Qadir, H.; Nanthakumar, K. Arrhythmogenic Ventricular Remodeling by Next-Generation Bruton's Tyrosine Kinase Inhibitor Acalabrutinib. *Int. J. Mol. Sci.* **2024**, *25*, 6207. https://doi.org/10.3390/ijms25116207

Academic Editors: Yosuke Okamoto and Kunichika Tsumoto

Received: 20 April 2024
Revised: 22 May 2024
Accepted: 24 May 2024
Published: 5 June 2024

1. Introduction

The introduction of Bruton's tyrosine kinase inhibitors (BTKi) is an important therapeutic advance in the management of B-cell lymphocyte malignancies [1]. This targeted therapeutic strategy avoids the cytotoxic side effects of conventional cancer chemotherapeutics. However, Ibrutinib (IBR)-associated cardiotoxicity is a significant concern, with cardiac arrhythmias being the most common cardiotoxic side effect [2,3]. Although atrial fibrillation (AF) is the most common cardiotoxicity associated with IBR, often prompting treatment cessation [3,4], ventricular arrhythmia (VA) also presents a substantial burden and is potentially fatal [5,6]. Off-target inhibition of other cellular kinases by IBR, a covalent BTKi, is thought to be responsible for cardiovascular side effects, including arrhythmias [7]. Acalabrutinib (ABR) is a next-generation selective BTKi with reduced off-target kinase inhibition [8]. In a head-to-head comparison of clinical efficacy in chronic lymphocytic leukemia, ABR was found to be non-inferior to IBR in progression-free survival, with a significantly lower incidence of AF [9]. A higher selectivity for BTK inhibition with a lesser degree of inhibition of off-target C-terminal Src Kinase (CSK), PI3K-Akt signaling, and Ca^{2+}/Calmodulin-dependent protein kinase II (CaMKII) may explain a lower incidence of AF with ABR [7,10–12].

Despite the small number of investigations that have drawn attention to VAs associated with ABR [13,14], its impact on VA remains inadequately assessed and the underlying mechanisms of ABR-induced VAs remain unclear. Our group has demonstrated that acute as well as chronic IBR treatments are associated with the aberration of membrane voltage and intracellular calcium (Ca^{2+}) dynamics in the heart, with a subsequent increase in VA vulnerability [15,16]. We hypothesized that ABR and IBR have distinct effects on membrane electrophysiology and calcium signaling, resulting in different levels of VA vulnerability and characteristics. We evaluated the effects of chronic ABR treatment on ventricular electrophysiology, Ca^{2+} dynamics, and arrhythmia susceptibility in a rodent model. We also compared the arrhythmogenic alteration in ventricular electrophysiology, VA susceptibility, VA characteristics, Ca^{2+} handling, and metabolic regulatory proteins between IBR and ABR treatments. This comparison may shed light on the differing clinical outcomes observed between IBR and ABR.

2. Results

2.1. Effects of IBR and ABR Treatment on VA Vulnerability and Characteristics

The effect of burst pacing is shown in Figure 1A. IBR treatment markedly increased VA vulnerability (62.0 ± 10.1% in IBR vs. 16.0 ± 6.5% in control; $p = 0.007$) and VA burden (148.7 ± 30.0 s in IBR vs. 47.6 ± 19.4 s in control; $p = 0.017$), compared to control hearts. The trends for VA vulnerability and VA burden with ABR treatment were higher than those of the control (52.0 ± 10.4% in ABR vs. 16.0 ± 6.5% in control for VA vulnerability; 110.5 ± 25.4 s in ABR vs. 47.6 ± 19.4 s in control for VA burden) but lower than seen with IBR treatment (52.0 ± 10.4% in ABR vs. 62.0 ± 10.1% in IBR for VA vulnerability; 110.5 ± 25.4 s in ABR vs. 148.7 ± 30.0 s IBR for VA burden). Although the ABR-treated hearts showed a tendency to higher VA vulnerability, as compared to control ($p = 0.056$), this did not achieve statistical significance. Similarly, statistical analysis did not demonstrate a significant difference in VA vulnerability between ABR-treated hearts and IBR-treated hearts ($p > 0.99$), and the VA burden was not statistically significant in ABR-treated hearts as compared to control ($p = 0.35$), or as compared to IBR-treated hearts ($p = 0.71$) (Figure 1B,C). Dominant frequency (DF) during induced VA was similar following IBR and ABR treatments (15.02 ± 0.55 Hz in ABR vs. 14.27 ± 0.74 Hz in IBR for VA; $p = 0.42$) (Figure 1D). However, ABR treatment was associated with a higher regularity index (RI) of VA compared to VAs in IBR-treated hearts (0.56 ± 0.03 in ABR vs. 0.41 ± 0.03 in IBR; $p = 0.002$) (Figure 1E), suggesting a more organized VA pattern with ABR treatment. Figure 1F shows voltage and calcium phase maps of the epicardial surface of rat hearts in the IBR and ABR groups, suggesting that hearts from IBR-treated rats exhibit a more spatiotemporal disorganized phase map during VA than those from ABR-treated rats (Videos S1 and S2). The number of voltage and calcium wavefronts for the first and last ten seconds during VA were measured. Both the mean numbers of voltage (3.47 ± 0.29/frame vs. 2.41 ± 0.31/frame in ABR, $p = 0.03$) and Ca^{2+} (2.93 ± 0.18/frame vs. 2.33 ± 0.19/frame in ABR, $p = 0.03$) wavefronts were higher in the IBR group (Figure 1G). The pseudo-electrogram (pseudo-ECG) also showed results consistent with the above observation; the VAs in the ABR group were primarily monomorphic, and this was significantly different from the predominant polymorphic VT of the IBR group ($p = 0.015$) (Figure S1).

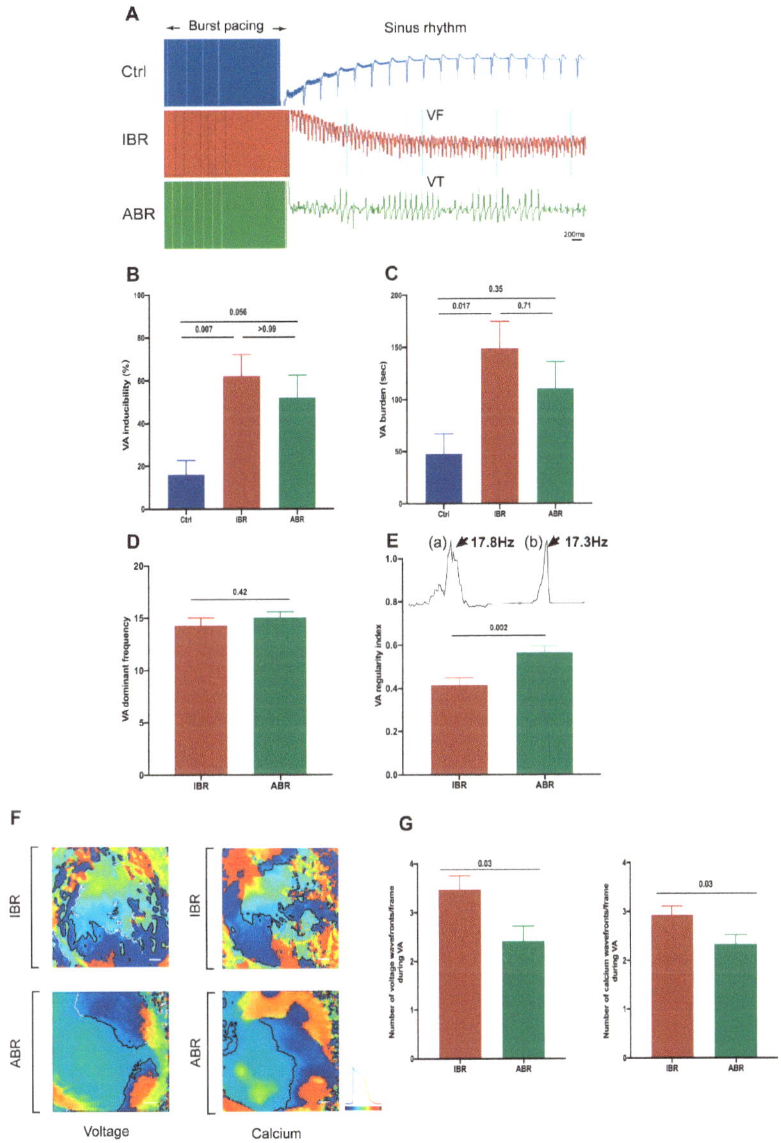

Figure 1. Effects of Ibrutinib (IBR) and Acalabrutinib (ABR) on ventricular arrhythmia (VA) vulnerability and characteristics. (**A**) Representative pseudo-electrogram (pseudo-ECGs) after VA inductions in the heart of control (Ctrl) and IBR- and ABR-treated rats. (**B**) VA inducibility and (**C**) VA burden in the ABR-treated group were higher compared to control but lower than those in the IBR-treated group (n = 10 in each group; p-value: Kruskal–Wallis test). (**D**) Dominant Frequencies (DF) of induced VA were similar in the IBR- and ABR-treated groups (n = 7 in each group; p-value: t-test). (**E**) The Regularity Index (RI) of induced VA was higher in the ABR-treated group compared to the IBR-treated group. Representative (a) IBR and (b) ABR frequency distribution during VA (n = 7 in each group; p-value: t-test). (**F**) Illustrative voltage and calcium phase mapping during VA in the hearts from IBR- and ABR-treated rats. Scale bar: 20 mm. (**G**) Number of membrane voltage and calcium wavefronts per frame during early (10 s) and late (10 s) VAs in IBR and ABR groups (n = 7–8 in each group; p-value: Kruskal–Wallis test or t-test).

2.2. Effects of IBR and ABR on Ventricular Electrophysiology

As shown in Figure 2, IBR treatment was associated with the shortening of the durations, from time 0 to 80%, relative to repolarization of action potential duration (APD_{80}) ($p = 0.008$) (Figure 2A,B), potentiated action potential duration (APD) alternans ($p = 0.046$), (Figure 2C,D) and significant APD spatially discordant alternans ($p = 0.026$) (Figure 2E,F). In contrast, ABR did not change the APD_{80} ($p = 0.71$), APD alternans ($p = 0.25$), or APD spatially discordant alternans ($p = 0.09$), compared to the control. Interestingly, both IBR and ABR treatments were associated with a significant reduction of conduction velocity (CV) ($p = 0.002$ for IBR and $p = 0.026$ for ABR) (Figure 2G,H) and an increase in CV dispersion ($p = 0.001$ for IBR and $p = 0.047$ for ABR) (Figure 2I,J), compared to controls.

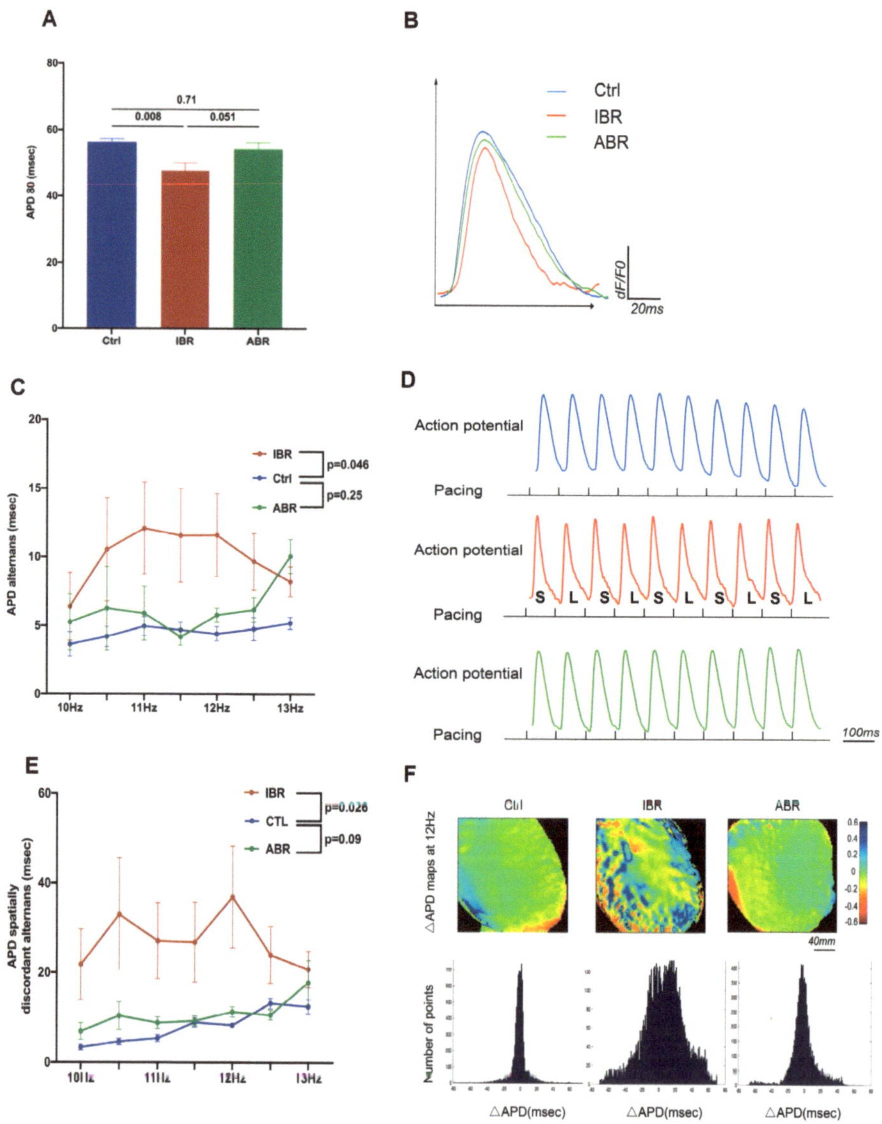

Figure 2. *Cont.*

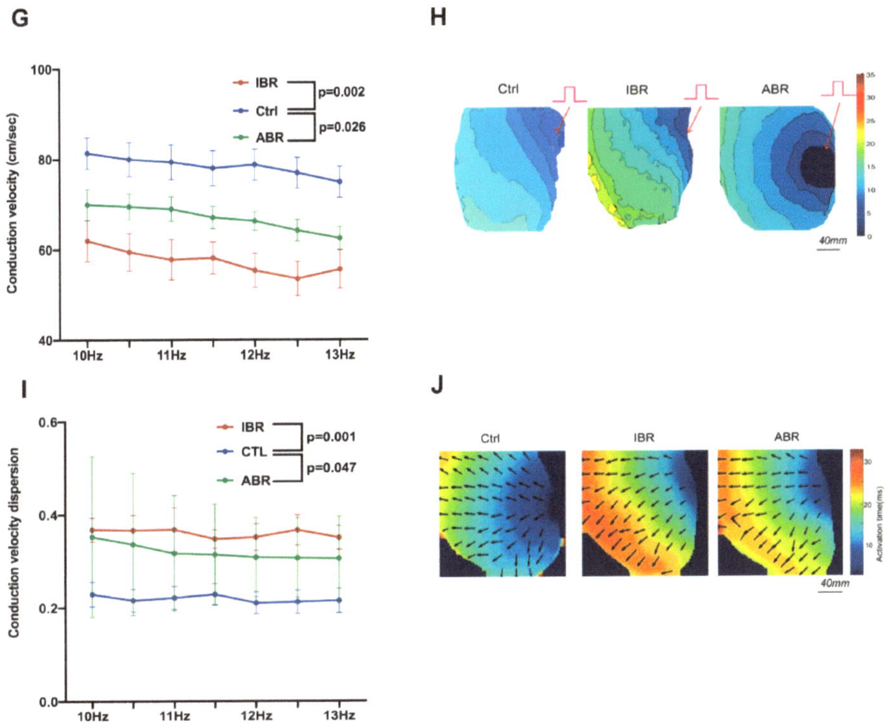

Figure 2. Effects of IBR and ABR on ventricular electrophysiology. (**A**) IBR, but not ABR, treatment was associated with reduced action potential duration (APD) at 80% repolarization (APD_{80}) ($n = 10$ at each group; p-value: t-test). (**B**) Representative traces of action potential at 10 Hz. (**C**) APD alternans was higher with IBR therapy, but not with ABR therapy. (**D**) Representative tracings with increased APD alternans from an IBR-treated heart and reduced APD alternans from Ctrl and ABR hearts at a pacing rate of 11.0 Hz. (**E**) IBR increased APD spatial discordance alternans. (**F**) Representative maps showing dispersion of APD_{80}, with color code, at 12 Hz. Histograms showing the distributions of ΔAPD across the LV epicardial mapping area, 12 Hz pacing (x-axis: ΔAPD [ms]; y-axis: the number of points from the ΔAPD map) (n = 8–10; p-value: two-way repeated ANOVA). (**G**)Epicardial conduction velocity (CV) was lower with IBR and ABR treatments. (**H**) Representative successive isochronal conduction maps with 10 Hz pacing. Red arrows show the pacing location. (**I**) CV dispersion was higher with IBR and ABR therapies. (**J**) Representative color maps showing CV dispersion in different groups during 10 Hz pacing. The arrows indicates the conduction direction. (**C**–**J**: n = 8–10 in each group; p-value: two-way repeated ANOVA).

2.3. Effects of IBR and ABR on Ventricular Calcium Dynamics

The effects of IBR and ABR treatments are shown in Figure 3. Neither IBR nor ABR treatment demonstrated a change in calcium transient (CaT) rise time compared to the control ($p = 0.97$ for IBR and $p = 0.31$ for ABR) (Figure 3A,B). However, CaT decay time (Tau value) was prolonged by IBR ($p = 0.014$) and ABR ($p = 0.046$), suggesting impaired diastolic Ca^{2+} clearance by both agents (Figure 3C,D). Spontaneous calcium elevation (SCaE), a marker of diastolic Ca^{2+} leak, was induced by IBR ($p = 0.044$), but not ABR ($p = 0.31$), treatment, and SCaE-induced spontaneous diastolic depolarization in IBR treated left ventricles (Figure 3E,F). CaT amplitude ratio was significantly reduced in IBR- ($p = 0.005$) and ABR-treated ($p = 0.039$) hearts, indicating potentiation of CaT alternans by these agents (Figure 3G,H).

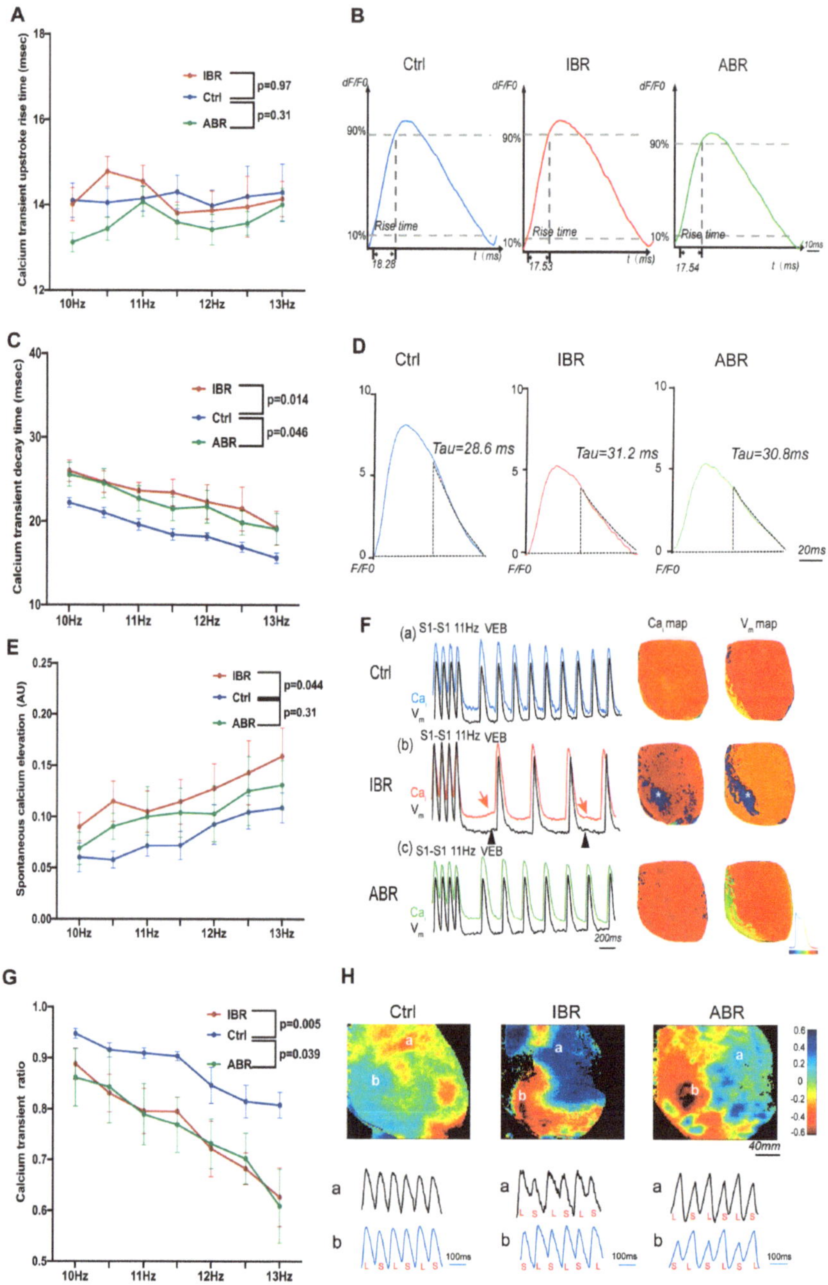

Figure 3. Effects of IBR and ABR on ventricular calcium cycling. (**A**) There were no differences in calcium transient (CaT) upstroke rise with IBR and ABR treatments. (**B**) Representative traces of CaT upstroke rise at 10 Hz. (**C**) The decay time constant of CaT was larger with IBR and ABR treatments. (**D**) Representative CaTs at 10 Hz. The decay portion of the CaT is marked as a black curve. (**E**) IBR, but not ABR, treatment was associated with significantly a greater spontaneous calcium elevation

(SCaE). (**F**) Representative traces of SCaE during 11 Hz pacing. (a–c) Indications of optical tracing and color maps of intracellular calcium (Ca_i) and membrane voltage (V_m), respectively. (b) SCaE (red arrows) and delayed after–depolarization (DAD) (black arrowhead) were elicited at the cessation of rapid pacing in the heart of IBR–treated rats. Optical images were captured from the sites labeled by asterisks in Cai and V_m maps. (**G**) More calcium alternans is observed with IBR and ABR treatments. (**H**) Color maps of calcium alternans and corresponding CaT traces at 12 Hz, a,b represent different heart regions of interest. (**A–H**: n = 9–10 in each group; p–value: two–way repeated ANOVA). VEB—ventricular escape beat.

2.4. Effects of IBR and ABR on Calcium-Handling and Metabolic Regulatory Proteins

To understand the mechanisms of abnormal calcium and voltage dynamics, we evaluate the abundance and phosphorylation of sarcoplasmic calcium-handling proteins and the metabolic regulator 5′-adenosine monophosphate-activated protein kinase (AMPK). There were no changes in the protein abundance of ryanodine receptor type 2 (RyR2) or Na^+-Ca^{2+} exchanger (NCX) with either IBR or ABR (Figure 4A). However, sarcoplasmic reticulum Ca^{2+}-ATPase 2a (SERCA2a) protein abundance was reduced by both agents ($p = 0.003$ for IBR and $p = 0.004$ for ABR) (Figure 4B). Protein abundance of phospholamban (PLB) was reduced by ABR ($p = 0.006$) and calcium/calmodulin-dependent kinase, type II (CaMKII) was reduced by IBR ($p = 0.025$) and ABR ($p = 0.021$); the degree of phosphorylation of these molecules was not altered by either of these agents (Figure 4C,D). Similarly, Ser 2814 and Ser 2808 phosphorylation of RyR2 were not changed by IBR ($p = 0.97$ for ser 2814; $p = 0.15$ for ser 2808) or ABR ($p = 0.80$ for ser 2814; $p = 0.089$ for ser 2808) (Figure 4E). IBR decreased AMPK phosphorylation ($p = 0.007$), whereas ABR reduced the protein ex abundance of AMPK ($p < 0.001$) and phosphorylated AMPK ($p < 0.001$). Hence, the p-AMPK/AMPK ratio was reduced by both IBR ($p = 0.007$) and ABR ($p = 0.002$) (Figure 4F).

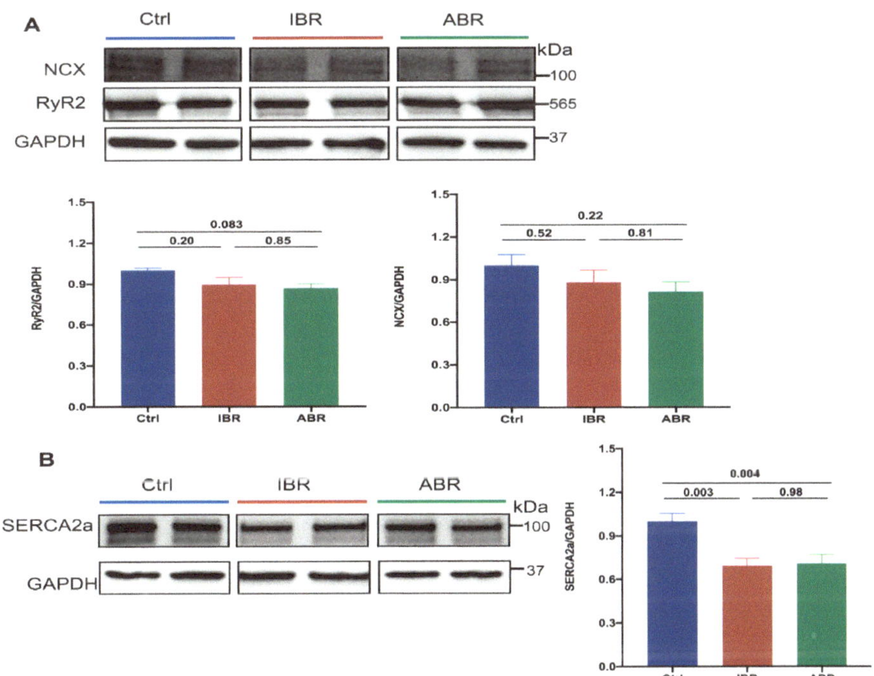

Figure 4. *Cont.*

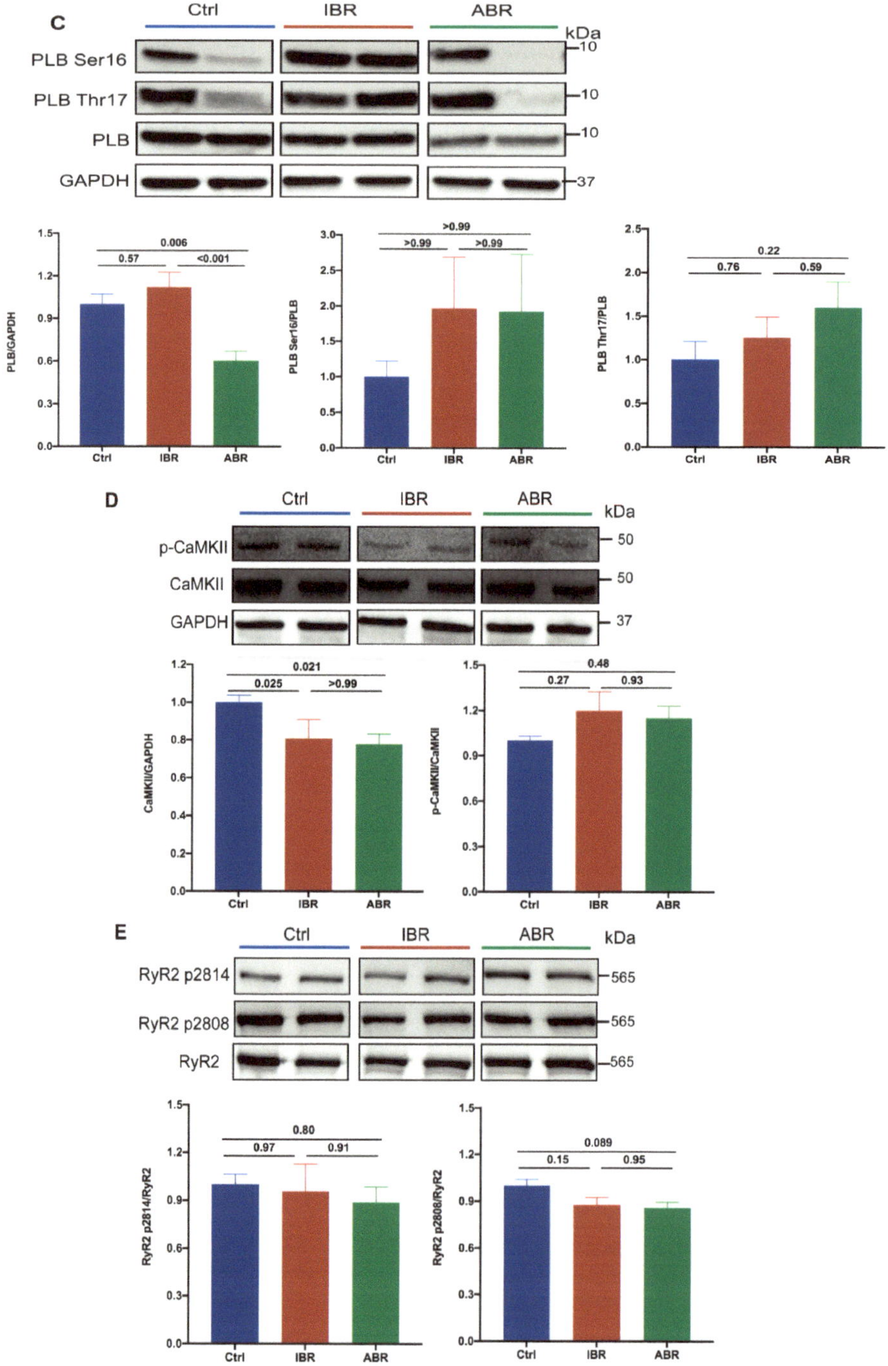

Figure 4. *Cont.*

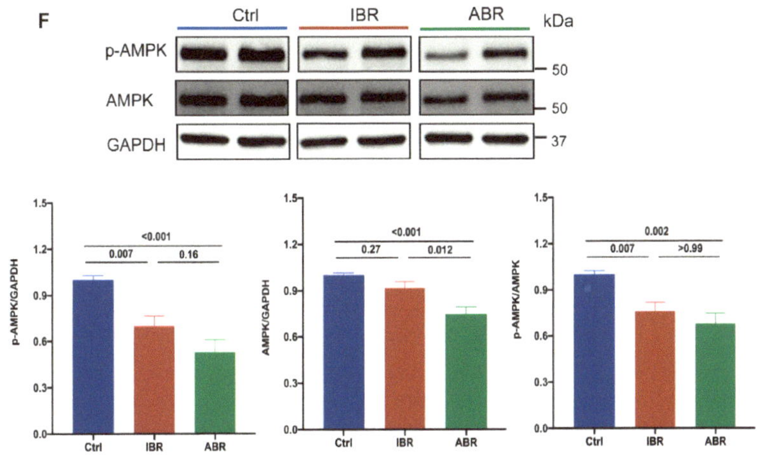

Figure 4. Effects of IBR and ABR on calcium-handling and metabolic regulatory proteins. Representative gel blots and normalized expression of (**A**) Na^+-Ca^{2+} exchanger (NCX) and total RyR2; n = 6–9 in each. (**B**) Sarcoplasmic reticulum Ca^{2+}-ATPase 2a (SERCA2a); n = 9–10 in each. (**C**)Phospholamban (PLB), and phosphorylation of PLB at Threonine-17 and Serine-16. (**D**) Calcium/calmodulin-dependent kinase, type II (CaMKII); phosphorylation of CaMKII at Thr 286 or 287. **C**,**D**: n = 9 in each. (**E**) Ryanodine receptor type 2 (RyR2) and phosphorylation of RyR2 at Ser2814 and Ser2808; n = 6 in Ctrl and n = 8 in IBR and ABR groups. (**F**) 5′-adenosine monophosphate-activated protein kinase (AMPK), phosphorylation of AMPK at Thr 172; n = 9 in each, between IBR, ABR, and Ctrl groups. The p value for (**A**–**F**): one-way ANOVA or Kruskal–Wallis test. GAPDH—glyceraldehyde 3-phosphate dehydrogenase.

We further explored the impacts of ABR on the levels of C-terminal Src kinase (CSK) and the phosphoinositol-3 kinase (PI3K)-Akt signaling pathway. Our analysis demonstrated no significant differences in CSK expression (p = 0.98; Figure S2A) or PI3K expression (p = 0.59; Figure S2B) between the control and ABR-treated groups. Additionally, there were no significant differences in Akt phosphorylation between the control and ABR groups (p = 0.70; Figure S2B).

3. Discussion

In our study, IBR treatment was associated with increased VA vulnerability and VA burden. The trend of VA vulnerability in ABR-treated hearts was higher than in controls but lower than in IBR-treated hearts, although the results did not reach statistical significance. Notably, the ABR-induced VAs were more spatially organized, compared to those with IBR. IBR treatment resulted in abnormalities in ventricular repolarization, as indicated by abbreviated APD and increased APD dispersion, as well as conduction abnormalities characterized by decreased CV and higher CV dispersion. In contrast, ABR treatment was associated only with abnormal ventricular conduction properties, including reduced ventricular CV and higher CV dispersion, albeit with a lower magnitude of CV dispersion abnormalities compared to IBR treatment. Both IBR and ABR treatments were associated with pro-arrhythmic calcium cycling abnormalities, including diastolic Ca^{2+} clearance and Ca^{2+} alternans.

3.1. Acalabrutinib and Ventricular Arrhythmia

Abnormal ventricular electrophysiology and increased VA risk caused by IBR have been demonstrated by experimental and clinical studies [6,17,18]. As a more selective BTKi, ABR is supposed to be less arrhythmogenic. However, there are reports of symptomatic VAs associated with next-generation BTKi [19]. Additionally, a follow-up study has demonstrated that while the VA risk with ABR is higher than that of the control, it remains lower than rates observed with IBR therapy [13]. Our study also suggests that although ABR treatment is associated with a higher risk of inducible VA and VA burden, the risk is lower compared to that with IBR. Furthermore, our study showed that induced VA in the heart of IBR-treated rats exhibited greater spatial disorganization than seen in hearts from the ABR group, making them more likely to cause insurmountable hemodynamic instability [20]. This finding aligns with the existing published literature on the characteristics of VA in patients treated with IBR and ABR [13,21–23]. The different impacts of IBR and ABR on ventricular membrane voltage, as demonstrated in our study, may underlie the varying vulnerability to VA and the distinct VA characteristics observed between these two agents [9].

3.2. *Acalabrutinib and Ventricular Membrane Voltage*

In our study, IBR, but not ABR, was associated with the abbreviation of APD, increased beat-to-beat APD alternans and seriously spatially discordant APD alternans. An abbreviated APD correlates with a shortened ventricular refractory period. On the other hand, spatial dispersion of APD (APD alternans) is known to be arrhythmogenic, and the risk is more with spatially discordant APD alternans [24,25]. The spatial dispersion of APD, especially the plateau phase, produces dynamic areas of unidirectional conduction block, facilitating the formation of functional reentry substrate [26–28]. On top of that, the diastolic Ca^{2+} overload generates the early after-depolarizations which trigger the induction of circus movement reentry and polymorphic ventricular tachycardia (PVT)/VF [26]. Depolarization heterogeneity from increased CV dispersion potentiates the functional reentry substrate of VA's incremental effect in creating the lines of block [29]. Both decreased CV and the reduced refractory period from abbreviated APD further promote the propagation of the reentrant PVT/VF by reducing the wavelength of the circuit [27].

ABR treatment was associated with impaired conduction properties, including decreased CV and increased CV dispersion without any repolarization abnormalities, compared to IBR. Isolated regional CV heterogeneity produces areas of functional unidirectional block, especially in presence of a premature ventricular beat [30]. The unidirectional block created by conduction heterogeneity, along with slow conduction, creates functional reentry circuits for VA, leading to the generation of monomorphic ventricular tachycardia (VT) [29,30]. The reentry circuits, in this setting, are commonly stable and fixed, leading to the generation of more organized VA [30,31].

3.3. *Acalabrutinib and Ventricular Calcium Cycling*

Both IBR and ABR therapies resulted in impaired diastolic Ca^{2+} clearance, as evidenced by an increase in diastolic Ca^{2+} decay time. Additionally, IBR treatment induced SCaE, which reflects a diastolic Ca^{2+} leak. These findings suggest that both IBR and ABR may elevate intracellular Ca^{2+} levels during diastole ($\uparrow$ $[Ca^{2+}]_i$). Abnormal ventricular Ca^{2+} signaling was previously reported following administration of IBR [15,16]. In the presence of higher diastolic $[Ca^{2+}]_i$, excess Ca^{2+} is extruded out of the cell in the exchange due to the intrusion of 3 sodium ions (Na^+) by electrogenic sodium-calcium exchange protein (NCX), leading to diastolic depolarization [32]. The diastolic depolarization subsequently produces a premature ventricular beat, which acts as a trigger for VA. Higher intracellular Ca^{2+} may also contribute to IBR- and ABR-induced reduction in CV [33]. In our study, CaT alternans was potentiated by both IBR and ABR. CaT alternans, the beat-to-beat alteration of CaT amplitude, is another manifestation of malfunction in sarcoplasmic Ca^{2+} handling. Enhanced CaT alternans is linked to ventricular arrhythmogenesis. Intracellular Ca^{2+} is shown to modulate APD by electrogenic feedback, and increased CaT alternans is associated with arrhythmogenic APD alternans [34]. To understand the mechanisms of aberrant Ca^{2+} signaling, we evaluated the expression and phosphorylation status of important Ca^{2+} handling proteins. Increased phosphorylation and activity of RyR2 is an important mechanism of abnormal Ca^{2+} signaling in a variety of cardiac conditions [32].

IBR, as well as ABR, did not modify the phosphorylation of Ser2808 and Ser2814 of RyR2 or CaMKII, ruling out CaMKII activation by both agents [32]. SERCA2a is largely responsible for sarcoplasmic Ca^{2+} uptake during diastole [32]. PLB, when associated with SERCA2a, inhibits SERCA2a activity. However, phosphorylation of PLB by protein kinase A (PKA) and CaMKII promotes dissociation of PLB from SERCA2a and increases SERCA2a activity. PLB phosphorylation was also not modified by IBR or ABR. Interestingly, SERCA2a expression was reduced by IBR and ABR. Importantly, reduced SERCA2a expression is associated with reduced SR Ca^{2+} load, diastolic Ca^{2+} overload, Ca^{2+} leak, and CaT alternans [35]. Overexpression of SERCA2a by genetic manipulation has been reported to improve cytosolic Ca^{2+} dynamics and attenuate VA [35,36]. Hence, BTKi-induced reduction in SERCA2a expression may be responsible for proarrhythmic remodeling of intracellular Ca^{2+} signaling in our study.

3.4. Acalabrutinib and Cardiac Metabolic Regulation

In our study, both steady-state protein levels and phosphorylation of AMPK were reduced by ABR, whereas IBR reduced AMPK phosphorylation. This finding aligns with observations in the context of the tyrosine kinase inhibitor sunitinib [37], suggesting a plausible class effect of constrained AMPK activation attributable to these two covalent BTKis [9]. AMPK is an important cellular-stress-sensing molecule that is activated by phosphorylation during metabolic challenges and helps to maintain cellular energy homeostasis by promoting catabolic pathways and improving mitochondrial biogenesis and function [38]. Other than metabolic control, AMPK also plays an important role in sarcoplasmic Ca^{2+} handling and membrane ion homeostasis, and reduced AMPK activity is linked to atrial and ventricular arrhythmias [38,39]. The reduced expression and phosphorylation of AMPK are associated with decreased AMPK activity and abnormal Ca^{2+} dynamics by BTKi molecules in our study, which may have resulted from decreased AMPK activity. The AMPK reserve is diminished with aging, and further AMPK inhibition is reported to exacerbate the aging-associated impairment of diastolic Ca^{2+} clearance [40]. Similarly to our study, Turdi et al. also demonstrated that decreased AMPK function is primarily associated with reduced SERCA2a expression without any modification of RyR2, PLB phosphorylation, or NCX, Voltage-Dependent Calcium Channel (VDCC) expression [40]. In our previous study, VA vulnerability following IBR exposure demonstrated a strong correlation with reduced AMPK phosphorylation. Importantly, concurrent treatment with 5-Aminoimidazole-4-carboxamide ribonucleotide (AICAR), an AMPK activator, was associated with improvement of ventricular Ca^{2+} and voltage parameters as well as mitigation of VA vulnerability [16].

3.5. Possible Mechanisms Contributing to the Different Arrhythmogenic Effects of IBR and ABR

It can be speculated that the inhibition of sodium current (I_{Na}), as well as the cytosolic Ca^{2+} overload, associated with BTKi may be responsible for reduced CV [33,41]. Moreover, the differential impacts of IBR and ABR as to depolarizing Na^+ and repolarizing K^+ current may explain the neutral effect of ABR on APD, as well as the reduced magnitude of changes in CV dispersion by ABR, as compared to IBR [41]. Although previous studies demonstrated inhibition of the PI3K-Akt pathway [11] and alterations in CSK activity [7] by IBR treatment, the above pathways were not influenced by ABR in our study. ABR, a next-generation BTKi with higher selectivity through binding to Cys-481, exhibits fewer off-target effects on other kinases and pathways than does IBR [8,42,43]. Coupled with our experimental findings, the reduced abundance of PLB in ABR-treated hearts diminishes the inhibition of SERCA2a activity [44]. These factors may underlie the distinct effects on ventricular electrophysiology and arrhythmogenesis observed between IBR and ABR. Nevertheless, further direct head-to-head clinical trials and precise mechanistic explorations, particularly those focusing on ventricular myocytes, are warranted.

3.6. *Study Limitations*

Although we have investigated the effects of IBR and ABR on ventricular conduction, repolarization, and Ca^{2+} cycling, the impacts of these agents on the individual membrane currents responsible for changes in action-potential characteristics have yet to be evaluated in our current study. However, we have summarized potential mechanisms of IBR's association with AF based on previous experiments, which may aid in predicting the proarrhythmic potency of BTKi in the ventricle (Table S1). Nonetheless, our study provides an important insight into the electrophysiological and metabolic mechanisms behind increased VA risk associated with second-generation BTKi therapy. We acknowledge that pathophysiology is more complicated in cancer patients with co-morbidities, and our model is a healthy rodent model. Furthermore, considering the differences in ventricular ion homeostasis between species, the results from this rodent model should be translated with caution to humans, although several studies have been conducted on murine and rodent models to evaluate the cardiac electrophysiological effects of BTKi [7,12,15,16,41]. Our study used only male rats, and it is important to address sex dimorphisms in BTKi research. Our recent data suggest repolarization shortening as one of IBR's major electrophysiological effects, and this may translate into QT shortening in a patient's electrocardiogram (EKG). Men are associated with shorter QTc than are women, and this, we believe, contributes to sexual dimorphism in IBR-induced VA. Consistent with this concept, male patients have been shown to be more susceptible to IBR-induced cardiotoxicities, including VA [17]. In a follow-up study by Bhat et al., all of the patients with ABR-induced VA were male [13]. Both action-potential characteristics and cardiac metabolism are subjected to modulation by sex hormones. Hence, the effects may be different in male and female rats; thus, we acknowledge the limitation of including only male rats. However, given male sex as a risk factor of BTKi-induced VA, we used that fact to maximize our understanding here, and we plan to study the mechanism of the impact of sex on the arrhythmia burden in future experiments, as it is very relevant to our experiments.

4. Materials and Methods

The experimental protocol was approved by the University Health Network Animal Care Committee, in accordance with the rules and regulations of the Canadian Council of Animal Care (Animal Use Protocol #5969).

4.1. *Animal Model*

A total of 30 male Sprague–Dawley (SD) rats aged 10–14 months (450 $\pm$ 50 g) were treated with an oral gavage of Ibrutinib (10 mg/day) (IBR group, n = 10), acalabrutinib (10 mg/day) (ABR group, n = 10), or an equal volume of vehicle (5% dimethyl sulfoxide) (Ctrl group, n = 10) for 4 weeks. Animal group sizes were as low as possible and empirically chosen. The dose of IBR was chosen based on our previous study on IBR-induced ventricular arrhythmia [16]. The dose of Acalabrutinib was primarily determined based on the dosing regimen used to evaluate its BTK inhibitory effect in two murine models of chronic lymphocytic leukemia (CLL) [45]. Subsequently, the rodent-appropriate dose was calculated using the dosage conversion relationship between rats and mice, with adjustments made for weight. Throughout the study period, the body weight of all rats was monitored every week, with any rats experiencing weight loss exceeding 10% being excluded from the follow-up experiment. There were no exclusions of animals attributed to illness/excessive weight loss during the experiments. The experimental protocol is described in Figure 5.

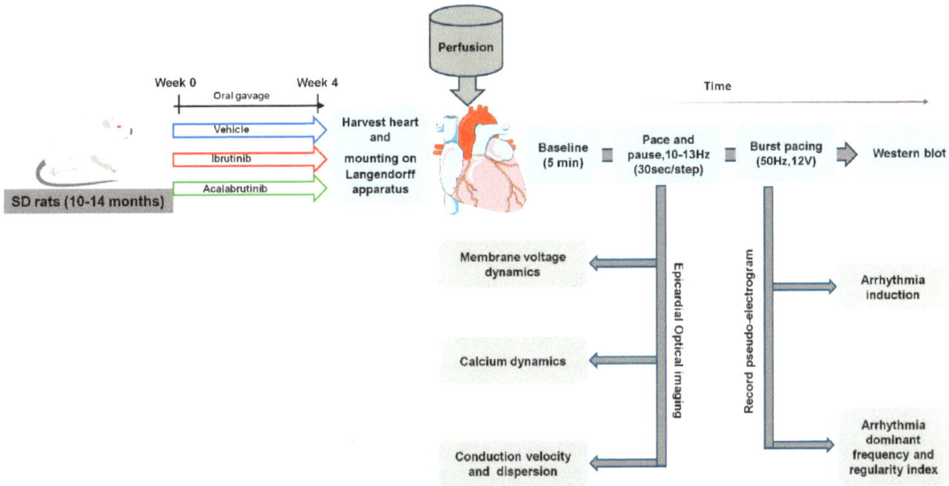

Figure 5. Experimental protocol and setup. Schematic diagram of the rat study protocol, groupings, treatment, and electrophysiology measurements at different steps.

4.2. Langendorff-Perfused Rat Hearts

After 4 weeks, each rat was anesthetized with isoflurane (2–5%), and the explanted heart, after harvesting through midline thoracotomy, was cannulated and mounted to a Langendorff apparatus for retrograde perfusion with perfusate containing NaCl (130 mM), $NaHCO_3$ (24 mM), KCl (4.4 mM), $MgSO_4$ (0.3 mM), $CaCl_2$ (2.2 mM), KH_2PO_4 (1.2 mM), and glucose (6 mM), and equilibrated with carbogen gas (95% O_2 and 5% CO_2), maintained at 37 °C, and kept under constant pressure (~70 mmHg). A pair of silver electrodes were mounted on a custom-made chair, positioned behind the heart from the camera perspective, to record pseudo-ECG.

4.3. Epicardial Optical Imaging

Fluorescence-based intracellular Ca^{2+} and membrane voltage dynamics were recorded simultaneously from the epicardial surface of the left ventricle with a dual optical camera. A detailed protocol for optical mapping has been previously published [15]. In brief, voltage-sensitive dye RH237 (0.25 µmol, Biotium, Inc., Fremont, CA, USA) and Ca^{2+}-sensitive dye Rhod-2 AM (0.1 µmol, Biotium, Inc., Fremont, CA, USA) were slowly infused into the perfusate. Blebbistatin (6 µmol, Enzo Life Sciences, Inc., Farmingdale, NY, USA) was also added to the perfusate to suppress motion artifacts from cardiac contractions. For epicardial fluorescence recording, a xenon light source (Moritek, Saitama, Japan) and a 530 nm green filter (Semrock, Rochester, NY, USA) were utilized to illuminate the left ventricle and excite dye fluorescence. A dichroic mirror at 665 nm divided the emission light beam. Light wavelengths above 665 nm were filtered through a pass with a length of 715 nm for action potential (AP) measurement, while those below 665 nm were filtered through a 585/40 nm bandpass filter for the measurement of Ca^{2+} transients. Voltage and Ca^{2+} signals were simultaneously recorded using different high-speed CMOS cameras (Ultima-L, Scimedia, Costa Mesa, CA, USA), each equipped with 10,000 pixels arranged in a 100 × 100 matrix on a 1-cm^2 sensor, and operating at a frequency of 500 frames per second. The optical arrangement included a Leica Plan APO 0.63× lens at the objective and a 1.0× on the condensing side, resulting in a spatial resolution of 160 µm/pixel. Electrical stimulation of the ventricle was delivered by a Grass Instruments S88X pulse stimulator through a pair of silver electrodes 1 mm apart connected to the epicardial surface of the ventricle; the electrodes were mounted on a homemade support made of syringe needles. Voltage and Ca^{2+} parameters were recorded with incremental pacing according to a "pace-&-pause"

protocol [15]. Hearts were paced for 30 s with burst pacing in a frequency range from 10.0 to 13.0 Hz to achieve a steady-state (6 V output, pulse width of 4 ms), and AP and Ca^{2+} signals were recorded during the last 2 s of pacing and the first 2 s of post-pacing intrinsic rhythm.

Voltage and Ca^{2+} signals obtained from optical mapping were imported into MATLAB for subsequent signal processing and calculation of AP and CaT parameters. APD was measured from the onset of AP to 80% of repolarization (APD_{80}). Beat-to-beat variation of APD_{80} (APD alternans) was used as a measure of the dispersion of repolarization. APD spatially discordant alternans was quantified as the width of the half-peak value of the APD alternations distribution histogram. The ΔAPD represented the average APD from even and odd beats. CaT rise time was calculated as the time required for fluorescence to rise from 10% to 90% during systolic release [46]. CaT decay time and spontaneous calcium elevation (SCaE) were assessed as markers of the diastolic Ca^{2+} load [46,47]. CaT amplitude ratio was calculated by dividing the smallest Ca^{2+} signals by the largest Ca^{2+} signals. A smaller ratio represents higher Ca^{2+} alternans, and vice versa [47].

Other than AP and CaT parameters, CV and CV dispersion were also measured. CV was determined by analyzing the propagation of the voltage signal during electrical stimulation using optical mapping data, and isochrone maps were constructed for each stimulated heartbeat, with the maximum positive time derivative serving as the reference point for activation time. Each map was divided into a 10×10 grid, and CV estimation was computed in the middle of each of the 81 squares constituting this grid. For each square, the activation time measured at the 4 corners was fitted into a polynomial surface, and CV was computed using a method similar to that used by Bayly and colleagues [48], giving, for each map, a 9×9 CV vector field. The CV dispersion at each stimulation frequency was calculated by dividing the standard deviation of CV at each point in the mapping area by the mean CV [30].

4.4. Determination of Ventricular Arrhythmia Vulnerability and Characteristics

At the end of optical mapping, hearts were subjected five episodes of VA induction by direct electrical stimulation with burst-pacing at 50 Hz and 12 V for 10 sec, with an interval of 2–3 min for recovery between inductions [15]. All VA episodes induced by electrical stimulation and persisting for $\geq$10 s were considered to be successfully induced VA and were included in the analyses of VA vulnerability and VA burden, as well as the calculation of VA dominant frequency (DF) and the regularity index (RI). Successful induction of VA was calculated from the percentage of successfully induced VF episodes in each heart during five inductions and was used as a measure of VF inducibility [15]. The mean duration of successfully induced VA (in seconds) for each heart was used as a parameter to assess the VA burden. The ECG signals of all VAs $\geq$ 10 s were subjected to power spectral density (PSD) analysis between 0–20 Hz. The frequency associated with the highest energy component was extracted as the DF [49,50]. The VA phenotypes were classified according to methods described in the previous literature [20]. The ratio of the power at the DF and the adjacent 1 Hz width to the total power (the sums over the range of 0–20 Hz) was defined as RI, an effective method used to directly quantify temporal organization [49,50]. The RI values range from 0 to 1, where higher values denote organized VA and lower values signify disorganized VA. We also performed optical mapping during the sustained VAs (VA $\geq$ 60 s). An analysis of the phase mapping of voltage recordings was conducted as a surrogate for VA spatial organization, similarly to that described in our previously published methodology. The analysis was performed on the total count of waves from voltage and calcium data observed during the two 10 s segments of VA [51,52].

4.5. Immunoblotting

Proteins extracted from the stored LV tissue were analyzed with standard Western blot protocols [15]. Equal amounts of proteins were separated on SDS-PAGE, transferred to PVDF membranes, and probed with the following primary antibodies: total and phospho-PLB (Badrilla), SERCA2a (Badrilla), NCX (biorbyt), phospho-RyR2 (Badrilla), total RyR2 (Thermo Fisher), CaMKII (pThr286 or 287) (Badrilla), CaMKII delta (abcam), total and phospho-AMPK(pThr172) (Cell signaling), GAPDH (Fitzgerald), and horseradish peroxidase-conjugated goat anti-rabbit or rabbit anti-mouse secondary antibodies. The proteins were detected with Image Lab Touch Software (version 2.4.0.03, Chemidoc Imaging System from Bio-Rad, Hercules, CA, USA), and Western blot bands were quantified by Image J (version 2.1.0); the results are presented in bar graphs.

4.6. Statistical Analysis

Experimental data are expressed as mean$\pm$ standard error of the mean. Data distribution was evaluated by the Shapiro–Wilk normality test. Student's *t*-test was performed to compare VA dominant frequency and VA regularity index between the IBR and ABR groups. Arrhythmia morphologies (monomorphic vs. polymorphic) were analyzed by X^2 test. Two-way repeated-measures ANOVA was performed for continuous variables in the analysis of APD_{80}, APD alternans, CV, CV dispersion, CaT rise time, CaT decay time, SCaE, and CaT ratio. One-way ANOVA (for normally distributed data with equal variance) or the Kruskal–Wallis test (for non-normally distributed data) were performed to compare the differences between the three groups. Statistical significance was defined as $p < 0.05$. Data analysis was performed using the Prism 9.00 (GraphPad, San Diego, CA, USA) software.

5. Conclusions

As previously observed with Ibrutinib, chronic Acalabrutinib treatment also resulted in increased risk of ventricular arrhythmia in our rodent model. Although reduced AMPK phosphorylation and abnormal Ca^{2+} dynamics are common to both generations of BTKi, the differential impacts of these agents on membrane electrophysiology may explain the relatively lower risk and more spatially organized ventricular arrhythmia following Acalabrutinib treatment, compared to Ibrutinib. In our model, a more organized ventricular arrhythmia with Acalabrutinib may particularly result from its neutral effect on repolarization and repolarization alternans. However, further clinical studies are necessary to determine whether Acalabrutinib will lead to more monomorphic ventricular tachycardia in humans. Additionally, there might be differences in responses to various therapies associated with Acalabrutinib-induced and Ibrutinib-induced ventricular arrhythmias.

Supplementary Materials: The following supporting information can be downloaded at: https://www.mdpi.com/article/10.3390/ijms25116207/s1.

Author Contributions: Writing—original draft preparation, Y.Z., P.C. and J.T.; Methodology, Y.Z., P.C. and S.M.; Software, S.M.; Investigation, Y.Z., J.T., T.S. and P.F.H.L.; Formal analysis, Y.Z., P.C. and T.S.; Visualization, Y.Z.; Writing—review and editing, P.C., P.T., F.B., H.A.-Q. and K.N.; Conceptualization, supervision and funding acquisition, K.N. All authors have read and agreed to the published version of the manuscript.

Funding: This work was supported by the Canadian Institutes of Health Research (CIHR), Project Grant PJT 190317, and a generous contribution by the Crispino family and the Hang Tough Initiative.

Institutional Review Board Statement: The experimental protocol followed the Guide to the Care and Use of Experimental Animals, and was approved by the University Health Network Animal Care Committee, in accordance with the rules and regulations of the Canadian Council of Animal Care (Animal Use Protocol #5969).

Informed Consent Statement: Not applicable.

Data Availability Statement: The data presented in this study are available on request from the corresponding author.

Acknowledgments: Y.Z. was partially supported by a grant from the China Scholarship Council.

Conflicts of Interest: The authors declare no conflicts of interest.

References

1. Hillmen, P.; Pitchford, A.; Bloor, A.; Broom, A.; Young, M.; Kennedy, B.; Walewska, R.; Furtado, M.; Preston, G.; Neilson, J.R.; et al. Ibrutinib and rituximab versus fludarabine, cyclophosphamide, and rituximab for patients with previously untreated chronic lymphocytic leukaemia (FLAIR): Interim analysis of a multicentre, open-label, randomised, phase 3 trial. *Lancet Oncol.* **2023**, *24*, 535–552. [CrossRef] [PubMed]
2. Dong, R.; Yan, Y.; Zeng, X.; Lin, N.; Tan, B. Ibrutinib-Associated Cardiotoxicity: From the Pharmaceutical to the Clinical. *Drug Des. Dev. Ther.* **2022**, *16*, 3225–3239. [CrossRef] [PubMed]
3. Quartermaine, C.; Ghazi, S.M.; Yasin, A.; Awan, F.T.; Fradley, M.; Wiczer, T.; Kalathoor, S.; Ferdousi, M.; Krishan, S.; Habib, A.; et al. Cardiovascular Toxicities of BTK Inhibitors in Chronic Lymphocytic Leukemia: JACC: CardioOncology State-of-the-Art Review. *JACC Cardio Oncol.* **2023**, *5*, 570–590. [CrossRef] [PubMed]
4. Mato, A.R.; Nabhan, C.; Thompson, M.C.; Lamanna, N.; Brander, D.M.; Hill, B.; Howlett, C.; Skarbnik, A.; Cheson, B.D.; Zent, C.; et al. Toxicities and outcomes of 616 ibrutinib-treated patients in the United States: A real-world analysis. *Haematologica* **2018**, *103*, 874–879. [CrossRef] [PubMed]
5. Lampson, B.L.; Yu, L.; Glynn, R.J.; Barrientos, J.C.; Jacobsen, E.D.; Banerji, V.; Jones, J.A.; Walewska, R.; Savage, K.J.; Michaud, G.F.; et al. Ventricular arrhythmias and sudden death in patients taking ibrutinib. *Blood* **2017**, *129*, 2581–2584. [CrossRef] [PubMed]
6. Guha, A.; Derbala, M.H.; Zhao, Q.; Wiczer, T.E.; Woyach, J.A.; Byrd, J.C.; Awan, F.T.; Addison, D. Ventricular Arrhythmias Following Ibrutinib Initiation for Lymphoid Malignancies. *J. Am. Coll. Cardiol.* **2018**, *72*, 697–698. [CrossRef] [PubMed]
7. Xiao, L.; Salem, J.E.; Clauss, S.; Hanley, A.; Bapat, A.; Hulsmans, M.; Iwamoto, Y.; Wojtkiewicz, G.; Cetinbas, M.; Schloss, M.J.; et al. Ibrutinib-Mediated Atrial Fibrillation Attributable to Inhibition of C-Terminal Src Kinase. *Circulation* **2020**, *142*, 2443–2455. [CrossRef]
8. Barf, T.; Covey, T.; Izumi, R.; van de Kar, B.; Gulrajani, M.; van Lith, B.; van Hoek, M.; de Zwart, E.; Mittag, D.; Demont, D.; et al. Acalabrutinib (ACP-196): A Covalent Bruton Tyrosine Kinase Inhibitor with a Differentiated Selectivity and In Vivo Potency Profile. *J. Pharmacol. Exp. Ther.* **2017**, *363*, 240–252. [CrossRef] [PubMed]
9. Byrd, J.C.; Hillmen, P.; Ghia, P.; Kater, A.P.; Chanan-Khan, A.; Furman, R.R.; O'Brien, S.; Yenerel, M.N.; Illés, A.; Kay, N.; et al. Acalabrutinib Versus Ibrutinib in Previously Treated Chronic Lymphocytic Leukemia: Results of the First Randomized Phase III Trial. *J. Clin. Oncol.* **2021**, *39*, 3441–3452. [CrossRef] [PubMed]
10. Patel, V.; Balakrishnan, K.; Bibikova, E.; Ayres, M.; Keating, M.J.; Wierda, W.G.; Gandhi, V. Comparison of Acalabrutinib, a Selective Bruton Tyrosine Kinase Inhibitor, with Ibrutinib in Chronic Lymphocytic Leukemia Cells. *Clin. Cancer Res.* **2017**, *23*, 3734–3743. [CrossRef] [PubMed]
11. McMullen, J.R.; Boey, E.J.; Ooi, J.Y.; Seymour, J.F.; Keating, M.J.; Tam, C.S. Ibrutinib increases the risk of atrial fibrillation, potentially through inhibition of cardiac PI3K-Akt signaling. *Blood* **2014**, *124*, 3829–3830. [CrossRef] [PubMed]
12. Jiang, L.; Li, L.; Ruan, Y.; Zuo, S.; Wu, X.; Zhao, Q.; Xing, Y.; Zhao, X.; Xia, S.; Bai, R.; et al. Ibrutinib promotes atrial fibrillation by inducing structural remodeling and calcium dysregulation in the atrium. *Heart Rhythm.* **2019**, *16*, 1374–1382. [CrossRef] [PubMed]
13. Bhat, S.A.; Gambril, J.A.; Azali, L.; Chen, S.T.; Rosen, L.; Palettas, M.; Wiczer, T.; Kalathoor, S.; Zhao, Q.; Rogers, K.A.; et al. Ventricular Arrhythmias and Sudden Death Events following Acalabrutinib initiation. *Blood* **2022**, *140*, 2142–2145. [CrossRef] [PubMed]
14. Boriani, G.; Menna, P.; Morgagni, R.; Minotti, G.; Vitolo, M. Ibrutinib and Bruton's tyrosine kinase inhibitors in chronic lymphocytic leukemia: Focus on atrial fibrillation and ventricular tachyarrhythmias/sudden cardiac death. *Chemotherapy* **2023**, *68*, 61–72. [CrossRef] [PubMed]
15. Du, B.; Chakraborty, P.; Azam, M.A.; Massé, S.; Lai, P.F.H.; Niri, A.; Si, D.; Thavendiranathan, P.; Abdel-Qadir, H.; Billia, F.; et al. Acute Effects of Ibrutinib on Ventricular Arrhythmia in Spontaneously Hypertensive Rats. *JACC Cardio Oncol.* **2020**, *2*, 614–629. [CrossRef] [PubMed]
16. Zhao, Y.; Du, B.; Chakraborty, P.; Denham, N.; Massé, S.; Lai, P.F.H.; Azam, M.A.; Billia, F.; Thavendiranathan, P.; Abdel-Qadir, H.; et al. Impaired Cardiac AMPK and Ca^{2+}-Handling, and Action Potential Duration Heterogeneity in Ibrutinib-Induced Ventricular Arrhythmia Vulnerability. *J. Am. Heart Assoc.*. *in press*.
17. Salem, J.E.; Manouchehri, A.; Bretagne, M.; Lebrun-Vignes, B.; Groarke, J.D.; Johnson, D.B.; Yang, T.; Reddy, N.M.; Funck-Brentano, C.; Brown, J.R.; et al. Cardiovascular Toxicities Associated with Ibrutinib. *J. Am. Coll. Cardiol.* **2019**, *74*, 1667–1678. [CrossRef] [PubMed]
18. Tuomi, J.M.; Xenocostas, A.; Jones, D.L. Increased Susceptibility for Atrial and Ventricular Cardiac Arrhythmias in Mice Treated with a Single High Dose of Ibrutinib. *Can. J. Cardiol.* **2018**, *34*, 337–341. [CrossRef] [PubMed]
19. Fazal, M.; Gomez, S.; Cheng, P.; Rhee, J.-W.; Baykaner, T. Tyrosine Kinase Inhibitor Associated Polymorphic Ventricular Tachycardia. *JACC Cardio Oncol.* **2022**, *4* (Suppl. S1), S4–S5. [CrossRef]
20. Chinyere, I.R.; Moukabary, T.; Hutchinson, M.D.; Lancaster, J.J.; Juneman, E.; Goldman, S. Progression of infarct-mediated arrhythmogenesis in a rodent model of heart failure. *Am. J. Physiol. Heart Circ. Physiol.* **2021**, *320*, H108–H116. [CrossRef] [PubMed]
21. Tomcsányi, J.; Nényei, Z.; Mátrai, Z.; Bózsik, B. Ibrutinib, an Approved Tyrosine Kinase Inhibitor as a Potential Cause of Recurrent Polymorphic Ventricular Tachycardia. *JACC. Clin. Electrophysiol.* **2016**, *2*, 847–849. [CrossRef] [PubMed]
22. Beyer, A.; Ganti, B.; Majkrzak, A.; Theyyunni, N. A Perfect Storm: Tyrosine Kinase Inhibitor-Associated Polymorphic Ventricular Tachycardia. *J. Emerg. Med.* **2017**, *52*, e123–e127. [CrossRef] [PubMed]

23. Fazal, M.; Wei, C.; Chuy, K.L.; Hussain, K.; Gomez, S.E.; Ba, S.S.; Pietrasik, G.; Yadav, N.; Ghazizadeh, Z.; Kapoor, R.; et al. Tyrosine kinase inhibitor-associated ventricular arrhythmias: A case series and review of literature. *J. Interv. Card. Electrophysiol.* **2023**, *66*, 1165–1175. [CrossRef] [PubMed]
24. Qu, Z.; Weiss, J.N. Cardiac Alternans: From Bedside to Bench and Back. *Circ. Res.* **2023**, *132*, 127–149. [CrossRef] [PubMed]
25. Liu, W.; Kim, T.Y.; Huang, X.; Liu, M.B.; Koren, G.; Choi, B.R.; Qu, Z. Mechanisms linking T-wave alternans to spontaneous initiation of ventricular arrhythmias in rabbit models of long QT syndrome. *J. Physiol.* **2018**, *596*, 1341–1355. [CrossRef] [PubMed]
26. Lukas, A.; Antzelevitch, C. Phase 2 reentry as a mechanism of initiation of circus movement reentry in canine epicardium exposed to simulated ischemia. *Cardiovasc. Res.* **1996**, *32*, 593–603. [CrossRef] [PubMed]
27. Tse, G.; Chan, Y.W.; Keung, W.; Yan, B.P. Electrophysiological mechanisms of long and short QT syndromes. *Int. J. Cardiol. Heart Vasc.* **2017**, *14*, 8–13. [CrossRef] [PubMed]
28. Extramiana, F.; Antzelevitch, C. Amplified transmural dispersion of repolarization as the basis for arrhythmogenesis in a canine ventricular-wedge model of short-QT syndrome. *Circulation* **2004**, *110*, 3661–3666. [CrossRef] [PubMed]
29. Schalij, M.J.; Boersma, L.; Huijberts, M.; Allessie, M.A. Anisotropic reentry in a perfused 2-dimensional layer of rabbit ventricular myocardium. *Circulation* **2000**, *102*, 2650–2658. [CrossRef] [PubMed]
30. Xu, L.; Zahid, S.; Khoshknab, M.; Moss, J.; Berger, R.D.; Chrispin, J.; Callans, D.; Marchlinski, F.E.; Zimmerman, S.L.; Han, Y.; et al. Conduction Velocity Dispersion Predicts Postinfarct Ventricular Tachycardia Circuit Sites and Associates with Lipomatous Metaplasia. *JACC Clin. Electrophysiol.* **2023**, *9*, 1464–1474. [CrossRef] [PubMed]
31. Anter, E.; Tschabrunn, C.M.; Buxton, A.E.; Josephson, M.E. High-Resolution Mapping of Postinfarction Reentrant Ventricular Tachycardia: Electrophysiological Characterization of the Circuit. *Circulation* **2016**, *134*, 314–327. [CrossRef] [PubMed]
32. Landstrom, A.P.; Dobrev, D.; Wehrens, X.H.T. Calcium Signaling and Cardiac Arrhythmias. *Circ. Res.* **2017**, *120*, 1969–1993. [CrossRef] [PubMed]
33. Shaw, R.M.; Rudy, Y. Ionic Mechanisms of Propagation in Cardiac Tissue. *Circ. Res.* **1997**, *81*, 727–741. [CrossRef] [PubMed]
34. Edwards, J.N.; Blatter, L.A. Cardiac alternans and intracellular calcium cycling. *Clin. Exp. Pharmacol. Physiol.* **2014**, *41*, 524–532. [CrossRef] [PubMed]
35. Sikkel, M.B.; Hayward, C.; MacLeod, K.T.; Harding, S.E.; Lyon, A.R. SERCA2a gene therapy in heart failure: An anti-arrhythmic positive inotrope. *Br. J. Pharmacol.* **2014**, *171*, 38–54. [CrossRef] [PubMed]
36. Prunier, F.; Kawase, Y.; Gianni, D.; Scapin, C.; Danik, S.B.; Ellinor, P.T.; Hajjar, R.J.; del Monte, F. Prevention of Ventricular Arrhythmias with Sarcoplasmic Reticulum Ca^{2+} ATPase Pump Overexpression in a Porcine Model of Ischemia Reperfusion. *Circulation* **2008**, *118*, 614–624. [CrossRef]
37. Kerkela, R.; Woulfe, K.C.; Durand, J.B.; Vagnozzi, R.; Kramer, D.; Chu, T.F.; Beahm, C.; Chen, M.H.; Force, T. Sunitinib-induced cardiotoxicity is mediated by off-target inhibition of AMP-activated protein kinase. *Clin. Transl. Sci.* **2009**, *2*, 15–25. [CrossRef] [PubMed]
38. Harada, M.; Nattel, S.N.; Nattel, S. AMP-activated protein kinase: Potential role in cardiac electrophysiology and arrhythmias. *Circ. Arrhythm. Electrophysiol.* **2012**, *5*, 860–867. [CrossRef] [PubMed]
39. Harada, M.; Tadevosyan, A.; Qi, X.; Xiao, J.; Liu, T.; Voigt, N.; Karck, M.; Kamler, M.; Kodama, I.; Murohara, T.; et al. Atrial Fibrillation Activates AMP-Dependent Protein Kinase and its Regulation of Cellular Calcium Handling: Potential Role in Metabolic Adaptation and Prevention of Progression. *J. Am. Coll. Cardiol.* **2015**, *66*, 47–58. [CrossRef] [PubMed]
40. Turdi, S.; Fan, X.; Li, J.; Zhao, J.; Huff, A.F.; Du, M.; Ren, J. AMP-activated protein kinase deficiency exacerbates aging-induced myocardial contractile dysfunction. *Aging Cell* **2010**, *9*, 592–606. [CrossRef] [PubMed]
41. Tuomi, J.M.; Bohne, L.J.; Dorey, T.W.; Jansen, H.J.; Liu, Y.; Jones, D.L.; Rose, R.A. Distinct Effects of Ibrutinib and Acalabrutinib on Mouse Atrial and Sinoatrial Node Electrophysiology and Arrhythmogenesis. *J. Am. Heart Assoc.* **2021**, *10*, e022369. [CrossRef] [PubMed]
42. Wang, L.; Sun, Y.; Liu, X.; Li, H.; Lu, C.; Yang, R.; Yang, C.; Li, B. SY-1530, a highly selective BTK inhibitor, effectively treats B-cell malignancies by blocking B-cell activation. *Cancer Biol. Med.* **2021**, *19*, 995–1007. [CrossRef] [PubMed]
43. Kapoor, I.; Li, Y.; Sharma, A.; Zhu, H.; Bodo, J.; Xu, W.; Hsi, E.D.; Hill, B.T.; Almasan, A. Resistance to BTK inhibition by ibrutinib can be overcome by preventing FOXO3a nuclear export and PI3K/AKT activation in B-cell lymphoid malignancies. *Cell Death Dis.* **2019**, *10*, 924. [CrossRef] [PubMed]
44. Dong, Q.; Wu, Z.; Li, X.; Yan, J.; Zhao, L.; Yang, C.; Lu, J.; Deng, J.; Chen, M. Resveratrol ameliorates cardiac dysfunction induced by pressure overload in rats via structural protection and modulation of Ca^{2+} cycling proteins. *J. Transl. Med.* **2014**, *12*, 323. [CrossRef] [PubMed]
45. Herman, S.E.M.; Montraveta, A.; Niemann, C.U.; Mora-Jensen, H.; Gulrajani, M.; Krantz, F.; Mantel, R.; Smith, L.L.; McClanahan, F.; Harrington, B.K.; et al. The Bruton Tyrosine Kinase (BTK) Inhibitor Acalabrutinib Demonstrates Potent On-Target Effects and Efficacy in Two Mouse Models of Chronic Lymphocytic Leukemia. *Clin. Cancer Res.* **2017**, *23*, 2831–2841. [CrossRef] [PubMed]
46. Jaimes, R.; Walton, R.D.; Pasdois, P.; Bernus, O.; Efimov, I.R.; Kay, M.W. A technical review of optical mapping of intracellular calcium within myocardial tissue. *Am. J. Physiol. Heart Circ. Physiol.* **2016**, *310*, H1388–H1401. [CrossRef] [PubMed]
47. Si, D.; Azam, M.A.; Lai, P.F.H.; Zamiri, N.; Kichigina, G.; Asta, J.; Massé, S.; Bokhari, M.M.; Porta-Sánchez, A.; Labos, C.; et al. Essential role of ryanodine receptor 2 phosphorylation in the effect of azumolene on ventricular arrhythmia vulnerability in a rabbit heart model. *J. Cardiovasc. Electrophysiol.* **2018**, *29*, 1707–1715. [CrossRef] [PubMed]

48. Bayly, P.V.; KenKnight, B.H.; Rogers, J.M.; Hillsley, R.E.; Ideker, R.E.; Smith, W.M. Estimation of conduction velocity vector fields from epicardial mapping data. *IEEE Trans. Biomed. Eng.* **1998**, *45*, 563–571. [CrossRef]
49. Massé, S.; Farid, T.; Dorian, P.; Umapathy, K.; Nair, K.; Asta, J.; Ross, H.; Rao, V.; Sevaptsidis, E.; Nanthakumar, K. Effect of global ischemia and reperfusion during ventricular fibrillation in myopathic human hearts. *Am. J. Physiol. Heart Circ. Physiol.* **2009**, *297*, H1984–H1991. [CrossRef] [PubMed]
50. Umapathy, K.; Masse, S.; Sevaptsidis, E.; Asta, J.; Ross, H.; Thavandiran, N.; Nair, K.; Farid, T.; Cusimano, R.; Rogers, J.; et al. Regional frequency variation during human ventricular fibrillation. *Med. Eng. Phys.* **2009**, *31*, 964–970. [CrossRef] [PubMed]
51. Azam, M.A.; Zamiri, N.; Massé, S.; Kusha, M.; Lai, P.F.; Nair, G.K.; Tan, N.S.; Labos, C.; Nanthakumar, K. Effects of Late Sodium Current Blockade on Ventricular Refibrillation in a Rabbit Model. *Circ. Arrhythm. Electrophysiol.* **2017**, *10*, e004331. [CrossRef] [PubMed]
52. Zamiri, N.; Massé, S.; Ramadeen, A.; Kusha, M.; Hu, X.; Azam, M.A.; Liu, J.; Lai, P.F.; Vigmond, E.J.; Boyle, P.M.; et al. Dantrolene improves survival after ventricular fibrillation by mitigating impaired calcium handling in animal models. *Circulation* **2014**, *129*, 875–885. [CrossRef] [PubMed]

International Journal of *Molecular Sciences*

MDPI

Review

Facilitation of hERG Activation by Its Blocker: A Mechanism to Reduce Drug-Induced Proarrhythmic Risk

Kazuharu Furutani

Department of Pharmacology, Faculty of Pharmaceutical Sciences, Tokushima Bunri University, 180 Nishihama-Boji, Yamashiro-cho, Tokushima 770-8514, Japan; furutani@ph.bunri-u.ac.jp

Abstract: Modulation of the human Ether-à-go-go-Related Gene (hERG) channel, a crucial voltage-gated potassium channel in the repolarization of action potentials in ventricular myocytes of the heart, has significant implications on cardiac electrophysiology and can be either antiarrhythmic or proarrhythmic. For example, hERG channel blockade is a leading cause of long QT syndrome and potentially life-threatening arrhythmias, such as *torsades de pointes*. Conversely, hERG channel blockade is the mechanism of action of Class III antiarrhythmic agents in terminating ventricular tachycardia and fibrillation. In recent years, it has been recognized that less proarrhythmic hERG blockers with clinical potential or Class III antiarrhythmic agents exhibit, in addition to their hERG-blocking activity, a second action that facilitates the voltage-dependent activation of the hERG channel. This facilitation is believed to reduce the proarrhythmic potential by supporting the final repolarizing of action potentials. This review covers the pharmacological characteristics of hERG blockers/facilitators, the molecular mechanisms underlying facilitation, and their clinical significance, as well as unresolved issues and requirements for research in the fields of ion channel pharmacology and drug-induced arrhythmias.

Keywords: hERG channel; I_{Kr}; drug-induced arrhythmias; molecular pharmacology; structural model

check for updates

Citation: Furutani, K. Facilitation of hERG Activation by Its Blocker: A Mechanism to Reduce Drug-Induced Proarrhythmic Risk. *Int. J. Mol. Sci.* **2023**, *24*, 16261. https://doi.org/10.3390/ijms242216261

Academic Editors: Yosuke Okamoto

Received: 24 October 2023
Revised: 8 November 2023
Accepted: 12 November 2023
Published: 13 November 2023

1. Introduction

The human *Ether-à-go-go-Related Gene* (hERG or KCNH2) encodes the alpha subunit of a time- and voltage-dependent potassium channel [1–4]. Its function is best understood in the heart, where it plays a crucial role in the robust repolarization of the cardiac action potential by mediating the I_{Kr} current [1–6]. The pharmacological blockade of hERG channels results in reduced I_{Kr} currents in ventricular myocytes, leading to delayed repolarization and prolonged action potential duration. This manifests as QT interval prolongation on electrocardiography, which is known as drug-induced long QT syndrome (LQT), potentially increasing the risk of life-threatening ventricular fibrillation and tachycardia, such as *torsades de pointes* [1,2,7–13].

Drug-induced LQT results from off-target inhibition of the hERG channel by structurally varied cardiac and non-cardiac drugs, and can pose a high risk of life-threatening arrhythmias [1,4,12]. Several drugs, including terfenadine, astemizole (antihistamine), sertindole (antipsychotics), grepafloxacin (antibacterial), cisapride (gastroprokinetic agents), and levomethadyl (opioid), have been withdrawn from the market due to this problem. However, assessing the risk of drug-induced lethal arrhythmias in humans before drug approval is ethically difficult. Therefore, the in vitro inhibition of hERG channels by drugs in cell cultures is considered a surrogate marker for drug-induced arrhythmias. In accordance with the drug safety guidelines developed by the International Council for Harmonization of Technical Requirements for Pharmaceuticals for Human Use (ICH), ICH S7B [14], all pharmaceutical companies currently conduct in vitro testing, known as the hERG test, to determine whether their drug candidates exhibit inhibitory activity on hERG channels. If a drug inhibits the hERG channels, it carries the risk of proarrhythmia. Notably, since the release of the ICH S7B guidelines in 2005, no approved drugs have been withdrawn from the market owing to hazardous arrhythmia risks, indicating that early testing of hERG channel block is extremely effective in eliminating the risk of approving potentially torsadogenic drugs.

However, the hERG test is an inadequate marker of true proarrhythmic risk, as several studies have demonstrated low proarrhythmic proclivities for hERG blockers [15–18]. Additionally, the hERG block in cell culture does not always correlate with arrhythmogenicity in humans, which may be attributed to the difference between the free therapeutic plasma concentration of drugs and the concentration of the hERG block [19]. Even if a drug exhibits hERG channel inhibitory activity, it may not pose a clinical problem if inhibition does not occur at the concentrations used for the treatment. Another reason is the absence of action potential duration or QT interval prolongation, which is typically expected with hERG block. This may be due to other effects of the drug, particularly on cardiac ion channels or transporters [20,21]. A good example is when a drug inhibits both hERG and calcium channels, resulting in no prolongation of the action potential. Considering these factors, the hERG test is not sufficiently selective for the assessment of life-threatening arrhythmias and may produce false positives [15,16].

Another concern is the high sensitivity of the hERG test. Reports indicate that 50–70% of the structurally diverse drugs tested exhibit some degree of hERG inhibitory activity [22–24]. The high sensitivity of the hERG test poses a challenge during drug development, as it complicates risk assessment.

Consequently, the identification of hazardous drugs that block the hERG channel remains an ongoing challenge and requires continued effort [15,16]. Addressing this challenge has the potential to enhance drug development efficiency, remove significant impediments to the approval of new drugs, and may incentivize the development and clinical application of safe hERG blockers for various diseases. Reports have also suggested the potential therapeutic utilization of hERG channel blockade or chaperone modulation in the treatment of cancer [25–29] and mental disorders [30], respectively, further underscoring the importance of improving our understanding of the function of this channel.

hERG blockers, such as amiodarone and nifekalant, are Class III antiarrhythmic agents that are used in clinical practice to terminate ventricular tachycardia and fibrillation [31–35]. These drugs are considered a last resort for patients who are unresponsive to other antiarrhythmic drugs; despite their proarrhythmic risks, their effectiveness and importance in clinical practice outweigh these risks. Intense pharmacological experiments have been conducted to understand the efficacy and safety of Class III antiarrhythmic agents. Notably, these studies have revealed that their effects on hERG channels are not solely blocking [31–37], but instead, they can facilitate the voltage-dependent activation of hERG channels, thus enhancing their activity near the membrane potential threshold for channel activation [36–38]. This increase in hERG current is achieved through a mechanism distinct from that of hERG channel activators [39–42]. This agonistic effect of hERG blockers, known as facilitation, has been observed not only in Class III antiarrhythmic agents but also in various non-cardiac drugs with hERG channel inhibitory activity used in clinical practice [43]. Importantly, recent studies have suggested that this drug action contributes to reducing the proarrhythmic risk associated with hERG blockers [44].

This review article seeks to offer a thorough examination of the present knowledge regarding the effects and underlying mechanisms of hERG channel facilitation by "blockers", and to summarize the consequences of these effects on cardiac arrhythmias.

2. Facilitation of hERG Activation by Its Blocker

In response to membrane depolarization, hERG channels undergo slow activation followed by much more rapid inactivation [1,4], resulting in inward rectification in the current–voltage (IV) relationship with a maximal outward current at voltages between −10 and 0 mV. This gating property of hERG channels also results in a decreasing current with further depolarization, e.g., during the plateau phase of the ventricular myocyte action potential. The maintenance of this plateau is crucial for ensuring sufficient time for the entry of extracellular Ca^{2+} into the myocyte and Ca^{2+} release from the sarcoplasmic reticulum to enable cardiac contraction [1,4]. As the myocyte repolarizes, hERG channel conductance increases due to recovering from inactivation and deactivation. The I_{Kr} current during phase 3 repolarization of the ventricular action potential accelerates the repolarization and terminates the action potential. In experimental studies, the sigmoidal conductance–voltage (GV) relationship was analyzed by measuring tail currents at a negative voltage, where inactivation was weak. Many drugs interact with hERG channels and influence cardiac electrophysiology. Some drugs, known as blockers, reduce both the IV and GV relationships. Additionally, some hERG blockers not only reduce the IV and GV relationships but also shift them to the left, as shown in Figure 1 [36–38,43,45–47]. This leftward shift in the GV relationship is referred to as the "facilitation" of voltage-dependent activation by the drug. In cases where the leftward shift was significant, the IV relationships of the control and drug-treated conditions may intersect, resulting in an increase in drug-induced hERG currents from the control at membrane voltages near the activation threshold. However, hERG channel activators increase hERG channel currents through a mechanism different from that of hERG blockers/facilitators (referred to as "hERG facilitators" in this review). Specifically, known activators enhance hERG channel activity by inhibiting inactivation [39,41,48–50], whereas the mechanism of hERG facilitation, which does not affect the inactivation process, differs from that of its activation. Perry had effectively summarized the pharmacological differences between the drugs in their review article [42].

Numerous reports on hERG facilitators have been published [36–38,43,45–47]. Among these hERG facilitators, our group focused on Class III antiarrhythmic agents such as amiodarone and nifekalant [36,37,44,51]. These agents are highly effective in terminating refractory ventricular tachycardia and fibrillation and can induce voltage-dependent facilitation, as shown in Figure 1. After treatment with nifekalant, the hERG currents induced by the test pulses decreased at −50 mV, which is considered a block and is thought to be the antiarrhythmic mechanism of Class III antiarrhythmic agents by prolonging the action potential duration and relative refractory period of ventricular myocytes. However, when strong depolarization to +60 mV was applied, the subsequent test pulse to −50 mV induced a large hERG current. This large hERG current was only observed in the presence of nifekalant, indicating that the drug (nifekalant) is necessary to increase the current. As shown in Figure 1A$_i$, the effect of a single strong depolarization was transient. The response to the test pulses gradually decreased and eventually returned to the response observed before the application of strong depolarization. These transient changes in hERG currents were caused by a leftward shift in the GV relationship (Figure 1B). A detailed biophysical assessment of this drug-induced leftward shift of the GV relationship revealed that both block and facilitation occurred with similar concentration dependence (Figure 2). Additionally, the extent to which a drug shifts its GV relationship to the left is inherent in the drug itself. Notably, amiodarone shifted the curve by approximately 30 mV, whereas nifekalant shifted it by approximately 25 mV to the left (Figure 2C). Although experiments are usually conducted using cells solely expressing hERG channels, a shift in the GV relationship also occurs when co-expressing the auxiliary subunit KCNE1 [44,52]. The expression system used did not significantly affect facilitation, even when recordings were carried out at different extracellular K^+ concentrations. Unlike activators, facilitators had little effect on inactivation.

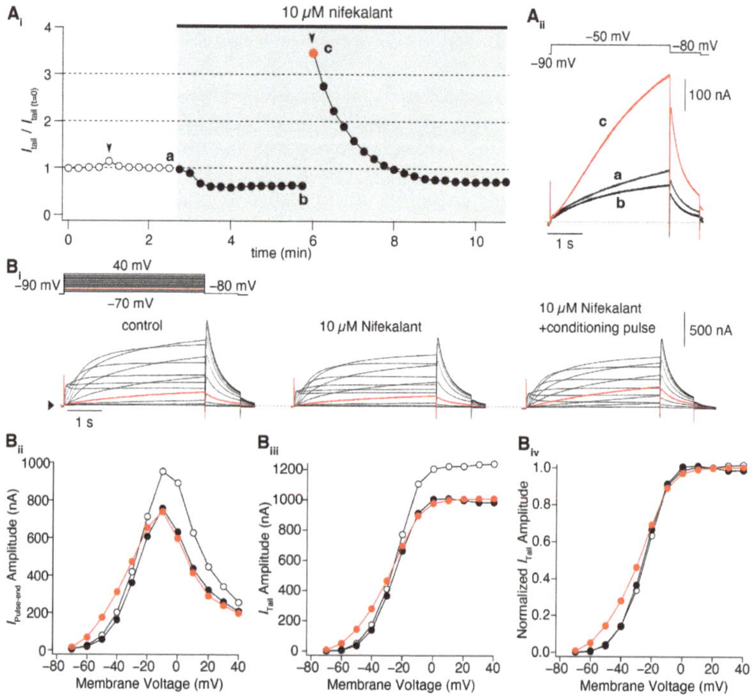

Figure 1. Nifekalant-induced facilitation of hERG activation. hERG channels were ectopically expressed in *Xenopus* oocytes, and the currents were recorded using a two-electrode voltage clamp method. (**A**) Depolarization-induced induction of facilitation. (**A$_i$**) Time course of changes in hERG tail current (recorded at −80 mV) evoked by repetitive test pulses to −50 mV every 15 s. Conditioning pulses (+60 mV, 4 s) were applied twice in this experiment (black arrowheads), first in the absence and then in the presence of nifekalant. (**A$_{ii}$**) Superimposed cell currents recorded in the same oocyte before (Time (a) in (**A$_i$**)) and after perfusion with 10 μM nifekalant, with (c) or without a conditioning pulse (b). Increased hERG current after the induction of facilitation effect by nifekalant is highlighted with red (red circle and trace in (**A$_i$**) and (**A$_{ii}$**), respectively). (**B**) Nifekalant plus conditioning pulse induced a shift in hERG activation curves. (**B$_i$**) Representative traces of hERG currents in the control (left), block (center), and block/facilitation (right) conditions. (**B$_{ii}$**) IV relationship, (**B$_{iii}$**) GV relationship, and (**B$_{iv}$**) normalized GV relationship. Open, filled black, and filled red circles represent the control, 10 μM nifekalant without a conditioned pulse (block), and 10 μM nifekalant with conditioned pulse (block/facilitation) conditions, respectively. Panel (**A**) was adapted with permission from Ref. [37].

Certain voltage-dependent properties of the drug-induced facilitation aid the investigation of the mechanisms by which a drug can exert both hERG block and facilitation. Recently, the mechanism by which depolarization induces hERG facilitation was revealed. The voltage dependency of the induction of facilitation is associated with the voltage dependency of the hERG channel activation, specifically the opening of the activation gate in the pore [51]. In this study, the D540K hERG mutant was utilized, which can be activated by both depolarization and hyperpolarization [53]. In the wild-type hERG channel, facilitation is induced only by depolarization, whereas in the D540K hERG channel, it is induced by both depolarization and hyperpolarization stimuli [51]. Furthermore, this study demonstrated that drugs can facilitate activation through hyperpolarization in the D540K hERG mutant [51]. While the GV relationship of depolarization-induced activation shifted leftward, the GV relationship of hyperpolarization-induced activation shifted rightward. Considering the difference in structural changes in the voltage sensor domain caused by membrane depolarization and hyperpolarization, it is anticipated that drugs affect structural changes in the pore domain, facilitating the opening of the hERG channel pore when structural changes occur in the voltage sensor.

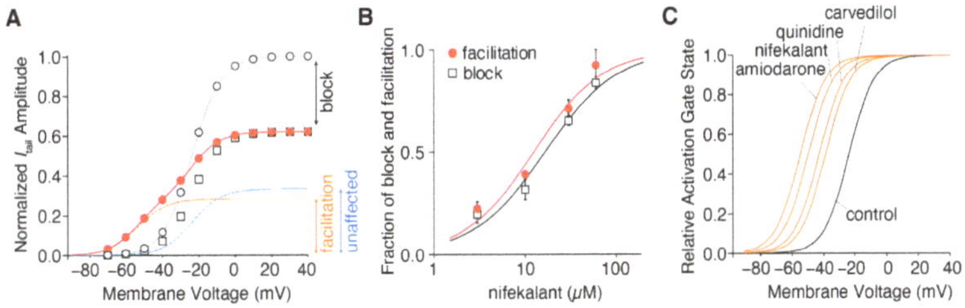

Figure 2. Drug-dependent negative shift in the hERG activation curve. (**A**) Voltage dependence of the hERG activation curves in the presence of 10 μM nifekalant. The tail currents of hERG in the absence (open circles) and presence of nifekalant with (filled circles) or without (open squares) the conditioning pulse were measured during the repolarizing pulse to −80 mV. The data were normalized to the current amplitude recorded following a voltage step of +10 mV in the absence of nifekalant. The model assumed two populations of channels with or without the facilitation effect of nifekalant (10 μM). The $V_{1/2}$ of activation for the facilitated fraction (orange dashed line) of channel was −50.7 mV, almost 28 mV negative to that of control channel (blue dashed lines). The red lines represent the double Boltzmann function (the sum of the Boltzmann functions for the facilitated fraction (orange dashed line) and the unaffected fraction (blue dashed lines)). (**B**) Concentration–response relationships for compound-induced block and facilitation by nifekalant. (**C**) Drug-dependent GV relationship in the facilitated fraction of hERG channels. Panel A was adopted with permission from Ref. [37].

Nifekalant and other facilitators act as open-channel blockers for hERG. It is important to note that despite this, some readers may still question why depolarization (channel opening) only affects facilitation and not inhibition. The experimental protocol illustrated in Figure 1A utilizes 4 s long test pulses that are commonly used to stimulate slow-activating hERG channels. While the block also requires pore opening, this is not evident in the steady state following the channel opening. Furthermore, the blocking effect of the drugs was assessed by comparing the magnitude of inhibition with the current in the absence of drugs, which essentially evaluates the effect of the drug solely on the opened channels. This evaluation method makes it challenging to discern differences in inhibition at different membrane potentials. Conversely, the facilitation induced by the pre-pulse was evaluated, and the subsequent reopening by the test pulse did not typically cause significant pore opening. This double-pulse protocol enables clear observation of the voltage-dependent induction of facilitation.

3. Structural Basis of hERG Facilitation

These studies have fielded our current understanding of the structural basis of facilitation. As previously mentioned, the facilitation effect is not limited to antiarrhythmic drugs, but can also be observed with non-cardiac hERG-blocking drugs [43]. The structural basis of hERG facilitators has been studied in detail. A three-dimensional quantitative structure–activity relationship (3D-QSAR) model was developed using EC50 values for facilitation and drug structures, consisting of one positively ionizable feature and three hydrophobic features [43]. The 3D-QSAR model was experimentally validated using test compounds, confirming the existence of common mechanisms among drugs for facilitation. Although this pharmacophore model of facilitation shares the same types and number of features as the block pharmacophore, their spatial arrangements differ slightly [43,54]. From this comparison, it was observed that there are drugs that satisfy both pharmacophore models for hERG channel block and facilitation, but the interaction with the channel may differ between block and facilitation.

The initial structure of the hERG channel was provided by the MacKinnon lab using single-particle cryo-electron microscopy [55]. The voltage sensor in this hERG structure had a non-domain-swapped architecture and appeared to be in a depolarized, activated conformation. The pores of the structure apparently opened. Subsequently, Asai et al. reported the structure of hERG complexed with astemizole, a representative hERG-blocking drug [56]. This structure clearly demonstrates that the drug interacts with several amino acids, including S624 and Y652 on the S6 helix, forming the central cavity of the hERG pore, thus supporting previous studies [1,57–66]. The involvement of F656 in direct interactions remains unclear, and it may play a role in mediating the access of the drug to the binding site [56,67]. Notably, astemizole is a classical hERG blocker. Therefore, these structures do not reveal the structural basis of the facilitation.

Hosaka et al. conducted alanine scanning of the pore and S6 helices of the hERG channel and analyzed the impact of mutations on both blockade and facilitation [36]. This study revealed that several mutants failed to exhibit both block and facilitation. For instance, substitutions of Y652 or F656 with alanine, which have been known to be crucial residues for the block, abolished the facilitation by nifekalant. Y652 is also crucial for the facilitation of phenanthrene [47]. These findings suggest that facilitation, similar to the block, occurs through interactions within residues that form the central cavity of the hERG pore. Importantly, mutants that had no effect on the block but affected facilitation were also identified. Some mutants, such as S624A, L646A, M651A, S654A, G657A, V659A, and L666A, inhibited facilitation, whereas others, such as I647A, S649A, S660A, I663A, R665A, and Y667A, enhanced it. It appears that residues around the central cavity of the hERG are specifically involved in facilitation and play a distinct role from that of the block. Structural models of open- and closed-state hERG channels were built to illustrate the amino acid residues associated with facilitation and putative drug interaction (Figure 3) [36]. We conducted a study using Rosetta to investigate hERG channel–drug interactions and found that certain drugs, known as hERG facilitators, are positioned in the hydrophobic pocket of the hERG central cavity [68]. This pocket is thought to play a role in drug interactions and is located in the open-state hERG channels, as identified by Wang and MacKinnon [55]. Our modeling study also suggested four potential fenestration regions in the pore, but hERG facilitators, including nifekalant and amiodarone, were not positioned in these fenestration regions [68]. In combination, these findings suggest that the hydrophobic pocket may be involved in the facilitation of hERG channels by these drugs. This intriguing possibility requires further validation and study.

Our working hypothesis for drug-induced facilitation (Figure 4) can be summarized as follows. Drugs enter the channel and inhibit channel function when the activation gate of the channel opens due to depolarization. It is then possible that drugs become trapped within the channel when the channel closes [69–72]. Drug interactions within the hydrophobic pocket may affect the closure of the S6 helix bundle, thereby altering the channel gating. Consistent with our results, experimental data suggest that facilitating drugs may act as a wedge to bias the hERG channel gating equilibrium toward an open state conformation. This can increase the hERG current amplitude in response to low-voltage depolarization [37,51]. Additional experimental measurements and molecular dynamics simulations are required to test structural model-based hypotheses.

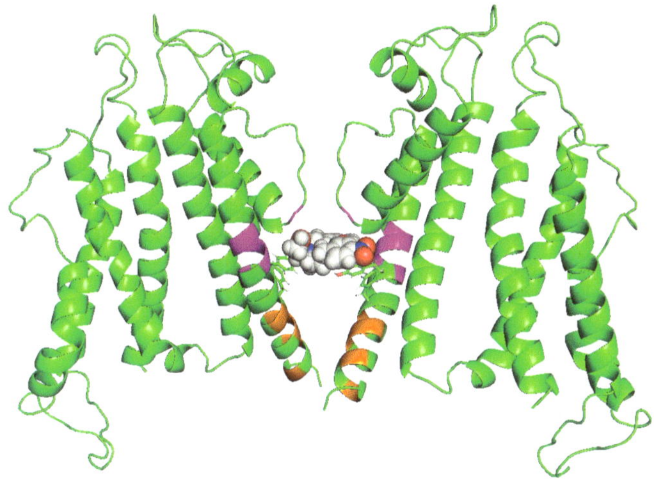

Figure 3. A possible hERG facilitator interaction site. Open hERG channel model (green). The side chains of Y652 and F656, crucial for the block, are shown as sticks. Magenta and orange indicate residues revealed experimentally by point mutagenesis influential for nifekalant's facilitation but not block. Magenta residues (L646, I647, S649, and M651) form hydrophobic pockets. In contrast, orange residues (G657, V659, S660, I663, and R665) may be located on the activation gate. The drug (nifekalant) may be positioned deep within the hydrophobic pocket [68]. The image was kindly provided by Dr. Igor Vorobyov (University of California, Davis).

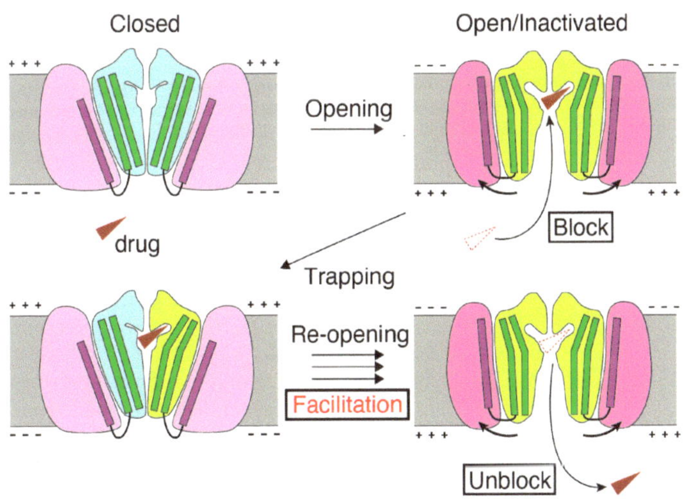

Figure 4. Schematic model of the mechanism underlying facilitation. (1) The hERG facilitator (red triangle) accesses the receptor site within the pore of the open or inactivated channels at depolarized potentials and "blocks" the channel. (2) Upon return to the resting potentials, channels close and trap the drug inside. (3) Trapped, the drug biases the open–closed equilibrium towards the open state and "facilitates". (4) The drug can escape from the re-opened hERG channels, creating open channels. Drug translocations to block and unblock are from red dotted to solid triangles. The cartoon of structural changes during hERG channel gating draws on the graphical abstract of Wang and MacKinnon's cell article [55].

4. A Possible Role of hERG Facilitation in hERG Block-Associated Arrhythmia

Yamakawa et al. investigated the facilitation effects of various non-cardiac drugs that block hERG channels [43]. The results revealed that drugs such as fluoxetine (an antidepressant), haloperidol (an antipsychotic), and chlorpheniramine (an antihistamine) exhibit both conventional blocking effects on hERG channels as well as facilitation effects similar to antiarrhythmic agents, such as amiodarone and nifekalant. This suggests that the facilitation effect is not exclusive to antiarrhythmic agents and may be commonly observed among clinically used hERG blockers. This study also reported that some classical hERG blockers, including atenolol, terfenadine, and sotalol, did not exhibit facilitation [43]. Terfenadine is an antihistamine that has been withdrawn from the market owing to its high risk of life-threatening arrhythmias. These findings suggest that the facilitation effect of hERG blockers may help reduce the risk of arrhythmias through their hERG-blocking mechanism. However, assessing the occurrence of facilitation in vivo presents technical challenges. In terms of the drug's concentration, nifekalant can exert block and facilitate hERG channels in almost the same concentration-dependent manner [37]. In addition, the repetitive excitation of ventricular myocytes may trigger facilitation. Experimental evidence has demonstrated that repeated stimulation with a voltage-clamp waveform that resembles action potentials can induce maximal facilitation [44]. This may suggest that facilitation occurs in living hearts.

An in vitro validation was performed to examine the effects of nifekalant, a hERG blocker with facilitation, and dofetilide, a hERG blocker without facilitation, on rat ventricular myocyte action potentials. Even at concentrations that inhibited I_{Kr} to a similar extent, dofetilide was more likely to induce early afterdepolarizations (EADs) than nifekalant, as shown in Figure 5A,B [44]. This suggests that facilitation may reduce the risk of arrhythmia. To further explore this concept, a theoretical study was conducted. We developed a mathematical model to simulate facilitation and its impact on action potential waveforms [44]. The facilitation model was formulated as a drug-induced shift in the GV relationship. By incorporating hERG block and facilitation models into a human ventricular myocyte action potential model (ORd human ventricular AP model [73]), we examined the influence of facilitation. Without a facilitation mechanism, increasing the concentration of a classical hERG blocker resulted in action potential prolongation and the development of EADs (Figure 5C,D). However, in the presence of a facilitation mechanism, the prolongation of the action potential was suppressed, and a higher concentration was required to induce EADs (Figure 5C,D). These observations theoretically illustrate that facilitation can lower the proarrhythmic risk associated with a drug.

Class III antiarrhythmic agents are hERG blockers used clinically to suppress ventricular tachyarrhythmias [1,31,32]. To suppress tachyarrhythmias without provoking *torsades de pointes*, the I_{Kr} block would ideally be use-dependent and prolong APD only in response to high-frequency stimulation [7,74]. However, reverse frequency-dependent action on APs is a property common to Class III antiarrhythmic agents [33,45,46,74–77], and the associated proarrhythmic risk limits their clinical usefulness [74,78]. The reverse frequency dependence of I_{Kr} block was first explained by an increase in the slowly activated delayed-rectifier K^+ current with rapid heart rate [75]. Thus, we theoretically tested whether the presence or absence of facilitation also influences the reverse frequency-dependent effect and suggested that facilitation reduces this dependence [44]. These effects are particularly prominent in models of heart failure [44], in which functional remodeling predisposes patients to arrhythmias [79].

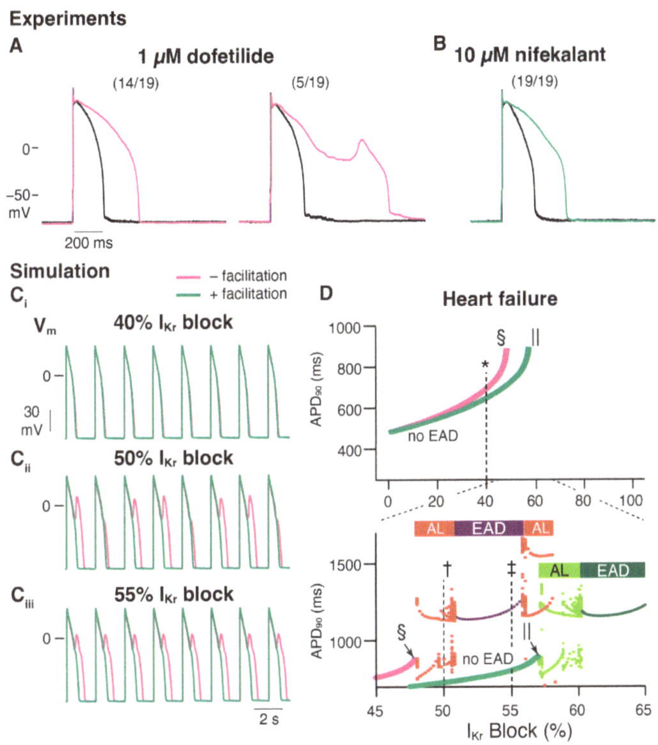

Figure 5. Facilitation suppresses the development of early afterdepolarization related to I_{Kr} blockade in ventricular cardiac myocytes. (**A**,**B**) Experimental study: Rabbit ventricular myocyte APs are more stable in nifekalant (**B**) than in dofetilide (**A**). AP responses in isolated rabbit ventricular myocytes were stimulated by minimal current injection (0.5 Hz) in whole-cell current clamp mode at 37 °C. Black line in (**A**,**B**) are control AP responses before the treatment. (**A**) Representative AP responses without (left) or with (right) EAD in 1 μM dofetilide. Of the 19 cells treated with 1 μM dofetilide (magenta), five showed EAD responses (26%). Fourteen cells showed prolonged APD at 1 μM but did not show EAD responses. (**B**) Representative AP responses to 10 μM nifekalant. All cells treated with 10 μM nifekalant (green) showed prolongation of APD upon treatment with 10 μM nifekalant but did not show EAD responses. (**C**,**D**) Simulation study: The effect of I_{Kr} facilitation on APD prolongation and EAD development by I_{Kr} block. (**C**) Steady-state AP trains with 40% (**C$_i$**), 50% (**C$_{ii}$**), and 55% (**C$_{iii}$**) I_{Kr} block in a heart failure model with and without facilitation. (**D**) Effect of I_{Kr} blockade and facilitation on APD and development of EADs in the heart failure model. Green and magenta lines indicate APD90 of AP (without EAD) for block with and without facilitation, respectively. The asterisk, dagger, and double-dagger indicate the conditions in (**C**), respectively. The sections and pipes indicate the upper limits of the I_{Kr} block, where APs are normally terminated. When EAD was observed, it was classified as either alternating or periodic EAD. In the bottom panel of D, red and deep purple dots indicate APD90 of AP with EAD for the block without facilitation, while light and deep green dots indicate APD90 of AP with EAD for the block with facilitation. Horizontal bars above the dots indicate alternating EAD, AL, periodic EAD, or EAD. This figure has been adapted with permission from Ref. [44].

The mechanisms underlying these effects have also been analyzed [44]. It is important to recall that the presence of blockers, such as nifekalant, can result in the crossing of the GV and IV relationships compared to their absence (Figures 1 and 2) [36,37]. When considering the trajectory of the ventricular action potential, depolarization occurs during phase 0 of the action potential, and this moves the myocyte membrane potential through the range of membrane potentials where facilitation increases the I_{Kr}/hERG current. Consequently, the effects of facilitation were minimal during phases 0 and 2. As the action potential repolarization begins and the myocyte voltage returns to the range of membrane potentials where facilitation occurs, facilitation increases the I_{Kr}/hERG currents. As a consequence, in this time- and voltage-window, the repolarizing currents are larger than in the absence of the drug. During a single ventricular action potential, hERG channel facilitators change their attributes to I_{Kr}. It decreases the current first and then increases it later, which prevents excessive prolongation of the action potential and repolarization impairments (Figure 6).

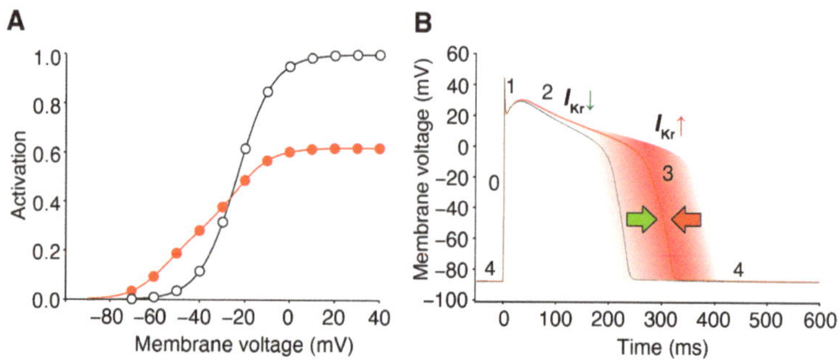

Figure 6. Block and facilitation impact on cardiac electrophysiology. (**A**) GV relationships of the control (open circles) and drug-treated conditions (red circles) intersect, resulting in drug-induced hERG/I_{Kr} currents decreasing at depolarized voltages and hERG/I_{Kr} currents increasing from the control at membrane voltages near the activation threshold. (**B**) The hERG channel blocker with a facilitation effect changes the attribute to I_{Kr} during the ventricular action potential. It decreases the current first and then increases it later, preventing excessive prolongation of the action potential and repolarization impairments. The different phases of the ventricular action potential (phases 0–4) are labeled. Green and red arrow indicate the dual drug actions on AP duration; I_{Kr} decrease by block prolongs (green), whereas I_{Kr} increase by facilitation prevents the prolongation of AP duration (red).

This facilitation-induced increase in the I_{Kr}/hERG current during phase 3 can become even more pronounced when repolarization is delayed [44]. This can be explained by the fact that, as the action potential duration is prolonged, the duration of the membrane potential within the range where facilitation occurs is also prolonged. This property can significantly affect the concentration-dependent and reverse frequency-dependent effects of hERG-blocking drugs on action potential duration. At low concentrations of these drugs, which only slightly prolong the action potential, the facilitation effect has a minimal impact. However, at higher concentrations, the action potential duration is further prolonged and the impact of facilitation can become more significant. Similarly, at high excitation frequencies, where the action potential is relatively short, the facilitation effect is less influential. However, at low excitation frequencies, where the action potential is relatively longer, it becomes more susceptible to facilitation. This mechanism allows the facilitation effect to prevent further prolongation of the action potential when it has already been extended by a drug-induced blockade.

Class III antiarrhythmic drugs are believed to exert their antiarrhythmic effects by extending the refractory period through hERG blockade, thereby prolonging the duration of action potential. There is some concern that the facilitation effect of hERG blockade on I_{Kr} current might also affect the antiarrhythmic effects of Class III antiarrhythmics. However, simulations have indicated that this impact is minimal. The increase in I_{Kr} current during phase 3 reduces excitability, offsetting the shortening effect on action potential duration. In simulations, prolongation of the relative refractory period by a hERG blocker with a facilitation effect was comparable to that of a classical hERG blocker [44].

Additionally, there is no report demonstrating that the facilitation effect in other Kv channels, such as I_{Ks} channels, contributes to repolarization in myocytes.

Further experimental validation is required; however, the notion that the proarrhythmic risk of hERG-blocking drugs may be mitigated by the facilitation effect is worth pursuing. In addition, to fully evaluate the risks associated with hERG-blocking drugs accurately, the presence or absence of facilitation effects must be considered.

5. Conclusions and Future Direction

In this review, we present an overview of the properties of hERG channel facilitation, its critical role in pharmacotherapy, and the current understanding of its functional and structural aspects. This material serves as the initial basis for the mechanism by which drugs facilitate hERG channel activation and suggests its clinical implications.

Important remaining questions include the following.

(i) How do drugs facilitate hERG activation? Which process(es) of hERG opening/closing are affected by the drug? It is highly likely that the drug is located within the central cavity of the hERG pore when it exerts facilitation [36,47,51]. By further investigating hERG channel gating and modulation, we can acquire a more profound understanding of its behavior. Advanced methods, such as single-channel recording and (single) molecular imaging, can be used to detect precise modifications in the hERG gating process triggered by the drug.

(ii) Where do the drugs bind exactly? To ascertain the structural location of the drug at the binding site for hERG activation, is it the hydrophobic pocket, or are other sites responsible for facilitation? Additionally, can the binding sites vary in size or shape between the closed and open states? If so, this may explain why certain drugs are capable of facilitating channel activation. Furthermore, it is conceivable that drugs within the channel pore can obstruct or eliminate hERG channel inactivation. Molecular dynamics simulations [80–86] would be particularly useful in revealing any modifications in the binding site resulting from drug binding in both closed and open states.

(iii) What are the structural differences between a hERG blocker/facilitator and a classical hERG blocker? A wide range of structurally diverse compounds interact with the hERG channel and inhibit its function. Is it possible to predict the facilitation effects of drugs based on their structures without experimental data? Although previous attempts have been made, the predictive capability of these methods has been limited to only a few selected drugs [43]. The facilitation effect, characterized by a leftward shift in the GV relationship, is unique to each drug. As such, the structure of the drug is likely related not only to its affinity but also to the strength of facilitation (modulation of hERG activation). Detailed exploration of pure hERG facilitators has not yet been conducted, and further research is necessary to understand the structural basis of the facilitation effect of drugs. Ultimately, this knowledge is crucial for the rational control and design of the facilitation effect of drugs.

(iv) How can we apply our understanding of the hERG facilitation to advance drug discovery and treatment? In the field of medical healthcare, improving the effectiveness and safety of pharmacotherapy remains an ongoing goal, requiring collaboration between industry, government, and academia. The hERG channel has garnered significant attention because of its association with arrhythmias, and an integrated approach incorporating multiscale, multiphysics biological simulations is essential to comprehensively understand and predict complex phenomena such as cardiac arrhythmias. To advance drug discovery and treatment, further testing of the hypothesis that "facilitation reduces the proarrhythmic risk of hERG blockers" [44] through theoretical research and proof-of-concept experiments is crucial. In principle, this knowledge could then be utilized to inform decision-making regarding drug therapy. Understanding the structure–activity relationship of drugs also opens up possibilities for theoretical drug design and modification, contributing to drug discovery. Although caution must continue to be exerted in the targeting of the hERG channel in drug discovery, if safety is confirmed, the conventional belief that "hERG blockers are dangerous" may evolve, possibly allowing vast resources of hERG-active compounds to be utilized for drug development.

Funding: This study was supported by Scientific Research (C) 21K06812 (K.F.) of the Ministry of Education, Culture, Sports, Science and Technology of Japan.

Institutional Review Board Statement: Not applicable.

Informed Consent Statement: Not applicable.

Data Availability Statement: The research data described in this paper are available upon request.

Acknowledgments: I thank Igor Vorobyov (University of California, Davis) for his support to make Figure 3.

Conflicts of Interest: The authors declare no competing financial interest.

References

1. Sanguinetti, M.C.; Tristani-Firouzi, M. hERG potassium channels and cardiac arrhythmia. *Nature* **2006**, *440*, 463–469. [CrossRef]
2. Sanguinetti, M.C.; Jiang, C.; Curran, M.E.; Keating, M.T. A mechanistic link between an inherited and an acquired cardiac arrhythmia: HERG encodes the IKr potassium channel. *Cell* **1995**, *81*, 299–307. [CrossRef]
3. Trudeau, M.C.; Warmke, J.W.; Ganetzky, B.; Robertson, G.A. HERG, a human inward rectifier in the voltage-gated potassium channel family. *Science* **1995**, *269*, 92–95. [CrossRef]
4. Vandenberg, J.I.; Perry, M.D.; Perrin, M.J.; Mann, S.A.; Ke, Y.; Hill, A.P. hERG K(+) channels: Structure, function, and clinical significance. *Physiol. Rev.* **2012**, *92*, 1393–1478. [CrossRef]
5. Clancy, C.E.; Kurokawa, J.; Tateyama, M.; Wehrens, X.H.; Kass, R.S. K+ channel structure-activity relationships and mechanisms of drug-induced QT prolongation. *Annu. Rev. Pharmacol. Toxicol.* **2003**, *43*, 441–461. [CrossRef]
6. Nerbonne, J.M.; Kass, R.S. Molecular physiology of cardiac repolarization. *Physiol. Rev.* **2005**, *85*, 1205–1253. [CrossRef]
7. Surawicz, B. Electrophysiologic substrate of torsade de pointes: Dispersion of repolarization or early afterdepolarizations? *J. Am. Coll. Cardiol.* **1989**, *14*, 172–184. [CrossRef]
8. Curran, M.E.; Splawski, I.; Timothy, K.W.; Vincent, G.M.; Green, E.D.; Keating, M.T. A molecular basis for cardiac arrhythmia: HERG mutations cause long QT syndrome. *Cell* **1995**, *80*, 795–803. [CrossRef]
9. Roden, D.M. Acquired long QT syndromes and the risk of proarrhythmia. *J. Cardiovasc. Electrophysiol.* **2000**, *11*, 938–940. [CrossRef]
10. Haverkamp, W.; Breithardt, G.; Camm, A.J.; Janse, M.J.; Rosen, M.R.; Antzelevitch, C.; Escande, D.; Franz, M.; Malik, M.; Moss, A.; et al. The potential for QT prolongation and proarrhythmia by non-antiarrhythmic drugs: Clinical and regulatory implications. Report on a policy conference of the European Society of Cardiology. *Eur. Heart J.* **2000**, *21*, 1216–1231. [CrossRef]
11. Roden, D.M. Cellular basis of drug-induced torsades de pointes. *Br. J. Pharmacol.* **2008**, *154*, 1502–1507. [CrossRef]
12. Kannankeril, P.; Roden, D.M.; Darbar, D. Drug-induced long QT syndrome. *Pharmacol. Rev.* **2010**, *62*, 760–781. [CrossRef] [PubMed]
13. Nachimuthu, S.; Assar, M.D.; Schussler, J.M. Drug-induced QT interval prolongation: Mechanisms and clinical management. *Ther. Adv. Drug Saf.* **2012**, *3*, 241–253. [CrossRef]
14. International Council for Harmonization of Technical Requirements for Pharmaceuticals for Human Use (ICH). The Non-Clinical Evaluation of the Potential for Delayed Ventricular Repolarization (QT Interval Prolongation) by Human Pharmaceuticals. Guidance on S7B. 2005. Available online: http://www.ich.org/products/guidelines/safety/safety-single/article/the-non-clinical-evaluation-of-the-potential-for-delayed-ventricular-repolarization-qt-interval-pro.html (accessed on 23 October 2023).
15. Sager, P.T.; Gintant, G.; Turner, J.R.; Pettit, S.; Stockbridge, N. Rechanneling the cardiac proarrhythmia safety paradigm: A meeting report from the Cardiac Safety Research Consortium. *Am. Heart J.* **2014**, *167*, 292–300. [CrossRef]
16. Gintant, G.; Sager, P.T.; Stockbridge, N. Evolution of strategies to improve preclinical cardiac safety testing. *Nat. Rev. Drug Discov.* **2016**, *15*, 457–471. [CrossRef]
17. Huang, H.; Pugsley, M.K.; Fermini, B.; Curtis, M.J.; Koerner, J.; Accardi, M.; Authier, S. Cardiac voltage-gated ion channels in safety pharmacology: Review of the landscape leading to the CiPA initiative. *J. Pharmacol. Toxicol. Methods* **2017**, *87*, 11–23. [CrossRef]
18. European Medicines Agency. *Topics ICH S7B Non-Clinical Evaluation of the Potential for Delayed Ventricular Repolarization (QT Interval Prolongation) by Human Pharmaceuticals—Scientific Guideline*; European Medicines Agency: Amsterdam, The Netherlands, 2022.
19. Redfern, W.S.; Carlsson, L.; Davis, A.S.; Lynch, W.G.; MacKenzie, I.; Palethorpe, S.; Siegl, P.K.; Strang, I.; Sullivan, A.T.; Wallis, R.; et al. Relationships between preclinical cardiac electrophysiology, clinical QT interval prolongation and torsade de pointes for a broad range of drugs: Evidence for a provisional safety margin in drug development. *Cardiovasc. Res.* **2003**, *58*, 32–45. [CrossRef]

20. Mirams, G.R.; Cui, Y.; Sher, A.; Fink, M.; Cooper, J.; Heath, B.M.; McMahon, N.C.; Gavaghan, D.J.; Noble, D. Simulation of multiple ion channel block provides improved early prediction of compounds' clinical torsadogenic risk. *Cardiovasc. Res.* **2011**, *91*, 53–61. [CrossRef]
21. Martin, R.L.; McDermott, J.S.; Salmen, H.J.; Palmatier, J.; Cox, B.F.; Gintant, G.A. The utility of hERG and repolarization assays in evaluating delayed cardiac repolarization: Influence of multi-channel block. *J. Cardiovasc. Pharmacol.* **2004**, *43*, 369–379. [CrossRef]
22. Finlayson, K.; Witchel, H.J.; McCulloch, J.; Sharkey, J. Acquired QT interval prolongation and HERG: Implications for drug discovery and development. *Eur. J. Pharmacol.* **2004**, *500*, 129–142. [CrossRef]
23. Shah, R.R. Drugs, QTc interval prolongation and final ICH E14 guideline: An important milestone with challenges ahead. *Drug Saf.* **2005**, *28*, 1009–1028. [CrossRef] [PubMed]
24. Raschi, E.; Vasina, V.; Poluzzi, E.; De Ponti, F. The hERG K+ channel: Target and antitarget strategies in drug development. *Pharmacol. Res.* **2008**, *57*, 181–195. [CrossRef] [PubMed]
25. Arcangeli, A. Expression and role of hERG channels in cancer cells. *Novartis Found. Symp.* **2005**, *266*, 225–232; discussion 232–234. [PubMed]
26. Masi, A.; Becchetti, A.; Restano-Cassulini, R.; Polvani, S.; Hofmann, G.; Buccoliero, A.M.; Paglierani, M.; Pollo, B.; Taddei, G.L.; Gallina, P.; et al. hERG1 channels are overexpressed in glioblastoma multiforme and modulate VEGF secretion in glioblastoma cell lines. *Br. J. Cancer* **2005**, *93*, 781–792. [CrossRef]
27. Pardo, L.A.; Contreras-Jurado, C.; Zientkowska, M.; Alves, F.; Stuhmer, W. Role of voltage-gated potassium channels in cancer. *J. Membr. Biol.* **2005**, *205*, 115–124. [CrossRef]
28. Wadhwa, S.; Wadhwa, P.; Dinda, A.K.; Gupta, N.P. Differential expression of potassium ion channels in human renal cell carcinoma. *Int. Urol. Nephrol.* **2009**, *41*, 251–257. [CrossRef]
29. Cicek, M.S.; Koestler, D.C.; Fridley, B.L.; Kalli, K.R.; Armasu, S.M.; Larson, M.C.; Wang, C.; Winham, S.J.; Vierkant, R.A.; Rider, D.N.; et al. Epigenome-wide ovarian cancer analysis identifies a methylation profile differentiating clear-cell histology with epigenetic silencing of the HERG K+ channel. *Hum. Mol. Genet.* **2013**, *22*, 3038–3047. [CrossRef]
30. Calcaterra, N.E.; Hoeppner, D.J.; Wei, H.; Jaffe, A.E.; Maher, B.J.; Barrow, J.C. Schizophrenia-Associated hERG channel Kv11.1-3.1 Exhibits a Unique Trafficking Deficit that is Rescued Through Proteasome Inhibition for High Throughput Screening. *Sci. Rep.* **2016**, *6*, 19976. [CrossRef]
31. Sanguinetti, M.C.; Jurkiewicz, N.K. Two components of cardiac delayed rectifier K+ current. Differential sensitivity to block by class III antiarrhythmic agents. *J. Gen. Physiol.* **1990**, *96*, 195–215. [CrossRef]
32. Vaughan Williams, E.M. Classifying antiarrhythmic actions: By facts or speculation. *J. Clin. Pharmacol.* **1992**, *32*, 964–977. [CrossRef]
33. Nakaya, H.; Tohse, N.; Takeda, Y.; Kanno, M. Effects of MS-551, a new class III antiarrhythmic drug, on action potential and membrane currents in rabbit ventricular myocytes. *Br. J. Pharmacol.* **1993**, *109*, 157–163. [CrossRef] [PubMed]
34. Zeng, J.; Laurita, K.R.; Rosenbaum, D.S.; Rudy, Y. Two components of the delayed rectifier K+ current in ventricular myocytes of the guinea pig type. Theoretical formulation and their role in repolarization. *Circ. Res.* **1995**, *77*, 140–152. [CrossRef] [PubMed]
35. Kodama, I.; Kamiya, K.; Toyama, J. Cellular electropharmacology of amiodarone. *Cardiovasc. Res.* **1997**, *35*, 13–29. [CrossRef] [PubMed]
36. Hosaka, Y.; Iwata, M.; Kamiya, N.; Yamada, M.; Kinoshita, K.; Fukunishi, Y.; Tsujimae, K.; Hibino, H.; Aizawa, Y.; Inanobe, A.; et al. Mutational analysis of block and facilitation of HERG current by a class III anti-arrhythmic agent, nifekalant. *Channels* **2007**, *1*, 198–208. [CrossRef]
37. Furutani, K.; Yamakawa, Y.; Inanobe, A.; Iwata, M.; Ohno, Y.; Kurachi, Y. A mechanism underlying compound-induced voltage shift in the current activation of hERG by antiarrhythmic agents. *Biochem. Biophys. Res. Commun.* **2011**, *415*, 141–146. [CrossRef]
38. Lo, Y.C.; Kuo, C.C. Temperature Dependence of the Biophysical Mechanisms Underlying the Inhibition and Enhancement Effect of Amiodarone on hERG Channels. *Mol. Pharmacol.* **2019**, *96*, 330–344. [CrossRef]
39. Perry, M.; Sachse, F.B.; Sanguinetti, M.C. Structural basis of action for a human ether-a-go-go-related gene 1 potassium channel activator. *Proc. Natl. Acad. Sci. USA* **2007**, *104*, 13827–13832. [CrossRef]
40. Grunnet, M.; Hansen, R.S.; Olesen, S.P. hERG1 channel activators: A new anti-arrhythmic principle. *Prog. Biophys. Mol. Biol.* **2008**, *98*, 347–362. [CrossRef]
41. Perry, M.; Sachse, F.B.; Abbruzzese, J.; Sanguinetti, M.C. PD-118057 contacts the pore helix of hERG1 channels to attenuate inactivation and enhance K+ conductance. *Proc. Natl. Acad. Sci. USA* **2009**, *106*, 20075–20080. [CrossRef]
42. Perry, M.; Sanguinetti, M.; Mitcheson, J. Revealing the structural basis of action of hERG potassium channel activators and blockers. *J. Physiol.* **2010**, *588*, 3157–3167. [CrossRef]
43. Yamakawa, Y.; Furutani, K.; Inanobe, A.; Ohno, Y.; Kurachi, Y. Pharmacophore modeling for hERG channel facilitation. *Biochem. Biophys. Res. Commun.* **2012**, *418*, 161–166. [CrossRef] [PubMed]
44. Furutani, K.; Tsumoto, K.; Chen, I.S.; Handa, K.; Yamakawa, Y.; Sack, J.T.; Kurachi, Y. Facilitation of I Kr current by some hERG channel blockers suppresses early afterdepolarizations. *J. Gen. Physiol.* **2019**, *151*, 214–230. [CrossRef] [PubMed]
45. Carmeliet, E. Use-dependent block and use-dependent unblock of the delayed rectifier K+ current by almokalant in rabbit ventricular myocytes. *Circ. Res.* **1993**, *73*, 857–868. [CrossRef]
46. Jiang, M.; Dun, W.; Fan, J.S.; Tseng, G.N. Use-dependent 'agonist' effect of azimilide on the HERG channel. *J. Pharmacol. Exp. Ther.* **1999**, *291*, 1324–1336. [PubMed]

47. Al-Moubarak, E.; Shiels, H.A.; Zhang, Y.; Du, C.; Hanington, O.; Harmer, S.C.; Dempsey, C.E.; Hancox, J.C. Inhibition of the hERG potassium channel by phenanthrene: A polycyclic aromatic hydrocarbon pollutant. *Cell. Mol. Life Sci.* **2021**, *78*, 7899–7914. [CrossRef]
48. Kang, J.; Chen, X.L.; Wang, H.; Ji, J.; Cheng, H.; Incardona, J.; Reynolds, W.; Viviani, F.; Tabart, M.; Rampe, D. Discovery of a small molecule activator of the human ether-a-go-go-related gene (HERG) cardiac K+ channel. *Mol. Pharmacol.* **2005**, *67*, 827–836. [CrossRef]
49. Zhou, J.; Augelli-Szafran, C.E.; Bradley, J.A.; Chen, X.; Koci, B.J.; Volberg, W.A.; Sun, Z.; Cordes, J.S. Novel potent human ether-a-go-go-related gene (hERG) potassium channel enhancers and their in vitro antiarrhythmic activity. *Mol. Pharmacol.* **2005**, *68*, 876–884. [CrossRef]
50. Hansen, R.S.; Diness, T.G.; Christ, T.; Wettwer, E.; Ravens, U.; Olesen, S.P.; Grunnet, M. Biophysical characterization of the new human ether-a-go-go-related gene channel opener NS3623 [N-(4-bromo-2-(1H-tetrazol-5-yl)-phenyl)-N′-(3′-trifluoromethylphenyl)urea]. *Mol. Pharmacol.* **2006**, *70*, 1319–1329. [CrossRef]
51. Furutani, K.; Kawano, R.; Ichiwara, M.; Adachi, R.; Clancy, C.E.; Sack, J.T.; Kita, S. Pore opening, not voltage sensor movement, underpins the voltage-dependence of facilitation by a hERG blocker. *Mol. Pharmacol.* **2022**, *102*, 223–233. [CrossRef]
52. Abbott, G.W.; Sesti, F.; Splawski, I.; Buck, M.E.; Lehmann, M.H.; Timothy, K.W.; Keating, M.T.; Goldstein, S.A. MiRP1 forms IKr potassium channels with HERG and is associated with cardiac arrhythmia. *Cell* **1999**, *97*, 175–187. [CrossRef]
53. Sanguinetti, M.C.; Xu, Q.P. Mutations of the S4-S5 linker alter activation properties of HERG potassium channels expressed in Xenopus oocytes. *J. Physiol.* **1999**, *514 Pt 3*, 667–675. [CrossRef] [PubMed]
54. Cavalli, A.; Poluzzi, E.; De Ponti, F.; Recanatini, M. Toward a pharmacophore for drugs inducing the long QT syndrome: Insights from a CoMFA study of HERG K(+) channel blockers. *J. Med. Chem.* **2002**, *45*, 3844–3853. [CrossRef] [PubMed]
55. Wang, W.; MacKinnon, R. Cryo-EM Structure of the Open Human Ether-a-go-go-Related K(+) Channel hERG. *Cell* **2017**, *169*, 422–430. [CrossRef]
56. Asai, T.; Adachi, N.; Moriya, T.; Oki, H.; Maru, T.; Kawasaki, M.; Suzuki, K.; Chen, S.; Ishii, R.; Yonemori, K.; et al. Cryo-EM Structure of K(+)-Bound hERG Channel Complexed with the Blocker Astemizole. *Structure* **2021**, *29*, 203–212.e204. [CrossRef] [PubMed]
57. Mitcheson, J.S.; Chen, J.; Lin, M.; Culberson, C.; Sanguinetti, M.C. A structural basis for drug-induced long QT syndrome. *Proc. Natl. Acad. Sci. USA* **2000**, *97*, 12329–12333. [CrossRef]
58. Lees-Miller, J.P.; Duan, Y.; Teng, G.Q.; Duff, H.J. Molecular determinant of high-affinity dofetilide binding to HERG1 expressed in Xenopus oocytes: Involvement of S6 sites. *Mol. Pharmacol.* **2000**, *57*, 367–374.
59. Kamiya, K.; Mitcheson, J.S.; Yasui, K.; Kodama, I.; Sanguinetti, M.C. Open channel block of HERG K(+) channels by vesnarinone. *Mol. Pharmacol.* **2001**, *60*, 244–253. [CrossRef]
60. Chen, J.; Seebohm, G.; Sanguinetti, M.C. Position of aromatic residues in the S6 domain, not inactivation, dictates cisapride sensitivity of HERG and eag potassium channels. *Proc. Natl. Acad. Sci. USA* **2002**, *99*, 12461–12466. [CrossRef]
61. Sanchez-Chapula, J.A.; Navarro-Polanco, R.A.; Culberson, C.; Chen, J.; Sanguinetti, M.C. Molecular determinants of voltage-dependent human ether-a-go-go related gene (HERG) K+ channel block. *J. Biol. Chem.* **2002**, *277*, 23587–23595. [CrossRef]
62. Fernandez, D.; Ghanta, A.; Kauffman, G.W.; Sanguinetti, M.C. Physicochemical features of the HERG channel drug binding site. *J. Biol. Chem.* **2004**, *279*, 10120–10127. [CrossRef]
63. Kamiya, K.; Niwa, R.; Mitcheson, J.S.; Sanguinetti, M.C. Molecular determinants of HERG channel block. *Mol. Pharmacol.* **2006**, *69*, 1709–1716. [CrossRef] [PubMed]
64. Milnes, J.T.; Witchel, H.J.; Leaney, J.L.; Leishman, D.J.; Hancox, J.C. hERG K+ channel blockade by the antipsychotic drug thioridazine: An obligatory role for the S6 helix residue F656. *Biochem. Biophys. Res. Commun.* **2006**, *351*, 273–280. [CrossRef] [PubMed]
65. Kamiya, K.; Niwa, R.; Morishima, M.; Honjo, H.; Sanguinetti, M.C. Molecular determinants of hERG channel block by terfenadine and cisapride. *J. Pharmacol. Sci.* **2008**, *108*, 301–307. [CrossRef] [PubMed]
66. Vandenberg, J.I.; Perozo, E.; Allen, T.W. Towards a Structural View of Drug Binding to hERG K(+) Channels. *Trends Pharmacol. Sci.* **2017**, *38*, 899–907. [CrossRef]
67. Mobley, D.L.; Dill, K.A. Binding of small-molecule ligands to proteins: "what you see" is not always "what you get". *Structure* **2009**, *17*, 489–498. [CrossRef]
68. Cortez, A.M.E.; DeMarco, K.R.; Furutani, K.; Bekker, S.; Sack, J.T.; Wulff, H.; Clancy, C.E.; Vorobyov, I.; Yarov-Yarovoy, V. Structural Modeling of hERG Channel -Drug Interactions Using Rosetta. *Front. Pharmacol.* **2023**, *14*, 1244166. [CrossRef]
69. Mitcheson, J.S.; Chen, J.; Sanguinetti, M.C. Trapping of a methanesulfonanilide by closure of the HERG potassium channel activation gate. *J. Gen. Physiol.* **2000**, *115*, 229–240. [CrossRef]
70. Windisch, A.; Timin, E.; Schwarz, T.; Stork-Riedler, D.; Erker, T.; Ecker, G.; Hering, S. Trapping and dissociation of propafenone derivatives in HERG channels. *Br. J. Pharmacol.* **2011**, *162*, 1542–1552. [CrossRef]
71. Linder, T.; Bernsteiner, H.; Saxena, P.; Bauer, F.; Erker, T.; Timin, E.; Hering, S.; Stary-Weinzinger, A. Drug trapping in hERG K(+) channels: (not) a matter of drug size? *Medchemcomm* **2016**, *7*, 512–518. [CrossRef]
72. Munawar, S.; Vandenberg, J.I.; Jabeen, I. Molecular Docking Guided Grid-Independent Descriptor Analysis to Probe the Impact of Water Molecules on Conformational Changes of hERG Inhibitors in Drug Trapping Phenomenon. *Int. J. Mol. Sci.* **2019**, *20*, 3385. [CrossRef]

73. O'Hara, T.; Virag, L.; Varro, A.; Rudy, Y. Simulation of the undiseased human cardiac ventricular action potential: Model formulation and experimental validation. *PLoS Comput. Biol.* **2011**, *7*, e1002061. [CrossRef] [PubMed]
74. Hondeghem, L.M.; Snyders, D.J. Class III antiarrhythmic agents have a lot of potential but a long way to go. Reduced effectiveness and dangers of reverse use dependence. *Circulation* **1990**, *81*, 686–690. [CrossRef] [PubMed]
75. Jurkiewicz, N.K.; Sanguinetti, M.C. Rate-dependent prolongation of cardiac action potentials by a methanesulfonanilide class III antiarrhythmic agent. Specific block of rapidly activating delayed rectifier K+ current by dofetilide. *Circ. Res.* **1993**, *72*, 75–83. [CrossRef]
76. Tande, P.M.; Bjornstad, H.; Yang, T.; Refsum, H. Rate-dependent class III antiarrhythmic action, negative chronotropy, and positive inotropy of a novel Ik blocking drug, UK-68,798: Potent in guinea pig but no effect in rat myocardium. *J. Cardiovasc. Pharmacol.* **1990**, *16*, 401–410. [CrossRef]
77. Hafner, D.; Berger, F.; Borchard, U.; Kullmann, A.; Scherlitz, A. Electrophysiological characterization of the class III activity of sotalol and its enantiomers. New interpretation of use-dependent effects. *Arzneimittelforschung* **1988**, *38*, 231–236.
78. Okada, Y.; Ogawa, S.; Sadanaga, T.; Mitamura, H. Assessment of reverse use-dependent blocking actions of class III antiarrhythmic drugs by 24-hour Holter electrocardiography. *J. Am. Coll. Cardiol.* **1996**, *27*, 84–89. [CrossRef] [PubMed]
79. Elshrif, M.M.; Shi, P.; Cherry, E.M. Representing variability and transmural differences in a model of human heart failure. *IEEE J. Biomed. Health Inform.* **2015**, *19*, 1308–1320. [CrossRef]
80. Dai, J.; Zhou, H.X. An NMDA receptor gating mechanism developed from MD simulations reveals molecular details underlying subunit-specific contributions. *Biophys. J.* **2013**, *104*, 2170–2181. [CrossRef]
81. Wang, Y.; Guo, J.; Perissinotti, L.L.; Lees-Miller, J.; Teng, G.; Durdagi, S.; Duff, H.J.; Noskov, S.Y. Role of the pH in state-dependent blockade of hERG currents. *Sci. Rep.* **2016**, *6*, 32536. [CrossRef]
82. Lev, B.; Murail, S.; Poitevin, F.; Cromer, B.A.; Baaden, M.; Delarue, M.; Allen, T.W. String method solution of the gating pathways for a pentameric ligand-gated ion channel. *Proc. Natl. Acad. Sci. USA* **2017**, *114*, E4158–E4167. [CrossRef]
83. Barbera, N.; Ayee, M.A.A.; Akpa, B.S.; Levitan, I. Molecular Dynamics Simulations of Kir2.2 Interactions with an Ensemble of Cholesterol Molecules. *Biophys. J.* **2018**, *115*, 1264–1280. [CrossRef] [PubMed]
84. Dietzen, N.M.; Arcario, M.J.; Chen, L.J.; Petroff, J.T., 2nd; Moreland, K.T.; Krishnan, K.; Brannigan, G.; Covey, D.F.; Cheng, W.W. Polyunsaturated fatty acids inhibit a pentameric ligand-gated ion channel through one of two binding sites. *elife* **2022**, *11*, e74306. [CrossRef]
85. Zhuang, Y.; Noviello, C.M.; Hibbs, R.E.; Howard, R.J.; Lindahl, E. Differential interactions of resting, activated, and desensitized states of the alpha7 nicotinic acetylcholine receptor with lipidic modulators. *Proc. Natl. Acad. Sci. USA* **2022**, *119*, e2208081119. [CrossRef] [PubMed]
86. Costa, F.; Ocello, R.; Guardiani, C.; Giacomello, A.; Masetti, M. Integrated Approach Including Docking, MD Simulations, and Network Analysis Highlights the Action Mechanism of the Cardiac hERG Activator RPR260243. *J. Chem. Inf. Model.* **2023**, *63*, 4888–4899. [CrossRef]

International Journal of *Molecular Sciences*

MDPI

Article

Ionic Mechanisms of Propagated Repolarization in a One-Dimensional Strand of Human Ventricular Myocyte Model

Yukiko Himeno *, Yixin Zhang, Suzuka Enomoto, Hiroto Nomura, Natsuki Yamamoto, Shotaro Kiyokawa, Mirei Ujihara, Yuttamol Muangkram, Akinori Noma and Akira Amano

Department of Bioinformatics, College of Life Sciences, Ritsumeikan University, Shiga 525-8577, Japan; sj0041xh@ed.ritsumei.ac.jp (Y.Z.); noma@sk.ritsumei.ac.jp (A.N.); a-amano@fc.ritsumei.ac.jp (A.A.)
* Correspondence: hime@fc.ritsumei.ac.jp

Abstract: Although repolarization has been suggested to propagate in cardiac tissue both theoretically and experimentally, it has been challenging to estimate how and to what extent the propagation of repolarization contributes to relaxation because repolarization only occurs in the course of membrane excitation in normal hearts. We established a mathematical model of a 1D strand of 600 myocytes stabilized at an equilibrium potential near the plateau potential level by introducing a sustained component of the late sodium current (I_{NaL}). By applying a hyperpolarizing stimulus to a small part of the strand, we succeeded in inducing repolarization which propagated along the strand at a velocity of 1~2 cm/s. The ionic mechanisms responsible for repolarization at the myocyte level, i.e., the deactivation of both the I_{NaL} and the L-type calcium current (I_{CaL}), and the activation of the rapid component of delayed rectifier potassium current (I_{Kr}) and the inward rectifier potassium channel (I_{K1}), were found to be important for the propagation of repolarization in the myocyte strand. Using an analogy with progressive activation of the sodium current (I_{Na}) in the propagation of excitation, regenerative activation of the predominant magnitude of I_{K1} makes the myocytes at the wave front start repolarization in succession through the electrical coupling via gap junction channels.

Keywords: repolarization propagation; mathematical 1D strand model; human ventricular myocyte; early afterdepolarization; late sodium current

Citation: Himeno, Y.; Zhang, Y.; Enomoto, S.; Nomura, H.; Yamamoto, N.; Kiyokawa, S.; Ujihara, M.; Muangkram, Y.; Noma, A.; Amano, A. Ionic Mechanisms of Propagated Repolarization in a One-Dimensional Strand of Human Ventricular Myocyte Model. *Int. J. Mol. Sci.* **2023**, *24*, 15378. https://doi.org/10.3390/ijms242015378

Academic Editor: Massimo Iacoviello

Received: 30 September 2023
Accepted: 13 October 2023
Published: 19 October 2023

1. Introduction

It has long been recognized that the repolarization of cardiac action potential (AP) can be regulated by both intrinsic ion channel properties, as well as cable properties conferred by the three-dimensional syncytium of cardiac myocytes [1–4]. However, the mechanisms of the propagation of repolarization in multicellular preparations of the cardiac muscle remain to be clarified based on ionic current systems. Initially, Weidman (1951) [5] observed all-or-none repolarizations in sheep's Purkinje fiber when he applied a hyperpolarizing current beyond a certain strength during the plateau phase. He observed that there was a definite 'threshold' for the all-or-none repolarization, and suggested that the wave of repolarization could propagate without decrement as in the case of excitation. Essentially the same threshold phenomenon was observed by Cranefield and Hoffman (1958) [6] when inducing repolarization in the papillary muscle preparation. However, attempts to determine the velocity of the propagation of repolarization were not entirely successful. In part, this were due to the fact that the velocity of the repolarization wave was much slower than that of the excitation, so the propagation of repolarization was expected to be overtaken by the normal process of repolarization after the beginning of the AP.

The first theoretical study of cardiac electrophysiology using detailed mathematical descriptors was presented by Noble in 1962 [1]. He modified the Hodgkin–Huxley equations of squid giant axon and combined these with the cable theory to quite accurately reproduce many of the electrical properties of Purkinje fiber APs and pacemaker potentials.

Purkinje fiber taken from the conduction system of juvenile sheep hearts had an exceptionally large diameter and length, which allowed extrapolation of the theory developed for the squid nerve to the cardiac muscle preparation after some modifications on parameters. After the epoch-making experimental milestones of the introduction of both dissociated cardiomyocyte preparations [7–9] and the patch clamp technique [10] in the next decades, the characteristics of ion channels and transporters were extensively clarified in the isolated cardiac myocytes from various cardiac tissues. These milestones made possible the development of various cardiac myocyte models based on the experimental data on isolated single myocytes [11]. Since then, cardiac myocyte models have become increasingly detailed and have been validated based on experimental data (for human ventricular myocyte models, ORd model [12], GPB model [13], Asakura et al., [14], Himeno et al. [15], etc.).

The aims of the present simulation study were (1) to clarify the ionic mechanisms of induced regenerative repolarization and the threshold phenomenon for the all-or-none repolarization in dissociated ventricular myocytes, (2) to evaluate the impacts of intercellular currents on the propagation of repolarization induced in the myocyte strand, and (3) to determine the velocity of the propagation of the repolarization wave. Accordingly, we first determined whether our comprehensive human ventricular myocyte (hVC) model [15] could reproduce the all-or-none repolarization as demonstrated by Trenor et al. (2017) [2] using the ORd model [12]. After confirming the fundamental behavior of all-or-none repolarization in the hVC model, we developed an in silico 1D strand of myocytes to analyze the effects of the intercellular currents on the regenerative repolarization as well as on the propagation of the repolarization wave. The clinical implications for the antiarrhythmic treatment of the present study will be described in the Section 3.

2. Results

2.1. Threshold Potentials for the All-or-None Repolarization in the hVC Model

Before examining the propagation of repolarization, we examined whether the all-or-none repolarization, i.e., the key experimental basis for suggesting the propagation of repolarization, can be observed in a single hVC model as has been proved in the ORd model by Trenor et al. (2017) [2]. In the simulation in Figure 1, the membrane potential (V_m) was clamped at various test potential (V_t) levels with 0.5 mV intervals for a duration of 5 ms.

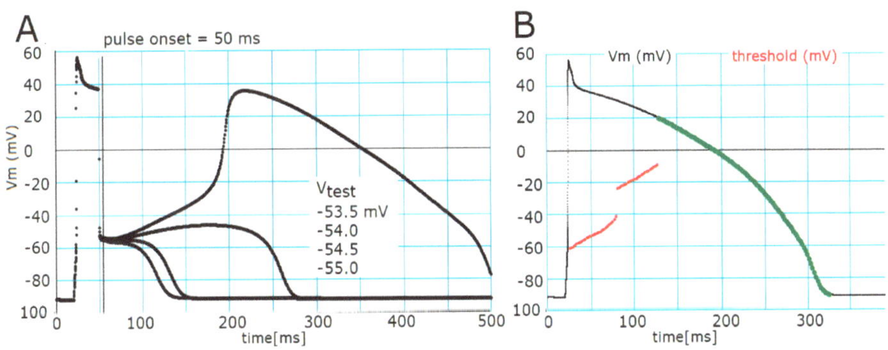

Figure 1. Threshold potentials detected by applying V_t of 5 ms in duration to different levels in the hVC model. (**A**) The application of V_t to −53.5, −54, −54.5, and −55 mV induced dynamic changes in the time course of plateau potential as displayed from top to bottom (four traces), respectively. The black vertical line in panel A indicates the off time (t = 55 ms) of the 5 ms clamp pulse. (**B**) The threshold potentials at which the given clamp pulse induced abolition of the plateau potential (red) were plotted against the off time of various test pulses. Green dots indicated there were no threshold potential. In this range colored in green, V_t did not induce all-or-none repolarization but led to monotonic repolarization.

After the membrane was released from the temporal voltage clamp, V_m changed automatically before progressive depolarization or hyperpolarization became obvious as seen in Figure 1A. The slope of this change reversed between −55 and −53.5 mV of the clamp pulse applied at 50 ms. By adjusting the V_t bit-by-bit, a relatively flat phase of V_m change, reflecting an unstable equilibrium, was obtained at around −50 mV. Above that level, delayed activation of an early after-depolarization (EAD)-like event started after the release time. A V_t far more positive than this threshold potential level induced the rising foot of an EAD-like event directly, while a more negative V_t induced monotonic repolarization. The threshold potentials, determined at an interval of 2 ms during the plateau phase, were plotted in Figure 1B (red dots). The threshold potential was time- and V_m-dependent; it shifted to more negative levels when the clamp pulse was applied earlier during the AP plateau. There were two apparent 'jumps' in the traces of the thresholds: at V_t = −42 and −23 mV. The green dots on the control time course of AP (black trace) indicated that the hyperpolarizing clamp pulse only accelerated the monotonic repolarization. All these findings were very similar to those reported in previous studies [2,3].

2.2. Ionic Mechanisms Responsible for Determining the Threshold Potentials for All-or-None Repolarization

As demonstrated in Figure 1A, the fate of all-or-none repolarization was determined by the initial slow V_m changes after switching off the clamp pulse (Figure 1A). To clarify the ionic mechanisms underlying the initial slow change, the changes in p(O) were plotted in Figure 2B. In both columns I and II, hyperpolarizing clamp pulses were applied at 50 ms. When the pulse induced repolarization to −55 mV, the p(O) of inward-going late sodium current (I_{NaL}), $p(O)_{INaL}$, gradually decreased, and p(O) of the rapid component of delayed rectifier potassium current (I_{Kr}), $p(O)_{IKr}$, also decreased. The magnitude of p(O) of the inward rectifier potassium channel (I_{K1}), $p(O)_{IK1}$, was slightly but significantly increased to enlarge I_{K1} (Figure 2C) by the negative jump in V_m from the plateau potential to −55 mV. The sum of the K^+ currents was larger, by about 0.3 pA/pF, than the sum of the major inward currents, which were mainly given by I_{NaL} and the background non-selective cation current, I_{bNSC}, to drive the gradual repolarization (sum of major I_{out} and total I_{in} were represented by gray lines in Figure 2C and D, respectively). Finally, the membrane quickly hyperpolarized to the resting potential via a positive feedback cycle; the rapid increase in $p(O)_{IK1}$ drives repolarization and more repolarization induces an increase in $p(O)_{IK1}$. Through this cycle, the amplitude of I_{K1} showed a sharp peak as large as 1.5 pA/pF at about −70 mV. From this fact, it may be concluded that the I_{K1} takes the central role in inducing regenerative repolarization toward the resting potential.

When the hyperpolarization from the plateau potential was increased only by 1 mV (to −54 mV), the slope of the slow depolarization was inverted from a negative (at −55 mV) to a positive direction (column II). As expected, no significant changes were obvious in the initial levels of p(O) or the amplitudes of I_{NaL}, I_{Kr}, I_{bNSC}, and I_{K1} immediately after the break in the voltage clamp pulse in comparison to the pulse to −54 mV. However, an evident exponential rise in the L-type calcium current (I_{CaL}) amplitude was observed 200 ms after the pulse (Figure 3D); the more the $p(O)_{ICaL}$ increased, the more depolarized the membrane became, and vice versa. The accompanying changes in I_{NaL}, I_{Kr}, I_{bNSC}, and I_{K1} may occur secondarily due to the membrane depolarization caused by I_{CaL}. The final discharge of an EAD-like event is caused by the much-enhanced positive feedback cycle of I_{CaL} activation.

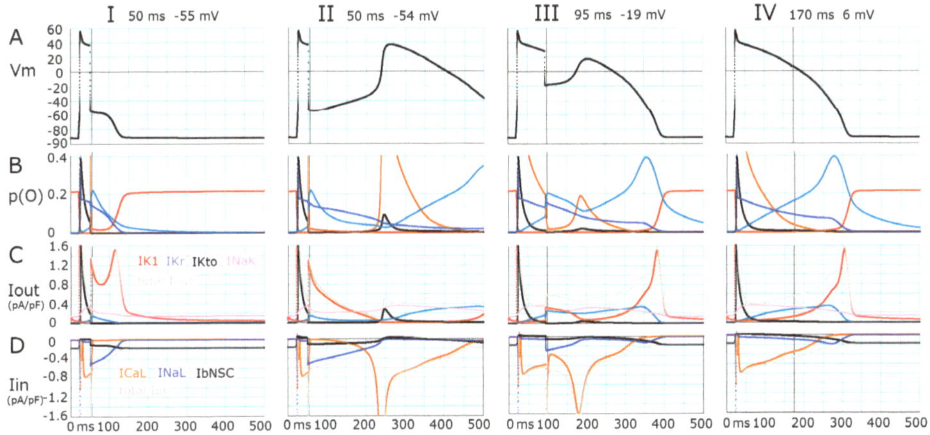

Figure 2. Ionic mechanisms underlying changes in V_m induced by applying 5 ms hyperpolarizing clamp pulses. Each time and voltage of the brief hyperpolarizing voltage clamp pulse applied were indicated at the top of each column. (**A**) indicates V_m, (**B**) indicates the open probabilities p(O) of currents, I_{K1} (red), I_{Kr} (steel blue), I_{Kto} (black), I_{CaL} multiplied by 2 (chocolate), and I_{NaL} multiplied by 100 (blue), (**C**,**D**) indicate amplitudes of outward (I_{out}) and inward (I_{in}) currents, respectively. The same colors were used as in (**B**). The sum of major outward and inward currents was calculated and illustrated in gray. The amplitude of I_{NaK} (pink) and I_{bNSC} (black) are also shown in (**C**) and (**D**), respectively. The vertical lines indicate the end of the clamp pulse.

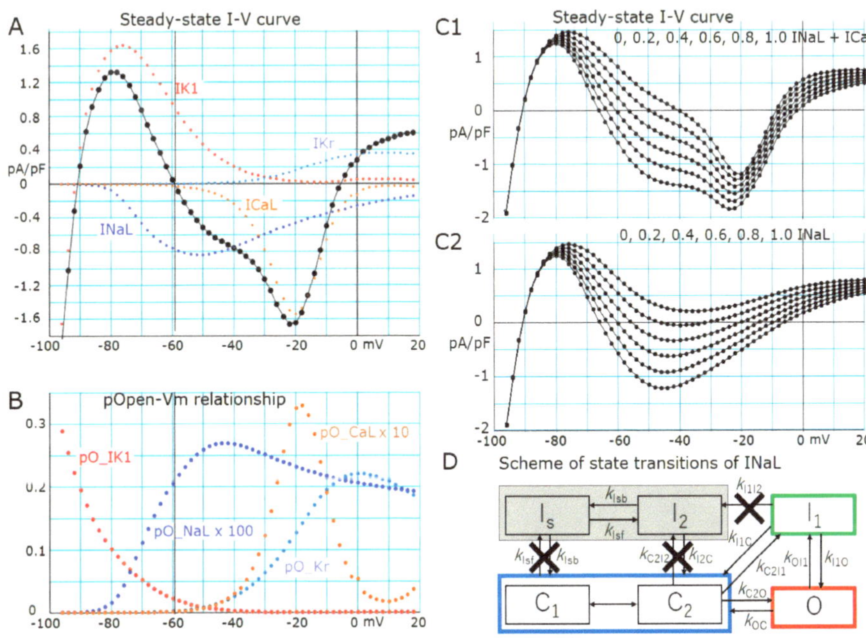

Figure 3. The N-shaped *I-V* curve with two stable equilibrium potentials and one unstable equilibrium potential of the hVC model. The *I-V* curves in (**A**,**C1**,**C2**) were measured by applying the voltage clamp test pulses of 4.9 s in duration given every 2 mV step from the holding potential of

−80 mV. Current magnitudes saturated within the duration of test pulses at the end of individual test pulses, and were plotted in these figures. Additionally, (**A**) shows total whole cell current (black) and current components I_{K1}, I_{Kr}, I_{NaL}, and I_{CaL} depicted in different colors as indicated. The steady-state p(O)s of major ion channels were plotted in (**B**). Note that $p(O)_{INaL}$ and $p(O)_{ICaL}$ were plotted after multiplication by 100 and 10, respectively. (**C1**,**C2**) shows the influence of increasing the amplitude of I_{NaL} on the appearance of equilibrium points in the presence and the absence of I_{CaL}, respectively. Increasing the amplitude of I_{NaL} shifted the N-shaped *I-V* curve negatively. All *I-V* and p(O)-V relationships share the same abscissa of V_m. In (**D**), the state of slow inactivation, I_s and the transitional inactivated state, I_2 of I_{NaL}, were shaded by gray color to indicate that these states were frozen, namely, the state transition from other states was totally prevented as indicated by cross marks. Thereby, repetitive state transitions between I_1 and O produce continuous bursts of brief openings of the channel, and 'late scattered mode' of Na channel [16] (k_{I1O} is close to k_{OI1}) during a clamp pulse to depolarized potentials. If integrated over bursts with numerous numbers of brief openings, the whole-cell I_{NaL} maintains a significant amplitude during depolarization. With an intact transition scheme, I_{NaL} gradually decreases in amplitude through the state transition from O to (I_s + I_2) when the membrane is depolarized. The state transition between the two closed states (**C1**,**C2**) was assumed to be instantaneous on the V_m jump; the probability of (**C2**) increases with membrane depolarization. In the state transition of transient sodium current (I_{NaT}) (see Supplementary Materials) the I_1 state is absent, and the inactivation occurs through a state transition from O to I_2 at transition rate k_{OI2}, which has the same magnitude as k_{OI1}.

In column III, the V_t was −19 mV, which gave a stable $p(O)_{ICaL}$ of about 0.02 to supply an I_{CaL} amplitude comparable to I_{NaL} at V_t = −54 mV (column II). On the other hand, the I_{NaL} amplitude was smaller if compared with that obtained by V_t at −54 mV because of V_m-dependent inactivation and the smaller driving force for the carrier ion Na^+. The I_{CaL} was also partially inactivated to give the decelerated rising phase of the EAD-like event. In column IV, the V_t (at 6 mV) was very close to the intrinsic plateau potential at 170 ms, and thereby the time courses of each current were almost the same as in the natural time course of final repolarization of the plateau. There was no threshold potential observed thereafter. It is important to note that the I_{Kto} is activated by the rapid depolarization, but is almost instantaneously deactivated by the hyperpolarization induced by the temporal voltage clamp pulse, thereby, I_{Kto} plays little role in initiating the full repolarization of the plateau (see Näbauer et al., 1996 [17]).

In summary, the I_{K1} takes the primary role in generating regenerative repolarization and the I_{CaL} plays a role in regenerative depolarization. Both I_{NaL} and I_{CaL} play modulatory roles in determining the threshold potential of repolarization.

2.3. Induction of a Stable Equilibrium Potential in the Current-Voltage Relationship of the hVC Model

In principle, if there is a second stable equilibrium potential in a more positive potential range than a resting membrane potential, as well as a negative conductance region in the steady-state *I-V* curve of a single myocyte model, it would be possible to induce a wave of repolarization in a multicellular preparation of myocyte model as predicted in the previous studies [1–4]. It is expected that the propagation of repolarization might be simply visualized in a 1D strand of such myocyte models whose V_m is set to another stable equilibrium potential. In brief, graded hyperpolarizing stimulus, which is strong enough to induce an all-or-none repolarization, is applied, it is expected that the repolarization would propagate as a wave based on the ionic mechanisms revealed for the threshold phenomenon at the single myocyte level (Figure 2): progressive increase in $p(O)_{IK1}$ supplemented by deactivation of I_{NaL} and I_{CaL} and activation of I_{Kr}.

A study by Chandler and Meves (1970) [18] used sodium fluoride to generate a long-lasting plateau in the AP of squid axons and reproduced the prolonged plateau potential mathematically by modifying the inactivation parameter of the sodium current (I_{Na}). In human ventricular myocytes, a slow inactivating sodium current component (I_{NaL}) has been described [16]. This I_{NaL} component in the human ventricle inactivates with a sufficiently fast time course to shape APD normally. However, it has been reported that the rate of slow inactivation of I_{NaL} is further delayed in heart failure, or by modulation of phospholipid associated with cardiovascular disease, to prolong the AP duration [19]. Based on these findings, we removed the inactivation of I_{NaL} to obtain a stable equilibrium potential near the plateau potential [14]. Figure 3D shows the removal of the slow inactivation steps to states (I_s + I_2), namely k_{Isf}, k_{Isb}, and k_{I1I2} were all set to zero. The fast inactivated state I_1 allows flickering openings of the channel to reconstruct the repetitive brief openings demonstrated in the single channel recordings [16] because the magnitude of k_{I1O} was set close to k_{OI1}.

The whole-cell current–voltage (*I-V*) relationships in Figure 3A were obtained by using the full set of current systems (Equation (3)) except for the slow inactivated states of I_{NaL} fixed at 0.45 for I_s and 0.001 for I_2, respectively. The *I-V* curve represented by black points connected by lines clearly demonstrated the stable equilibrium (zero-current) potential at approximately −5 mV. Indeed, the slope of the whole-cell *I-V* curve was positive on both sides of the stable equilibrium potential. It should be noted that a similar positive slope was also evident around the stable equilibrium (resting) potential at −90.5 mV. In contrast, the slope of the steady-state *I-V* curve is negative around the unstable zero-current potential at −59 mV (vertical line).

To get deeper insights into the ionic mechanisms of the abolition of the AP plateau phase as well as the propagation of repolarization, we examined the composition of the whole-cell *I-V* relationship. In Figure 3A, the major determinants of *I-V* relations, such as I_{K1} (red), I_{Kr} (steel blue), I_{NaL} (blue), and I_{CaL} (chocolate) were superimposed on the whole-cell current. Toward the negative potential side of the threshold potential (vertical line), a sharp voltage-dependent increase in both $p(O)_{IK1}$ and a decrease in $p(O)_{INaL}$ (blue dots in Figure 3B) were evident, indicating that resultant increase in outward I_{K1} and decrease in inward I_{NaL} were mainly responsible for the acceleration of repolarization. On the positive side of the threshold potential, the sharp V_m-dependent increase in $p(O)_{ICaL}$ and the resultant increase in I_{CaL} amplitude with depolarization promotes the depolarization. This influence of I_{CaL} activation was antagonized by the voltage-dependent increase in $p(O)_{IKr}$ to give the stable equilibrium potential at about −5 mV.

The I_{NaL} took a key role in inducing the second stable equilibrium potential at the voltage range of AP plateau as demonstrated in Figure 3C1,C2, where the amplitude of I_{NaL} was decreased from one to zero in steps of 0.2. In Figure 3C1, it is evident that the unstable equilibrium potential was largely shifted toward negative potentials from −40 mV to −60 mV with increasing magnitude of I_{NaL}. If the limiting conductance of I_{K1} was increased or decreased by +20%, the unstable equilibrium potential was shifted only by +2 mV or −3 mV, respectively (not shown). In Figure 3C2, the 'window component' of I_{CaL} was removed to reveal the role of I_{NaL} in inducing the unstable equilibrium potential. In the absence of I_{CaL}, two intersections of the *I-V* relation with the zero-current level disappeared when I_{NaL} was intact, shown in in Figure 3C2, leaving the stable equilibrium potential at the resting potential intact. The addition of I_{CaL} scarcely affected the location of unstable equilibrium potential when the relative magnitude of I_{NaL} remained in the physiological range (>0.2). If the gating kinetics of I_{NaL} (illustrated in Figure 3D) are intact under the physiological condition, the time-dependent slow inactivation accumulates the $p(I_s)_{INaL}$ to promote membrane repolarization to the resting potential.

2.4. Propagation of Repolarization from the Second Stable Equilibrium Potential at Around 0 mV

To separate the propagation of repolarization from the natural time course of AP repolarization, the V_m of the myocyte strand was stabilized to the equilibrium potential near the plateau potential by fixing 45.1% of I_{NaL} channels to the inactivated states (I_s and I_2). The stimulus current pulse (−45 pA/pF, 30 ms in duration) was applied to cell No. 1 (Figure 4A) at 10 ms after starting the V_m recording. The first AP of cell No. 1 (blue trace) was conducted sequentially from myocyte to myocyte in the 1D strand model of hVCs (the AP depicted in red is recorded from cell No. 17). With the modified I_{NaL} (shown in

Figure 3A,D), a train of EADs was initiated spontaneously during the repolarizing phase of the AP, but the EAD subsequently transformed into decremental oscillations in the myocyte strand to level off at a stable depolarization (Figure 4A) as predicted by the steady-state *I-V* curve.

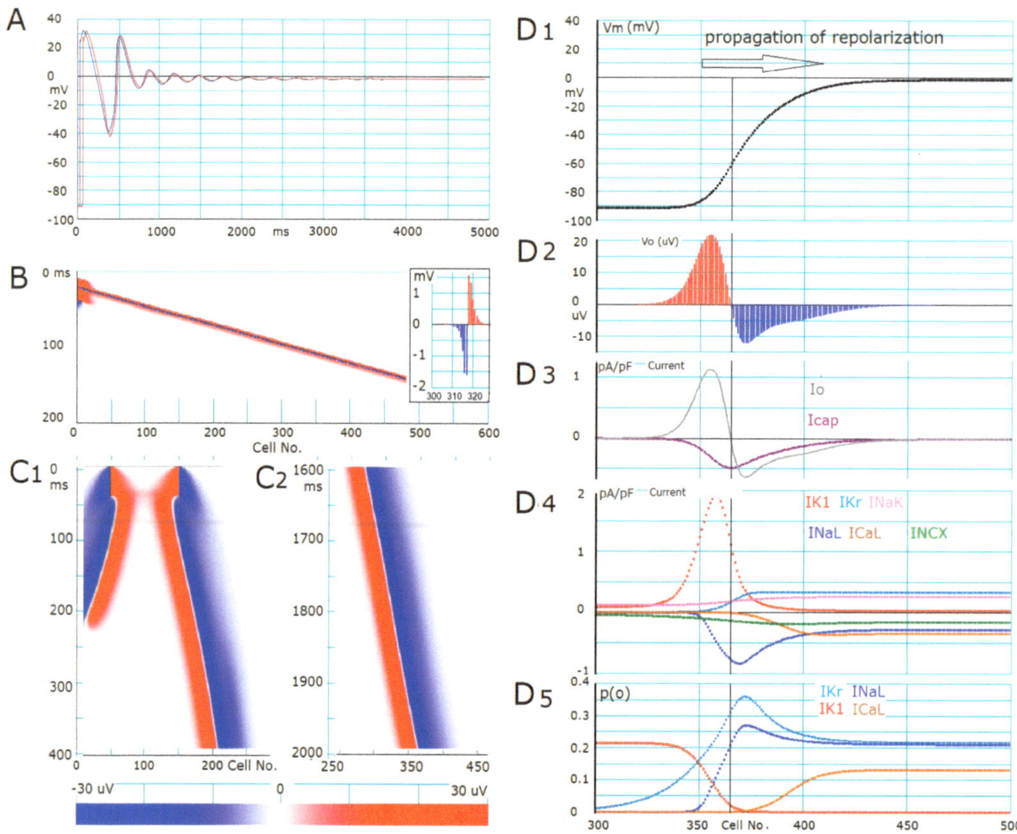

Figure 4. Propagated repolarization in a linear strand of the EAD-prone myocytes in comparison to that of excitation. Panel A shows the time course of V_m, moving from the resting membrane potential (−91 mV) to the second stable equilibrium potential (−2 mV) in the myocyte strand, in which the deeply inactivated states ($I_s + I_2$) of I_{NaL} were removed. Panel (**A**) indicates V_m recordings in cell No. 1 (blue) and 17 (red). The AP was evoked by applying a depolarizing pulse to cell No. 1 in the 1D strand of the hVC model. The V_m stabilized at about −3 mV at the end of the record in all myocytes within the strand. Panel (**B**) indicates the propagation of the O/I pattern (propagation of excitation) along the myocyte strand. The inset shows the stable O/I pattern of V_o; positive V_o evoked by outward whole-cell current, I_o (red), or negative V_o due to inward I_o (blue). Note, that the cell number on the abscissa of the inset is the same as in panel B, but the scale of the V_o is reduced by 100 times. Panels (**B**,**C**) indicate the movement of O/I patterns (propagation of repolarization) along the myocyte strand by using the color code of V_o shown at the bottom of panel (**C**). The O/I patterns illustrated in (**C1**,**C2**) were obtained at different recording times. See text for the explanation of panels (**D1**–**D5**). Note that $p(O)_{INaL}$ and $p(O)_{ICaL}$ were plotted after multiplication by 100 and 10, respectively, in panel (**D5**). The time scale shown in (**D5**) is common for all (**D1**–**D5**). The arrow head in (**D1**) indicates the direction of the propagation of repolarization. The scaling factor of I_{K1} = 1.2 was used in this simulation. Note that the scale of the V_o is reduced by 100 times in panel (**C**) if compared to that in panel (**B**).

The rate of AP conduction was measured from the slope of the profile of V_o along the 1D strand of in silico myocytes (Figure 4B). An example snapshot of the original V_o profile along cell No. on the *x* axis from cell No. 300~330 is demonstrated in the inset of Figure 4B. The V_o of a given myocyte within the myocyte strand is positive (here depicted in red) when the whole-cell membrane current is outward and vice versa (depicted in blue). It was evident that the direction of I_o, which corresponded to the sign of V_o, was inverted from negative to positive between cell No. 317 and 318. The wavefront of excitation propagation (in the right-ward direction) was defined by the location of this inversion of the sign of the I_o (defined as O/I center). In Figure 4B, the V_o of an individual myocyte was represented by the color code indicated at the bottom of Figure 4C, in which the ordinate is time, and the abscissa represents cell No. of the myocyte strand. The propagation rate of membrane excitation was determined to be 48.0 cm/s from the slope of the linear relationship between the propagation time and the number of the myocytes, whose length was 120 mm in the hVC model (Figure 4B).

The propagation of repolarization could be evoked in the 1D strand model with 600 hVCs which was depolarized continuously, as in Figure 4A. If compared with the propagation of excitation in panel B, the propagation of repolarization was characterized by a much smaller current density (by two orders of magnitude) of the major repolarizing current I_{K1} than that of the major excitatory current I_{Na}. To trigger the propagation of repolarization in the myocyte strand, the hyperpolarizing current should be applied either to a large fraction of myocytes within the strand, or at a large amplitude, or for a longer duration of current pulse to overcome the buffering action of the rest fraction of myocytes in the strand. This was indicated by the findings demonstrated in Figure 4C1, where a trigger current pulse of 400 pA/pF and 40 ms in duration was applied simultaneously to 100 myocytes (cell No. 50~150). Immediately after switching off the test pulse, the distance between the two O/I centers on the left and right sides at the borders of stimulated and unstimulated myocytes slightly decreased for ~10 ms before starting the propagation of repolarization. If the triggering pulse was smaller (<310 pA/pF), and/or shorter in duration, or applied to a smaller number of myocytes, the distance between the two O/I centers was shortened to push back the repolarization, which is consistent with the presence of a threshold phenomenon observed in the multicellular preparations.

The repolarizing electric stimulus applied to the 1D strand model of myocytes could be replaced by increasing the G_{K1}, for example. Increasing the $p(O)_{IK1}$ beyond 0.03 for 40 ms in the central 100-myocyte segment successfully triggered the propagation of repolarization. Similarly, the propagation of repolarization could be evoked by activation of the ATP-sensitive K^+ channel (I_{KATP}), which has a large maximum conductance (G_{KATP}). Since the maximum conductance of I_{Kr} (G_{Kr}) is much smaller than that of I_{K1}, increasing the $p(O)_{IKr}$ failed to evoke hyperpolarization to trigger the propagation of repolarization. Hyperpolarizing stimulus applied by nullifying the $p(O)$ of the major inward currents, I_{NaL} and I_{CaL}, during the AP plateau phase, was not effective in inducing propagation of repolarization.

The rate of the propagation of repolarization (the slope of the movement of the O/I center) slightly decreased with time and reached a steady rate of 1.7 cm/s after ~400 ms. The ionic mechanisms of repolarization propagation were analyzed in panels D1–D5 in Figure 4, which shows snapshots taken 800 ms after the initiation of the triggering pulse. Surprisingly, the V_o profiles in Figure 4D2 were distributed over a much wider range of ~100 myocytes, in which V_m continues to repolarize (Figure 4D1), in contrast to the narrower width of V_o profile over 10 myocytes in the case of propagation of membrane excitation (inset in Figure 4B). The location of the O/I center is indicated by the vertical line in Figure 4D1–D5. On the left side of the O/I center, the $p(O)_{IK1}$ increased and $p(O)_{INaL}$ decreased with a steep slope (Figure 4D5), and the amplitude of I_{K1} increased while the amplitude of I_{NaL} decreased (Figure 4D4). The profile of all variables shown in panels D1–D5 remained the same as far as the rate of propagation was constant, indicating that the propagation proceeded without decrement.

To compare the contribution of membrane ionic currents (I_m) and extracellular current (I_o) to varying V_m of individual myocytes during the propagation of repolarization, profiles of all related parameters are illustrated on a common *x*-axis of cell No. in the strand.

$$I_c = I_o - I_m \quad (1)$$

where I_c (or I_{cap} in Figure 4D3) is the current through the membrane capacitor to determine dV_m/dt, and I_m represents the sum of whole-cell ionic currents. Major ionic currents, such as I_{K1}, I_{Kr}, I_{NaL}, and I_{CaL}, and transporter currents I_{NaK} and I_{NCX} are depicted in different colors in Figure 4D4. It is evident that the peak amplitudes of these major currents are roughly in the same order of magnitude, indicating that each one of these major currents has the potential to affect the rate of propagation of+ repolarization. However, if the current amplitudes in Figure 4D4 were compared to the V_o or I_o profiles in Figure 4D2 or Figure 4D3, it would seem to indicate that I_{K1} and I_{NaL} took the central role in driving repolarization over a potential range more negative than −60 mV. On the right side of the O/I center, the increase in $p(O)_{IKr}$ and decrease in $p(O)_{ICaL}$ over the segment of myocytes from No. 365 to 420 were mainly responsible for the initial repolarization over the V_m range from 0 to −60 mV. The currents generated by the Na/K pump (pink) and Na/Ca exchange (green) were relatively small and were mainly determined by the V_m change.

It is of note that the large increasing outward current component of I_{K1} and decreasing I_{NaL} in the myocytes on the left side of the O/I center with V_m below −60 mV (red and blue in Figure 4D4) not only contribute to repolarize themselves but also repolarize the myocytes on the right side of the O/I center (see I_{cap} in Figure 4D3) through gap junction channels (see I_o in Figure 4D3). The area of the positive and negative I_o in Figure 4D3 was always equal and was distributed electrotonically along the strand. This electrotonic current started to repolarize the myocytes at the propagation wavefront where the inward and outward currents were still almost balanced at the potential near the quasi-equilibrium around 0 mV. These findings on current dynamics observed near the unstable equilibrium point around −60 mV were quite comparable, in qualitative aspects, with the ionic mechanisms observed in both the steady-state *I-V* curve (Figure 3) and the abolition experiment of the AP plateau (Figures 1 and 2).

2.5. Effects of Varying the Conductance of I_{NaL}, I_{K1}, and Gap Junction Channel (G_g) on the Rate of Propagation of Repolarization

The velocities of propagation for the repolarization were measured at various conductances of I_{NaL} and I_{K1} as listed in Table 1. The limiting conductance of I_{NaL} (G_{NaL}) was increased by decreasing the magnitude of fixed ($p(I_2)$ + $p(I_s)$), and vice versa when decreasing G_{NaL}. The rate of propagation of repolarization increased with decreasing G_{NaL}, suggesting that the influence of I_{NaL} to interfere with the hyperpolarizing influence of I_o (= I_{cap} + I_m) in myocytes located in front of the O/I center is greater than the accelerating influence of I_{NaL} deactivation in myocytes located behind the O/I center. As expected, an increase in the conductance of outward I_{K1} (G_{K1}) increased the propagation of repolarization to support the central role of I_{K1} in promoting the propagation.

Table 1. The rate of propagation of repolarization in the strand model. The sign 'F' indicates that the stimulus failed to evoke the propagation of repolarization. Note that increasing the fraction of I_2 in (I_2 + I_s) by 10 or 20% decreases G_{NaL}.

	0.639 (−20%)	**0.594 (−10%)**	**Relative G_{NaL} 0.549 after Fixing (I_2 + I_s)**	**0.504 (+10%)**	**0.459 (+20%)**
−10%	F	F	F	F	F
control G_{K1}	F	F	F	1.69	2.11
+10%	F	1.52	1.94	2.38	2.94
+20%	1.79	2.114	2.40	2.69	3.04

In Table 1, the letter 'F' indicates that the triggering hyperpolarizing pulse failed to evoke the propagation of repolarization because of smaller G_{K1} and/or larger G_{NaL}. Under these conditions, EAD-like events occurred in myocytes located before the wavefront after switching off the hyperpolarizing pulse. Then, the EAD-like event subsequently triggered an AP in myocytes behind the wavefront that stayed at the resting potential, resulting in a retrograde propagation of AP in the 1D strand.

These findings, obtained by varying the gap junction conductance (G_g), are consistent with the theory that the magnitude of the intercellular current is one of the determinants for promoting the rate of AP propagation (Table 2).

Table 2. The rate of propagation of excitation and repolarization calculated by varying gap junction conductances (G_g) in the strand model.

G_g (nS)	1000	1500	2000	2500
rate excitation propagation (cm/s)	36.5	48.0	51.79	58.5
rate repolarization propagation (cm/s)	1.93	2.43	2.60	2.94

3. Discussion

3.1. Brief Summary of the Results

(1) The ionic mechanisms underlying the induced all-or-none repolarization at the myocyte level (Figure 2), and the propagation of repolarization in a 1D linear strand of myocytes (Figure 4), were examined. Unambiguous propagation of repolarization that did not overlap with the natural time course of the AP plateau was observed only when the 1D strand of myocytes was depolarized at the second stable equilibrium potential, which was introduced by freezing the fraction of slow inactivation of I_{NaL} to 0.451 in Figure 4.

(2) The key ionic currents were I_{K1}, I_{Kr}, I_{NaL}, and I_{CaL} in both the abolition of the AP plateau in a single ventricular myocyte and the propagation of repolarization in the in silico 1D strand of the myocyte model. Since the voltage range of the kinetic dynamics for I_{K1} and I_{NaL} are more negative compared to that for I_{Kr} and I_{CaL}, as demonstrated in Figure 3A, I_{K1} and I_{NaL} play their role during the late phase of repolarization, while I_{Kr} and I_{CaL} work in the early phase of repolarization. The positive feedback mechanisms accelerate the rate of V_m change caused by the kinetics dynamics of these currents.

(3) In the abolition experiments in Figures 1 and 2, the triggering pulse of hyperpolarization was supplied by the external current source, while the propagation of repolarization in the 1D strand was initiated by local current I_o and evoked by the myocytes located behind the wavefront of repolarization before the O/I center (Figure 4D2).

(4) The kinetic mechanisms of the four currents were most probably responsible for the stable propagation of repolarization.

(5) The O/I profile of V_o around the O/I center (Figure 4D2) is a mirror image of that in the propagation of excitation (Figure 4B, inset). The width of the O/I profile was much more elongated in the propagation of repolarization (Figure 4D1–D5) compared with that for the propagation of AP (Figure 4B), reflecting the relatively small current amplitude of I_{K1}, I_{NaL}, I_{Kr}, and I_{CaL} in repolarization. The merit of using multiple channel species may be attributed to its flexibility in adjusting the configuration of the AP plateau phase more precisely and stably.

(6) The rate of propagation of repolarization was dependent mainly on the amplitude of I_{K1} and I_{NaL} (Table 1). The rate increased with increasing the G_g within a limited range of G_g.

3.2. Clinical Implications

Noble et al. [1,3] demonstrated, for the first time, the principal ionic mechanisms underlying the induction of repolarization as well as the propagation of repolarization in cardiac myocytes, based on the computer model of cardiac membrane excitation. The present study used one of the detailed contemporary computer models (human ventricular myocyte models [12–15]) to clarify the ionic mechanisms of the propagation of repolarization in terms of individual ion channels as well as ion transporters to reconstruct the key experimental findings. The model was modified only by removing the very slow inactivation of I_{NaL}, leaving other fast mechanisms of I_{NaT}, I_{CaL}, I_{Kr}, the transient outward potassium current (I_{Kto}), the slow component of delayed rectifier potassium current (I_{Ks}), and I_{K1} intact to create the stable equilibrium potential near 0 mV. Since the kinetic time constant of the slow inactivation of I_{NaL} is in the order of several hundreds ms, fixing the I_{NaL} slow inactivation hardly interferes with the repolarizing process from the stable equilibrium potential. Thus, the ionic mechanisms of the repolarization demonstrated in the present study are all relevant to repolarization mechanisms under physiological conditions.

The ionic mechanisms underlying the propagation of repolarization might be examined using an in silico model of a simple 1D strand, in which hVCs were connected by gap junctions. Here, let us assume that the 1D strand is depolarized by preventing its repolarization by removing the slow inactivation step from the state transition scheme of I_{NaL} (Figure 3D). This strand will be connected to an intact 1D strand of a similar size. At the junction of the two strands, the I_{K1} channel in the latter strand will provide a hyperpolarizing current through gap junctions to the depolarized myocytes to repolarize toward the resting potential. This repolarization through the cell-to-cell interaction should propagate into the depolarized strand as demonstrated in Figure 4. Through this propagation of repolarization, the whole strand will recover to its resting potential. Thus, the propagation of repolarization might be relevant in clinical anti-arrhythmic treatment. Indeed, the present study found that the propagation of repolarization could be initiated by amplifying G_{K1} above 0.03, which is at the same level of G_{K1} at the resting potential, in a fraction of the in silico 1D strand of hVC model. As expected, it was found in our preliminary study that magnifying G_{KATP} was also useful in initiating the propagation of repolarization.

For the relevance of the present study on cardiac AP repolarization as well as the influences of these theoretical studies on clinical sciences, see the excellent review by Trenor et al. (2017) [2].

3.3. Ionic Mechanisms of Propagation of Repolarization in Comparison to the Propagation of Excitation

Here, a comprehensive view of the propagation of repolarization is discussed in comparison to that of excitation, including the threshold phenomena for the abolition of the AP plateau phase, the steady-state *I-V* relations, and positive feedback mechanisms. The mechanisms of the propagation of membrane excitation in the cardiac tissue were well established, as described in many textbooks. In essence, the depolarization-dependent activation of I_{Na} and the I_{Na}-dependent membrane depolarization cause the positive feedback cycle to promote the rapidly rising phase of AP. Depolarization of a myocyte within the 1D strand generates outward-going I_o in neighbor myocytes through gap junctions to cause the positive cycle of I_{Na} activation in the next myocyte. In contrast, the propagation of membrane repolarization takes place through the positive feedback cycle of repolarization-dependent activation of I_{K1} and I_{K1}-dependent repolarization in the presence of supplemental ionic currents. Repolarization of myocytes in the strand induces inward-going I_o to repolarize myocytes which are still depolarized at the repolarization wavefront of the strand through gap junctions.

In the intact hVC model, all the parameters of I_{NaL}, including the slow inactivation states (I_2 and I_s) of the model, continuously change throughout the time course of the AP plateau phase driven by the time- and voltage-dependent gating parameters to promote the membrane repolarization. On the other hand, the state transitions among the C, O, and I_1 states of I_{NaL} are much faster than the V_m change during the relatively slow AP plateau, and thereby these fast parameters roughly follow the steady-state values during the time course of the AP plateau phase. Thus, it is quite possible that the model may take nearly the same parameter set as that used in the steady-state *I-V* curve (Figure 3A) at an instant during the natural time course of the plateau phase.

It seems difficult to find any ionic current that can substitute the role of I_{NaL} in the simulation of propagated repolarization in the multicellular preparations. Especially as the unstable equilibrium (threshold) potential at around −60 mV allowed the sharp inversion of I_o from inward to outward current at the O/I center, as demonstrated in Figure 4D1–D5. Note that these equilibrium potentials are common for both excitation and repolarization under physiological conditions (Figure 3). The window current of I_{CaL}, which was depicted in chocolate in Figure 3A, can also provide the unstable equilibrium potential at about −40 mV in addition to the inducible stable equilibrium potential at −8.5 mV, even in the absence of I_{NaL} (Figure 3C1). However, around this threshold potential (−40 mV), the amplitude of the outward current (I_{K1}) is much smaller than that at −60 mV, so the rate of repolarization may be too slow to evoke the regenerative propagation of repolarization against the buffering effect of the multicellular strand. In contrast, the threshold potential at ~−60 mV allowed the smooth and swift transition from the initial half of repolarization (ahead of the O/I center) to the final (behind the O/I center) repolarization due to the massive magnification of I_{K1} as shown in Figure 4D1–D5. The rapid upstroke of AP also starts at about −60 mV when the AP is triggered from the resting potential.

Qu et al. [20] in their review paper, introduced a general framework to explain the mechanisms of EADs using the quasi-stable *I-V* relationships. The same framework is also useful in understanding how the second stable equilibrium potential was introduced in the present study. The description of a given current, I_X, is described using three parameters; the limiting conductance G_X, the open probability of the channel ($p(O)_X$), and the driving force ($V_m - E_X$), where E_X is the reversal potential of the ionic species of the current X.

$$I_X = G_X \cdot p(O)_X \cdot (V_m - E_X) \quad (2)$$

The steady-state *I-V* curve is measured in experiments (or is computed in simulation in Figure 4) by clamping V_m until the I_m reaches a stable level at different voltages along the V_m axis. While the quasi-steady-state *I-V* curve in the present study was obtained by the same protocol, the magnitude p(O) of a selected slow parameter was fixed during the natural time course of the AP plateau; for example, the $p(O)_{INaL}$ in the present study (Figure 4C). Thus, at a depolarized V_m during AP plateau, the quasi-steady-state whole-cell *I-V* relationship shown in Figure 4C moves upward smoothly with the continuous reduction of $p(O)_{INaL}$ accompanied by increasing inactivated fraction ($p(I_s)_{INaL} + p(I_2)_{INaL}$). At the plateau potential of the AP, the V_m traces the negative shift of the second stable equilibrium potential. Finally, the quasi-steady-state *I-V* relation fails to cross the X-axis over the depolarized potential range, and the membrane quickly repolarizes to the resting potential. Thus, the stable equilibrium potential could be induced in the 1D myocyte strand when the decrease of the inactivated fraction of I_{NaL} was frozen at an optimum level during the progress of the AP plateau with all the other kinetic parameters left intact.

Lastly, some of the reported values of the velocity of repolarization in previous studies are compared. In the classic study of Cranefield and Hoffmann [6], using papillary muscle fiber, they stated, "no value was given for conduction velocity under Results because of the uncertainty of measurement" and described a velocity of 20 cm/s estimated from the records of a propagated repolarization wave in their Discussion. In addition, they commented on the value as "this value may not be correct, but it is certain that the velocity is low compared with that of propagated depolarization" and referred to the original study of Weidemann [5]. Our simulation using a strand composed of intact hVC demonstrated that it was impossible to determine a unique value for the velocity of propagated repolarization because the velocity increased continuously and approached infinity as the repolarization propagated [21]. If the slope of the O/I center was roughly calculated to be t = 205~255 ms after the hyperpolarizing stimulation, a value of 17 cm/s was obtained, which was in the same order of magnitude as that reported by Cranefield and Hoffmann. However, it should be noted that this velocity is not a pure rate of the propagation of repolarization, instead it is a rate of the propagation of repolarization overlapping the intrinsic spontaneous repolarization of each myocyte. Previous work in this area is of interest. There has been a study on the conduction across the interface between two different cell groups using micropatterned cocultures of genetically engineered (GE) excitable cells and neonatal rat ventricular myocytes (NRVMs) [22]. The GE cell was a monoclonal HEK293 cell line stably expressing human I_{Na} ($\mathrm{Na_V}$ 1.5), I_{K1} (Kir2.1), and rat connexin-43 (Cx43) gap junctions and had shorter APD (31.9 ms) than NRVMs (153.2 ms). They stimulated the co-cultured strand from the GE cell side and mapped the conduction of activation and repolarization using voltage-sensitive dye optically. In this strand, intrinsic repolarization occurred spontaneously in individual cells, but the GE cells of the former half triggered repolarization of the NRVMs composing the latter half of the strand because the APD of the GE cells was much shorter than NRVMs. If we calculate the rate of propagation of repolarization focusing on the donor–host interface from their result, the rate was about 1.8 cm/s which was much slower than that of activation (22 cm/s). In addition, Sperelakis et al. [23] measured the propagation of repolarization by inducing repolarization from another stable potential level to a resting potential as we did in the present study. They use a five-cell array of an extremely simplified model of electrical activity (2005). They reported that the rate of repolarization propagation is considerably (2.3~15 times) slower than that of excitation propagation, and that it varies with the number of gap junction channels (50~200 cm/s for repolarization propagation compared to 115~3000 cm/s for excitation propagation in the strand with 100~10,000 gap junction channels of 100 pS). Although the values previously reported in various experimental settings varied widely, the conclusion that the velocity of propagation of repolarization was much smaller than that of excitation was consistent in the present study.

4. Materials and Methods

4.1. The hVC Model

A single ventricular myocyte having a cuboid shape (120 μm on the longitudinal axis, 37.6 μm on the transverse axis, and 8.4 μm in depth) was assumed to calculate the propagation rate of excitation and repolarization. The membrane ionic current (I_{m}) was given as a sum of 15 components as described in Equation (3).

$$I_{\mathrm{m}} = I_{\mathrm{NaT}} + I_{\mathrm{NaL}} + I_{\mathrm{CaL}} + I_{\mathrm{Kto}} + I_{\mathrm{Kr}} + I_{\mathrm{Ks}} + I_{\mathrm{K1}} + I_{\mathrm{Kpl}} + I_{\mathrm{bNSC}} + I_{\mathrm{Cab}} + I_{\mathrm{L(Ca)}} + I_{\mathrm{KATP}} + I_{\mathrm{NaK}} + I_{\mathrm{NaCa}} + I_{\mathrm{PMCA}} \quad (3)$$

As will be described in the present study, the I_{K1} takes a pivotal role in both the repolarization of the AP plateau as well as in the propagation of repolarization. In this simulation study, we used the I_{K1} model developed by the Ishihara group [24,25]. Fabbri et al. (2019) [26] compared six computer formulations of I_{K1} to overcome the deficiency of I_{K1} conductance in the human-induced pluripotent stem cells (hiPSC-CMs) through the dynamic clamp (DC) technique. They found that the optimum compensation of the AP waveform was satisfactory when the Ishihara I_{K1} model was used. The subendocardial type of I_{Kto} was used (I_{to1} in Näbauer et al., 1996 [17]). The open probabilities of individual currents are plotted in Figures 2–4. The open probability of whole cell I_{CaL} ($p(\mathrm{O})_{I\mathrm{CaL},X}$) was a weighted sum of those distributed in three $\mathrm{Ca^{2+}}$ spaces (Xs) in *jnc*, *iz*, and *blk* in the hVC model (Equation (4)).

$$p(\mathrm{O})_{\mathrm{I_{CaL}}} = frcCaL_{jnc} \cdot p(\mathrm{O})_{\mathrm{I_{CaL}},jnc} + frcCaL_{iz} \cdot p(\mathrm{O})_{\mathrm{I_{CaL}},iz} + frcCaL_{blk} \cdot p(\mathrm{O})_{\mathrm{I_{CaL}},blk} \quad (4)$$

where $frcCaL_{\mathrm{X}}$ represents the fraction of I_{CaL} channel distribution in X space (X: *jnc*, *iz*, and *blk*).

For calculating each current component in the hVC, abbreviations, physical constants, ionic composition of external solution, parameters, and initial values of the model components, see model equations in Supporting Materials and Himeno et al. (2015) [15] or Asakura et al. (2014) [14]. The dimensions of variables in the model equation were usually given as time ms, length μm, concentration mM (except $[Ca^{2+}]_i$ μM for contraction), electric potential mV, conductance nS, and current pA, unless otherwise described. In the figures for membrane ionic currents in the present paper, the standard pA/pF is used.

The ordinary differential equations (ODE) of the hVC model [15] were time-integrated using the Heun method or Euler method. Computer programs were coded in Visual Basic and run on Visual studio. The time step of integration was set at 0.01 ms, approximately half of the maximum interval, which gives results consistent with those obtained using smaller step sizes. In some preliminary simulations, the Adams–Bashforth and Adams–Moulton integration methods were used to validate the numerical integration using simple methods.

4.2. One-Dimensional Strand of the hVC Model

Figure 5 shows the electrical coupling between two neighboring electric circuits in the myocyte model in a one-dimensional strand of N_c (=400) myocytes. The intracellular potential (V_i) was defined as the sum of the extracellular potential (V_o) and the membrane potential (V_m) of each myocyte (Equation (5)).

$$V_{i,n} = V_{o,n} + V_{m,n} \tag{5}$$

The velocity of propagation along the fiber orientation was determined from the slope of the relationship between time and location of the point where the sign of the V_o was inverted.

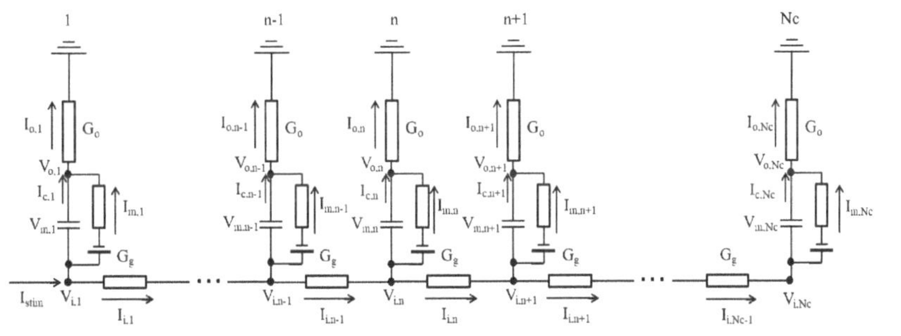

Figure 5. The electrically equivalent circuit of the myocyte strand. G_o; the extracellular conductance of 10 μS, V_o; extracellular potential near the cell surface, C_m; membrane capacitance of 192.5 pF, V_m; membrane potential, V_i; intracellular potential, I_o; extracellular current, I_c; capacitor current, I_m; membrane current, I_i; gap junction current, G_g; the gap junction conductance of 1.5 μS assumed for the end-to-end junction along the longitudinal axis of the cuboid shape myocyte.

4.3. Calculating V_o Outside the Myocyte Membrane in the Strand

The equivalent electric circuit of the myocyte strand is shown in Figure 5. The electrically equivalent circuit of the myocyte. Note that this myocyte strand model is different from the bidomain model because all myocytes within the strand have direct contact with the ground through an extracellular medium that has high conductance. This circuit represented experimental conditions of small multicellular preparations soaked in the bath solution, as in the classical experimental studies [5], rather than myocyte sheets tightly packed in the intact heart. In the presented model of the myocyte strand, the V_o of individual myocyte models was determined from individual V_m, G_o, and G_g according to the algorithm as described in Algorithm 1.

Algorithm 1. Summary of the calculation for the 1D myocyte strand model.

Step1: Calculate conductance matrix of A, B, C in Equation (9);
Step2: Calculate the inverse matrix of the conductance matrix of A, B, C;
Step3: Calculate the right-hand side of Equation (9) from $V_{m,n}^t$;
Step4: Calculate $V_{o,n}^t$ in Equation (9) by using the inverse matrix obtained in Step 2;
Step5: Calculate $V_{i,n}^t$ from $V_{o,n}^t$ and $V_{m,n}^t$ using Equation (5);
Step6: Calculate $I_{i,n}^t$ from $V_{i,n}^t$ and $V_{i,n+1}^t$ using Equation (6);
Step7: Calculate $I_{o,n}^t$ from $I_{i,n}^t$ and $I_{i,n-1}^t$ using Equation (7);
Step8: Determine $\frac{dV_{m,n}^t}{dt}$ using total membrane current $I_{m,n}^t$ and $I_{o,n}^t$
$\frac{dV_{m,n}^t}{dt} = \frac{I_{c,n}^t}{C_m} = \frac{-\left(I_{m,n}^t - I_{o,n}^t\right)}{C_m}$;
Step9: Integration of parameters for the next time step
$V_{m,n}^{t+dt} = V_{m,n}^t + \frac{dV_{m,n}^t}{dt} \cdot dt$;
Step10: Go to Step3.

The current along the longitudinal axis (across the gap junction, $I_{\mathrm{i,n}}$) was determined by Equation (6), which deals with the G_g and the voltage difference across the G_g.

$$I_{\mathrm{i,n}} = G_g \cdot (V_{\mathrm{i,n}} - V_{\mathrm{i,n+1}}) = G_g \cdot ((V_{\mathrm{o,n}} + V_{\mathrm{m,n}}) - (V_{\mathrm{o,n+1}} + V_{\mathrm{m,n+1}})) \quad (6)$$

The intercellular current along the transverse axis (I_o) is driven by the difference between $I_{\mathrm{i,n-1}}$ and $I_{\mathrm{i,n}}$ through G_o.

$$I_{\mathrm{o,n}} = I_{\mathrm{i,n-1}} - I_{\mathrm{i,n}} = G_o \cdot V_{\mathrm{o,n}} \quad (7)$$

Replacing I in Equation (7) with the potentials (V_o, V_m) provides an equation composed of V and G,

$$-G_g \cdot V_{\mathrm{o,n-1}} + (G_o + 2 \cdot G_g) \cdot V_{\mathrm{o,n}} - G_g \cdot V_{\mathrm{o,n+1}} = G_g \cdot ((V_{\mathrm{m,n-1}} - V_{\mathrm{m,n}}) - (V_{\mathrm{m,n}} - V_{\mathrm{m,n+1}})) \quad (8)$$

which is then rearranged for V_o of three sequential myocyte models ($V_{\mathrm{o,n-1}}$, $V_{\mathrm{o,n}}$, and $V_{\mathrm{o,n+1}}$) as a function of voltages (V_o and V_m) as follows;

$$A \cdot V_{\mathrm{o,n-1}} + B \cdot V_{\mathrm{o,n}} + C \cdot V_{\mathrm{o,n+1}} = D \cdot V_{\mathrm{m,n-1}} + E \cdot V_{\mathrm{m,n}} + F \cdot V_{\mathrm{m,n+1}} \quad (9)$$

here, the V_m of each unit is obtained beforehand by integrating the whole family of membrane currents to calculate the charge across the membrane capacitance. In brief, values of V_o for all myocytes are calculated by the parallel linear relations of all myocytes using a homemade solver every time after the numerical integration of the variables of individual myocytes. V_o was assigned a positive sign when I_o was outward whereas V_o was assigned a negative sign when I_o was inward and vice versa. As the G_o (= 10 μS) was larger than the G_g (= 1.5 μS), the conduction velocity was largely dependent on G_g. The G_o should be a scalable parameter, adjusted to the real size and dimension of the external electrode used for measurements. For a tentative G_o magnitude, the recordings of the QRS complex using the epicardial electrogram [27] were referred to. With a G_o of 5 μS, full-sized experimental V_o deflections of 5~10 mV were obtained with a standard maximum rate of rise of 250 V/s.

The gap junction current (I_{gap}) was calculated on the right side of the nth equivalent circuit.

$$I_{\mathrm{gap}} = I_{\mathrm{i,n}} = G_g \cdot [(V_{\mathrm{i,n}} - V_{\mathrm{i,n+1}}) + E_{\mathrm{K,n}}] \quad (10)$$

To calculate changes in $[K^+]_i$, I_{gap} was assumed to be carried purely by K^+. Thus, the Nernst equilibrium potential E_K was determined across the gap junction.

$$E_{\mathrm{K,n}} = \frac{RT}{F} \cdot \ln\left(\frac{[K^+]_{\mathrm{n}}}{[K^+]_{\mathrm{n+1}}}\right) \quad (11)$$

4.4. Electrical Stimulation to the Myocyte Model and the Myocyte Strand Model

The conditioning AP was evoked by applying a 5 ms current pulse of −12 pA/pF (K^+-mediated current) in the single myocyte experiment. In the 1D strand model experiment, the stimulus current was magnified to −35 pA/pF in amplitude and 30 ms in duration to evoke the conditioning AP, which propagated throughout the axis to the right end of the strand.

To confirm whether the threshold phenomenon is observed in the single hVC model, hyperpolarizing voltage clamp pulses of 5 ms in duration at various test potential (V_t) levels in hyperpolarization were applied during the AP plateau phase. The compensation current given by Equation (12) was applied at various selected time points during the AP plateau phase.

$$I_{\text{compensation}} = -(V_t - V_m)\cdot\frac{1}{fb_{\text{gain}}} \quad (12)$$

where fb_{gain} was $1/(10 \times dt)$, dt = 0.01 ms.

To trigger the propagation of repolarization in the 1D strand of myocytes, a trigger current pulse of 400 pA/pF and 40 ms in duration was applied simultaneously to 100 myocytes (cell No. 50~150).

These stimulus currents as well as $I_{compensation}$ were added to the total membrane current $I_{m,n}$ of the myocyte model, which is used to calculate dV_m/dt in each myocyte. The propagation of repolarization started from the border of the segment of the 100 myocytes. The movement of the wave of propagation was monitored by plotting the V_o in the 2D display (the myocyte number abscissa and the time ordinate).

4.5. Induction of the Second Stable Equilibrium V_m in Myocyte Model within the In Silico 1D Strand

The propagation of repolarization could be observed in a straightforward manner if the myocyte strand was depolarized to a stable equilibrium potential near the plateau potential around −5 mV. This was achieved by removing the slow V_m-dependent inactivation of I_{NaL} from the state transition scheme, described later in Figure 3D. The equilibrium potential was examined by recording the steady-state current–voltage relationship (*I-V* relationship) by applying long voltage clamp pulses of 4.9 s in duration. After confirming that the time-dependent changes were virtually saturated near the end of clamp pulses, the current amplitude was measured. Influences of varying the I_{NaL} were examined by changing the fractional amplitude of the fixed I_s state of I_{NaL}.

Supplementary Materials: The following supporting information can be downloaded at: https://www.mdpi.com/article/10.3390/ijms242015378/s1. Supplementary materials include model equations and computer program codes. References [14,15,24,28–37] are cited in the supplementary materials.

Author Contributions: Conceptualization, A.N. and Y.H.; methodology, A.N. and N.Y.; software, A.N., N.Y., M.U., H.N., Y.Z. and Y.M.; validation, S.E., Y.H., S.K., M.U., Y.Z. and A.A.; formal analysis, S.K. and S.E.; investigation, N.Y. and Y.Z.; writing—original draft preparation, A.N., Y.M. and Y.H.; writing—review and editing, Y.H.; visualization, N.Y., S.E. and H.N.; supervision, A.A. and Y.H.; project administration, Y.H.; funding acquisition, Y.H. and A.A. All authors have read and agreed to the published version of the manuscript.

Funding: This research was funded by The Research Promotion Program for Acquiring Grants in-Aid for Scientific Research (KAKENHI) at Ritsumeikan University to Y.H. and A.A.

Institutional Review Board Statement: Not applicable.

Informed Consent Statement: Not applicable.

Data Availability Statement: The data that supports the findings of this study are available in the Supplementary Material (Tables S1–S8 and Schemes S1–S6) of this article.

Acknowledgments: We would like to thank Haruka Hirata for her secretarial support.

Conflicts of Interest: The authors declare no conflict of interest.

References and Note

1. Noble, D. A modification of the Hodgkin-Huxley equations applicable to Purkinje fibre action and pacemaker potentials. *J. Physiol.* **1962**, *160*, 317–352. [CrossRef] [PubMed]
2. Trenor, B.; Cardona, K.; Saiz, J.; Noble, D.; Giles, W. Cardiac action potential repolarization revisited: Early repolarization shows all-or-none behaviour. *J. Physiol.* **2017**, *595*, 6599–6612. [CrossRef] [PubMed]
3. Noble, D.; Hall, A.E. The Conditions for Initiating "All-or-Nothing" Repolarization in Cardiac Muscle. *Biophys. J.* **1963**, *3*, 261–274. [CrossRef] [PubMed]
4. Noble, D. The voltage dependence of the cardiac membrane conductance. *Biophys. J.* **1962**, *2*, 381–393. [CrossRef] [PubMed]
5. Weidmann, S. Effect of current flow on the membrane potential of cardiac muscle. *J. Physiol.* **1951**, *115*, 227–236. [CrossRef] [PubMed]
6. Cranefield, P.F.; Hoffman, B.F. Propagated repolarization in heart muscle. *J. Gen. Physiol.* **1958**, *41*, 633–649. [CrossRef] [PubMed]
7. Powell, T.; Steen, E.M.; Twist, V.W.; Woolf, N. Surface characteristics of cells isolated from adult rat myocardium. *J. Mol. Cell. Cardiol.* **1978**, *10*, 287–292. [CrossRef]
8. Powell, T.; Twist, V.W. A rapid technique for the isolation and purification of adult cardiac muscle cells having respiratory control and a tolerance to calcium. *Biochem. Biophys. Res. Commun.* **1976**, *72*, 327–333. [CrossRef]
9. Powell, T.; Terrar, D.A.; Twist, V.W. Electrical properties of individual cells isolated from adult rat ventricular myocardium. *J. Physiol.* **1980**, *302*, 131–153. [CrossRef]
10. Sakmann, B.; Neher, E. Patch clamp techniques for studying ionic channels in excitable membranes. *Annu. Rev. Physiol.* **1984**, *46*, 455–472. [CrossRef]
11. Fink, M.; Niederer, S.A.; Cherry, E.M.; Fenton, F.H.; Koivumäki, J.T.; Seemann, G.; Thul, R.; Zhang, H.; Sachse, F.B.; Beard, D.; et al. Cardiac cell modelling: Observations from the heart of the cardiac physiome project. *Prog. Biophys. Mol. Biol.* **2011**, *104*, 2–21. [CrossRef] [PubMed]
12. O'Hara, T.; Virág, L.; Varró, A.; Rudy, Y. Simulation of the undiseased human cardiac ventricular action potential: Model formulation and experimental validation. *PLoS Comput. Biol.* **2011**, *7*, e1002061. [CrossRef]
13. Grandi, E.; Pasqualini, F.S.; Bers, D.M. A novel computational model of the human ventricular action potential and Ca transient. *J. Mol. Cell. Cardiol.* **2010**, *48*, 112–121. [CrossRef] [PubMed]
14. Asakura, K.; Cha, C.Y.; Yamaoka, H.; Horikawa, Y.; Memida, H.; Powell, T.; Amano, A.; Noma, A. EAD and DAD mechanisms analyzed by developing a new human ventricular cell model. *Prog. Biophys. Mol. Biol.* **2014**, *116*, 11–24. [CrossRef]
15. Himeno, Y.; Asakura, K.; Cha, C.Y.; Memida, H.; Powell, T.; Amano, A.; Noma, A. A human ventricular myocyte model with a refined representation of excitation-contraction coupling. *Biophys. J.* **2015**, *109*, 415–427. [CrossRef] [PubMed]
16. Undrovinas, A.I.; Maltsev, V.A.; Kyle, J.W.; Silverman, N.; Sabbah, H.N. Gating of the late Na^+ channel in normal and failing human myocardium. *J. Mol. Cell. Cardiol.* **2002**, *34*, 1477–1489. [CrossRef]
17. Näbauer, M.; Beuckelmann, D.J.; Uberfuhr, P.; Steinbeck, G. Regional differences in current density and rate-dependent properties of the transient outward current in subepicardial and subendocardial myocytes of human left ventricle. *Circulation* **1996**, *93*, 168–177. [CrossRef]
18. Chandler, W.K.; Meves, H. Slow changes in membrane permeability and long-lasting action potentials in axons perfused with fluoride solutions. *J. Physiol.* **1970**, *211*, 707–728. [CrossRef]
19. Undrovinas, A.I.; Fleidervish, I.A.; Makielski, J.C. Inward sodium current at resting potentials in single cardiac myocytes induced by the ischemic metabolite lysophosphatidylcholine. *Circ. Res.* **1992**, *71*, 1231–1241. [CrossRef] [PubMed]
20. Qu, Z.; Xie, L.H.; Olcese, R.; Karagueuzian, H.S.; Chen, P.S.; Garfinkel, A.; Weiss, J.N. Early afterdepolarizations in cardiac myocytes: Beyond reduced repolarization reserve. *Cardiovasc. Res.* **2013**, *99*, 6–15. [CrossRef]
21. Himeno, Y. Ritsumeikan University, Shiga, Japan. Our simulation using a strand composed of intact hVC demonstrated that it was impossible to determine a unique value for the velocity of propagated repolarization because the velocity increased continuously and approached infinity as the repolarization propagated. 2023.
22. Kirkton, R.D.; Badie, N.; Bursac, N. Spatial Profiles of Electrical Mismatch Determine Vulnerability to Conduction Failure Across a Host–Donor Cell Interface. *Circ. Arrhythm. Electrophysiol.* **2013**, *6*, 1200–1207. [CrossRef] [PubMed]
23. Sperelakis, N.; Ramasamy, L.; Kalloor, B. Propagated repolarization of simulated action potentials in cardiac muscle and smooth muscle. *Theor. Biol. Med. Model* **2005**, *2*, 5. [CrossRef]
24. Yan, D.H.; Ishihara, K. Two Kir2.1 channel populations with different sensitivities to Mg^{2+} and polyamine block: A model for the cardiac strong inward rectifier K^+ channel. *J. Physiol.* **2005**, *563*, 725–744. [CrossRef] [PubMed]
25. Ishihara, K.; Sarai, N.; Asakura, K.; Noma, A.; Matsuoka, S. Role of Mg^{2+} block of the inward rectifier K^+ current in cardiac repolarization reserve: A quantitative simulation. *J. Mol. Cell. Cardiol.* **2009**, *47*, 76–84. [CrossRef]
26. Fabbri, A.; Goversen, B.; Vos, M.A.; van Veen, T.A.B.; de Boer, T.P. Required GK1 to Suppress Automaticity of iPSC-CMs Depends Strongly on IK1 Model Structure. *Biophys. J.* **2019**, *117*, 2303–2315. [CrossRef] [PubMed]

27. Taggart, P.; Sutton, P.; John, R.; Hayward, R.; Swanton, H. The epicardial electrogram: A quantitative assessment during balloon angioplasty incorporating monophasic action potential recordings. *Br. Heart J.* **1989**, *62*, 342–352. [CrossRef]
28. Shannon, T.R.; Wang, F.; Puglisi, J.; Weber, C.; Bers, D.M. A mathematical treatment of integrated Ca dynamics within the ventricular myocyte. *Biophys. J.* **2004**, *87*, 3351–3371. [CrossRef]
29. Negroni, J.A.; Lascano, E.C. Simulation of steady state and transient cardiac muscle response experiments with a Huxley-based contraction model. *J. Mol. Cell. Cardiol.* **2008**, *45*, 300–312. [CrossRef]
30. Rush, S.; Larsen, H. A practical algorithm for solving dynamic membrane equations. *IEEE Trans. Biomed. Eng.* **1978**, *25*, 389–392. [CrossRef]
31. Ishihara, K.; Yan, D.H. Low-affinity spermine block mediating outward currents through Kir2.1 and Kir2.2 inward rectifier potassium channels. *J. Physiol.* **2007**, *583*, 891–908. [CrossRef]
32. Takeuchi, A.; Tatsumi, S.; Sara, N.; Terashima, K.; Matsuoka, S.; Noma, A. Ionic mechanisms of cardiac cell swelling induced by blocking Na^+/K^+ pump as revealed by experiments and simulation. *J. Gen. Physiol.* **2006**, *128*, 495–507. [CrossRef]
33. Oka, C.; Cha, C.Y.; Noma, A. Characterization of the cardiac Na^+/K^+ pump by development of a comprehensive and mechanistic model. *J. Theor. Biol.* **2010**, *265*, 68–77. [CrossRef] [PubMed]
34. Smith, N.P.; Crampin, E.J. Development of models of active ion transport for whole-cell modelling: Cardiac sodium-potassium pump as a case study. *Prog. Biophys. Mol. Biol.* **2004**, *85*, 387–405. [CrossRef] [PubMed]
35. Grandi, E.; Pasqualini, F.S.; Pes, C.; Corsi, C.; Zaza, A.; Severi, S. Theoretical investigation of action potential duration dependence on extracellular Ca2+ in human cardiomyocytes. *J. Mol. Cell. Cardiol.* **2009**, *46*, 332–342. [CrossRef] [PubMed]
36. Hinch, R.; Greenstein, J.L.; Tanskanen, A.J.; Xu, L.; Winslow, R.L. A simplified local control model of calcium-induced calcium release in cardiac ventricular myocytes. *Biophys. J.* **2004**, *87*, 3723–3736. [CrossRef]
37. Tran, K.; Smith, N.P.; Loiselle, D.S.; Crampin, E.J. A thermodynamic model of the cardiac sarcoplasmic/endoplasmic Ca^{2+} (SERCA) pump. *Biophys. J.* **2009**, *96*, 2029–2042. [CrossRef]

International Journal of *Molecular Sciences*

MDPI

Article

The W101C *KCNJ5* Mutation Induces Slower Pacing by Constitutively Active GIRK Channels in hiPSC-Derived Cardiomyocytes

Anne Kayser [1], Sven Dittmann [1], Tomo Šarić [2], Giulia Mearini [3], Arie O. Verkerk [4,5,*,†] and Eric Schulze-Bahr [1,†,‡]

1 Institute for Genetics of Heart Diseases (IfGH), University Hospital Münster, 48149 Münster, Germany; sven.dittmann@ukmuenster.de (S.D.); eric.schulze-bahr@ukmuenster.de (E.S.-B.)
2 Center for Physiology and Pathophysiology, Institute for Neurophysiology, Faculty of Medicine and University Hospital Cologne, University of Cologne, 50931 Cologne, Germany; tomo.saric@uni-koeln.de
3 Institute of Experimental Pharmacology and Toxicology, University Medical Center Hamburg-Eppendorf, 20246 Hamburg, Germany
4 Department of Medical Biology, Amsterdam University Medical Centers, University of Amsterdam, 1105 AZ Amsterdam, The Netherlands
5 Department of Experimental Cardiology, Amsterdam University Medical Centers, University of Amsterdam, 1105 AZ Amsterdam, The Netherlands
* Correspondence: a.o.verkerk@amsterdamumc.nl
† These authors contributed equally to this work.
‡ Member of the European Reference Network for Rare, Low Prevalence and Complex Diseases of the Heart: ERN GUARD-Heart' (ERN GUARDHEART).

Abstract: Mutations in the *KCNJ5* gene, encoding one of the major subunits of cardiac G-protein-gated inwardly rectifying K$^+$ (GIRK) channels, have been recently linked to inherited forms of sinus node dysfunction. Here, the pathogenic mechanism of the W101C *KCNJ5* mutation underlying sinus bradycardia in a patient-derived cellular disease model of sinus node dysfunction (SND) was investigated. A human-induced pluripotent stem cell (hiPSCs) line of a mutation carrier was generated, and CRISPR/Cas9-based gene targeting was used to correct the familial mutation as a control line. Both cell lines were further differentiated into cardiomyocytes (hiPSC-CMs) that robustly expressed GIRK channels which underly the acetylcholine-regulated K$^+$ current ($I_{K,ACh}$). hiPSC-CMs with the W101C *KCNJ5* mutation (hiPSC^{W101C}-CM) had a constitutively active $I_{K,ACh}$ under baseline conditions; the application of carbachol was able to increase $I_{K,ACh}$, further indicating that not all available cardiac GIRK channels were open at baseline. Additionally, hiPSC^{W101C}-CM had a more negative maximal diastolic potential (MDP) and a slower pacing frequency confirming the bradycardic phenotype. Of note, the blockade of the constitutively active GIRK channel with XAF-1407 rescued the phenotype. These results provide further mechanistic insights and may pave the way for the treatment of SND patients with GIRK channel dysfunction.

Keywords: *KCNJ5*; sinus node dysfunction; hiPSC; cardiomyocytes; disease model; $I_{K,ACh}$ blocker XAF-1407

Citation: Kayser, A.; Dittmann, S.; Šarić, T.; Mearini, G.; Verkerk, A.O.; Schulze-Bahr, E. The W101C *KCNJ5* Mutation Induces Slower Pacing by Constitutively Active GIRK Channels in hiPSC-Derived Cardiomyocytes. *Int. J. Mol. Sci.* **2023**, *24*, 15290. https://doi.org/10.3390/ijms242015290

Academic Editors: Yosuke Okamoto and Kunichika Tsumoto

Received: 18 September 2023
Revised: 13 October 2023
Accepted: 15 October 2023
Published: 18 October 2023

1. Introduction

Sinus node dysfunction (SND) is clinically manifested in sinus bradycardia, chronotropic incompetence or sinus arrest as a cause of disturbed cardiac impulse generation, delayed or absent cardiac conduction between sinoatrial node (SAN) and atria [1]. Depending on the pathogenesis, SND can classified into primary or secondary SND [2]. Patients with SND show a broad range of clinical symptoms, e.g., dizziness, syncope, fall [3], and it may increase the risk of cardiac arrhythmias [4]. SND may be due to extrinsic or intrinsic factors and both exhibit variable manifestations, as recently reviewed by Sathnur and colleagues [5]. It was increasingly recognized that ion channel dysfunction and inherited forms of so-called channelopathies are an important cause for intrinsic and familial SND resulting in an upcoming list of ion-channel genes being involved in SND (for reviews, see [4,6–8]). Among them, the p.Trp101Cys (short: W101C) mutation in the *KCNJ5* gene has recently been identified in a family with SND [9]. *KCNJ5* encodes the G-protein-gated inwardly rectifying K^+ channel Kir3.4 (GIRK4), which is mainly expressed in the heart, pancreas, and in parts of the adrenal gland [10–12]. Kir3.4 forms heterotetramers with Kir3.1 (GIRK1, encoded by *KCNJ3*), which results in the acetylcholine-activated G-protein-gated inwardly rectifying K^+ (GIRK) channels which underly the K^+ current, $I_{K,ACh}$ [4,13], that is present in the human SAN and atria. The activation of GIRK channels, resulting in an increased $I_{K,ACh}$, elongates the intrinsic cycle length of SAN cells and hyperpolarizes the maximal diastolic membrane potential (MDP) in SAN and atrial cardiomyocytes [14–18]. In addition to the high-level protein expression of GIRK4 subunits in atrium, a higher expression was also detected in the hypothalamus [19]. Here, it was especially found in the paraventricular nucleus of the cardiac vagal neurons, which may play an additional role in heart-rate control and cardiac function [20].

Currently, 31 disease-causing variants in the *KCNJ5* gene are known to date. Potential heart disease variants in *KCNJ5* have been described and may be causative for long QT syndrome, Andersen-Tawil syndrome, atrial fibrillation, short QT syndrome, or unspecified events like sudden cardiac arrest or sudden unexpected death [21–29]. However, due to the small number of cases and mostly missing functional data, *KCNJ5* is not a proven gene for the respective diseases. In addition, mainly somatic variants have been associated with primary aldosteronism and hypertension. The *KCNJ5* variant W101C has been reported as a gain-of-function variant in familial SND [9]; functional data in heterologous systems, e.g., CHO cells and *Xenopus* oocytes, demonstrated a constitutively active $I_{K,ACh}$ [9]; thus far, data in human-derived cardiomyocytes disease models are lacking. Here, we generated patient-specific (from a *KCNJ5* W101C mutation carrier) induced pluripotent stem cells (hiPSC) and an isogenic control cell line via CRISPR/Cas9. Cardiomyocytes derived from these hiPSC lines (hiPSC-CMs) were electrophysiologically analyzed by patch-clamp techniques. We found that hiPSC-CMs with the mutant GIRK channel had a slowed pacemaking activity due to a constitutive channel activation and that $I_{K,ACh}$ channel inhibition might pave a way for potential targets for SND treatment.

2. Results

2.1. Generation and Genetic Correction of hiPSCs from a Patient with the W101C KCNJ5 Variant

For characterization of mutant and isogenic corrected $I_{K,ACh}$ channels, first patient-specific peripheral blood mononuclear cells (PBMCs) were isolated via density gradient centrifugation with Ficoll™ and frozen at −80 °C until further use. The reprogramming of these PBMCs into stem cells was carried out via Sendai-virus-based vectors, as set out in detail in Section 4.1. The cell clone KCNJ5 K8 was used for this study and registered in the Human Pluripotent Stem Cell Registry (hPSCreg® number: UKMi005-A). In addition, an isogenic control hiPSC line (hiPSC^{corr}) was generated with CRISPR/Cas9-mediated genome editing, as described in Section 4.3. Figure 1A shows a brightfield microscopy image of the control *KCNJ5* hiPSC line (hiPSC^{corr}), and Figure 1B demonstrates the successful correction of the pathogenic *KCNJ5* W101C variant as addressed by Sanger sequencing. The hiPSC^{corr} line maintained expression of NANOG, SOX2, and OCT4 on the RNA level and protein level (Figure 1C–H), and CRISPR/Cas9-mediated genome editing had no effect on the karyotype (Figure 1I). This cell line is also registered at hPSCreg® (number: UKMi006-A).

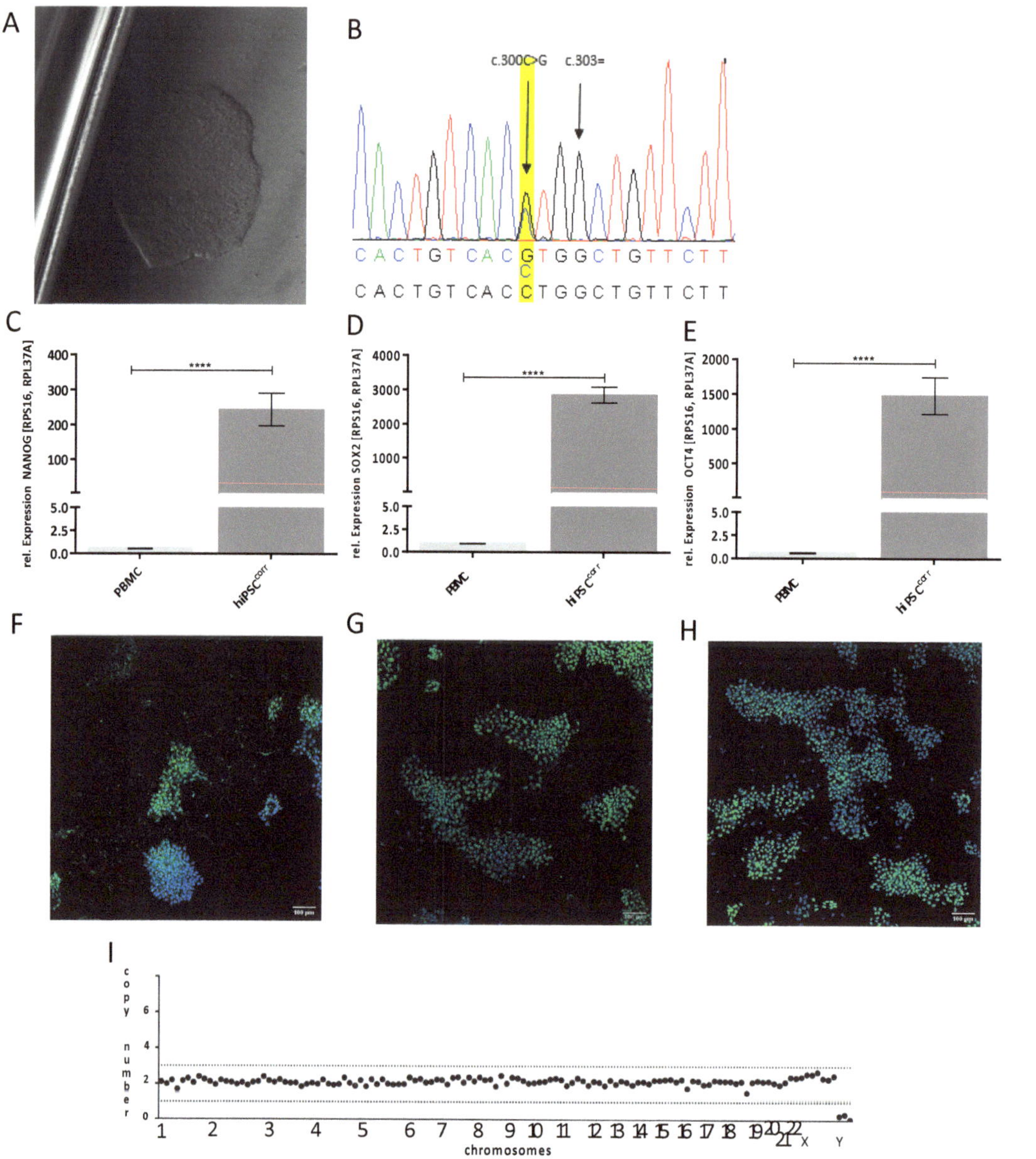

Figure 1. Characterization of the hiPSC^{corr} line. (**A**) Brightfield microscopy picture of a corrected *KCNJ5* W101C hiPSC clone (hiPSC^{corr}). (**B**) Sanger sequencing results after correction of the pathogenic *KCNJ5* W101C variant. The nucleotide exchange at position c.303 was successfully corrected and a silent mutation for protection of the HDR-repair template (c.300C>G) was inserted. (**C–E**) NANOG, SOX2, and OCT4 as pluripotency markers on RNA level via RT-qPCR. **** = $p < 0.0001$. (**F–H**) Immunofluorescence staining of NANOG (green), SOX2 (green), and OCT4 (green). The nucleus is stained with DAPI (blue). (**I**). Molecular karyogram of corrected *KCNJ5* W101C hiPSC (hiPSC^{corr}). No chromosomal abrasions were detected.

2.2. *Variant W101C KCNJ5 Results in a Constitutively Active $I_{K,ACh}$ Current after Retinoic Acid (RA)-Based Differentiation into Atrial-like hiPSC^{W101C}-CMs*

To study the effect of the W101C variant on $I_{K,ACh}$, isogenic hiPSC^{corr} and hiPSC^{W101C} lines were next differentiated into cardiomyocyte-like cells (hiPSC-CMs). Hence, a retinoic acid (RA)-based differentiation protocol was used, from which it is well-known that they promote the acquisition of atrial-like features and with robust expression of *KCNJ3* as an atrial marker (Supplement Figure S1C) [30–36]. Such hiPSC-CMs further enable electrophysiological studies of mutations involved in $I_{K,ACh}$ channel activation [32].

First, it was tested whether the W101C *KCNJ5* variant results in constitutively activated $I_{K,ACh}$ channels, similarly as previously shown in a heterologous cell expression system [9]. Therefore, we tested if the new developed highly-selective $I_{K,ACh}$ current inhibitor XAF-1407 [37–39] reduces $I_{K,ACh}$ currents under baseline conditions during a descending voltage-clamp ramp (see Section 4.9). Figure 2A,B shows typical recordings and average current–voltage (I–V) relationships, respectively, in hiPSC^{corr}-CMs and hiPSC^{W101C}-CMs. The results show that XAF-sensitive currents were virtually absent in hiPSC^{corr}-CMs (Figure 2A,B). However, in hiPSC^{W101C}-CMs, an XAF-sensitive current was clearly present at baseline conditions (Figure 2A,B). For example, the average current density of the XAF-sensitive current in hiPSC^{W101C}-CMs at −120 mV was -1.2 ± 0.2 pA/pF (n = 8), which is significantly larger than in hiPSC^{corr}-CMs (Figure 2B). Thus, the W101C *KCNJ5* mutation results in a constitutively active $I_{K,ACh}$ current in hiPSC-CMs.

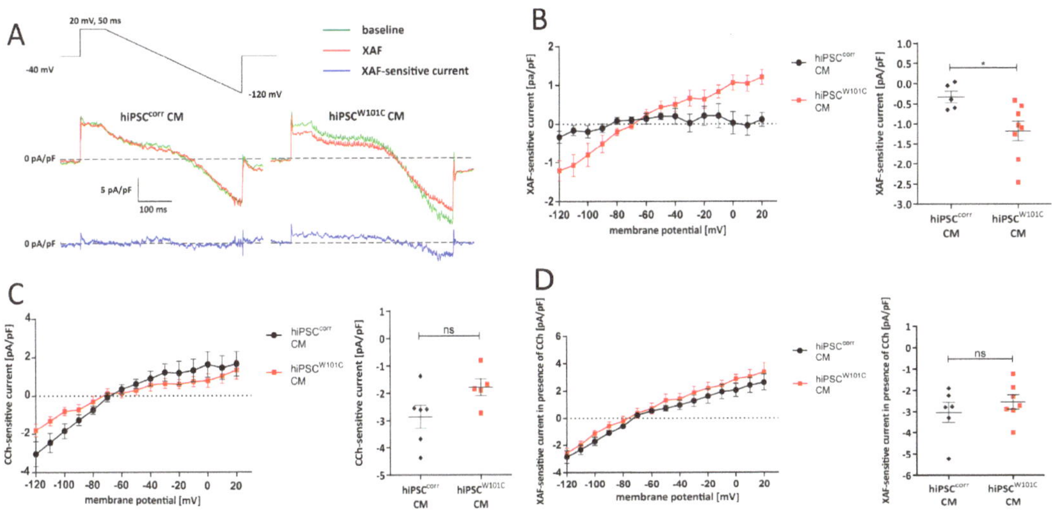

Figure 2. $I_{K,ACh}$ densities in isogenic RA-treated hiPSC^{corr}-CMs and hiPSC^{W101C}-CMs. (**A**) Typical ramp traces in the absence and presence of XAF-1407 (XAF), a selective $I_{K,ACh}$ channel locker, in hiPSC^{corr}-CMs and hiPSC^{W101C}-CMs. (**B**) Current–voltage (I–V) relationships of XAF-sensitive currents in hiPSC^{corr}-CMs (black) and hiPSC^{W101C}-CMs (red) under baseline conditions (**left panel**) and dot plots of the XAF-sensitive currents at −120 mV (**right panel**). * = $p < 0.05$. (**C**) I–V relationships of CCh-sensitive currents (**left panel**) and dot plots of CCh-sensitive currents at −120 mV (**right panel**). ns = not significant. (**D**) I–V relationships of XAF-sensitive currents in the presence of CCh (**left panel**) and dot plots of XAF-sensitive currents in the presence of CCh at −120 mV (**right panel**). ns = not significant.

Secondly, it was analyzed whether all mutant GIRK channels were constitutively open due to the presence of the W101C mutation under baseline conditions and whether carbachol (CCh), a structure analog of acetylcholine and agonist of the muscarinic receptor, could open additional channels. By applying 10 µM CCh, a robust $I_{K,ACh}$ current could be evoked in both hiPSC^{corr}-CMs and hiPSC^{W101C}-CMs (Figure 2C). There was a tendency of larger CCh-activated $I_{K,ACh}$ in hiPSC^{corr}-CMs (Figure 2C), but this was finally not significant ($p = 0.08$; unpaired *t*-test). In a third voltage clamp experiment, the total $I_{K,ACh}$ density, i.e., constitutively active $I_{K,ACh}$ plus CCh-activated $I_{K,ACh}$, was tested between hiPSC^{corr}-CMs and hiPSC^{W101C}-CMs. Therefore, we measured the XAF-sensitive current in the presence of CCh. The XAF-sensitive current densities in the presence of CCh were not significantly different between both hiPSC-CMs groups (Figure 2D). In addition, the I–V relationships of XAF-sensitive current densities in the presence of CCh in hiPSC^{corr}-CMs and hiPSC^{W101C}-CMs were virtually overlapping, indicating similar reversal potentials (E_{rev}; -77.3 ± 5.0 mV for hiPSC^{W101C}-CMs (n = 7) and -74.2 ± 4.1 mV for hiPSC^{Corr}-CMs (n = 6)) and amounts of rectification. Thus far, these series of voltage clamp experiments showed that the W101C *KCNJ5* mutation does not affect the total amount of $I_{K,ACh}$ channels or its I–V relationship in hiPSC-CMs but resulted in a more constitutively active $I_{K,ACh}$.

2.3. W101C KCNJ5 Results in a Lower Pacemaking Frequency in RA-Treated hiPSC-CMs

Subsequently, the effects of the W101C *KCNJ5* variant on spontaneous activity were evaluated. Figure 3A shows typical action potentials (APs) of a spontaneously beating RA-treated hiPSC^{corr}-CM and hiPSC^{W101C}-CM under baseline conditions. The average AP parameters are summarized in Figure 3B–F. hiPSC^{W101C}-CMs had a slower spontaneous beating activity as indicated by the significant lower frequency of spontaneous APs (Figure 3B). On average, the frequency in hiPSC^{corr}-CM and hiPSC^{W101C}-CM was 2.0 ± 0.2 Hz and 1.5 ± 0.1 Hz ($p < 0.05$, unpaired *t*-test), respectively. In addition, the maximal diastolic potential (MDP) was about $\approx$6 mV more negative in hiPSC^{W101C}-CMs (-75.1 ± 0.6 mV) compared to the hiPSC^{corr}-CMs (-69.3 ± 0.7 mV) (Figure 3C). Furthermore, the diastolic depolarization rate (DDR) was significantly slower in hiPSC^{W101C}-CMs (19.3 ± 2.9 mV/s) compared to the hiPSC^{corr}-CMs (49.6 ± 6.3 mV/s) (Figure 3D). Other AP parameters were unaffected by the W101C *KCNJ5* variant, except for a lower AP plateau amplitude in hiPSC^{W101C}-CMs (Figure 3E–G).

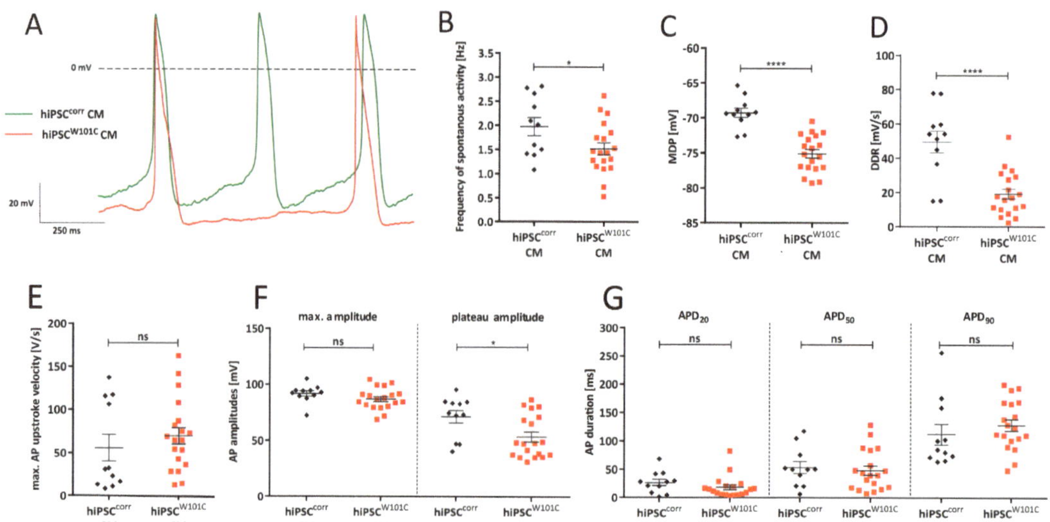

Figure 3. Action potential (AP) parameters under baseline conditions in RA-treated hiPSC^{corr}-CMs and hiPSC^{W101C}-CMs. (**A**) Typical APs of a hiPSC^{corr}-CM and hiPSC^{W101C}-CM. (**B–F**) Dot plots of spontaneous activity frequency: (**B**) maximal diastolic potential (MDP), (**C**) diastolic depolarization rate (DDR), (**D**) maximal AP upstroke velocity, (**E**) AP amplitudes, (**F**) and AP durations at 20, 50 and 90% of repolarization (APD_{20}, APD_{50}, and APD_{90}; (**G**)). * = $p < 0.05$; **** = $p < 0.0001$; ns = not significant.

2.4. $I_{K,ACh}$ Inhibition Restores Spontaneous Activity in RA-Treated hiPSC^{W101C}-CMs

Since hiPSC^{W101C}-CMs were shown to have a lower spontaneous beat frequency and a more hyperpolarized MDPs (Figure 3), it was tested whether this was due to the constitutively active channel as observed in hiPSC^{W101C}-CMs (Figure 2). Therefore, the effects of the $I_{K,ACh}$ blocker XAF-1407 on spontaneous activity in RA-treated hiPSC^{corr}-CMs and hiPSC^{W101C}-CMs were measured. Figure 4A shows typical APs in the absence and presence of XAF-1407 and the average AP parameters in hiPSC^{W101C}-CMs and hiPSC^{corr}-CMs are summarized in Supplemental Tables S1 and S2. XAF-1407 resulted in a faster spontaneous or pacemaking activity in hiPSC^{W101C}-CMs as indicated by the increased frequency of the spontaneous APs (1.2 ± 0.1 Hz (baseline) vs. 1.7 ± 0.1 Hz (XAF-1407); $p < 0.001$, paired t-test; Supplemental Table S1), while it did not affect the pacemaking activity of the isogenic control line hiPSC^{corr}-CMs (Figure 4A, Supplemental Table S2). In fact, the frequency of hiPSC^{W101C}-CMs in the presence of the channel blocker (1.7 ± 0.1 Hz, n = 9) is now similar to the frequency of the hiPSC^{corr}-CMs in the presence of XAF-1407 (1.7 ± 0.3 Hz, n = 3). This indicates that $I_{K,ACh}$ current inhibition in hiPSC^{W101C}-CMs restores the frequency of pacemaking. In addition, XAF-1407 also resulted in the depolarization of the MDP in hiPSC^{W101C}-CMs (−74.0 ± 0.9 mV (baseline) vs. −72.3 ± 1.0 mV (XAF); $p < 0.001$, paired t-test; Supplemental Table S1), without a significant effect on the MDP in hiPSC^{corr}-CMs (Figure 4A, Supplemental Table S2). XAF-1407 also significantly decreased the V_{max} in hiPSC^{W101C}-CMs (Supplemental Table S1), consistent with the reduced availability of sodium channels in response to MDP depolarization [40]. Other AP parameters were not affected by XAF-1407 in hiPSC^{W101C}-CMs and hiPSC^{corr}-CMs (Supplemental Tables S1 and S2). Thus, while AP parameters under baseline conditions showed a clear phenotype in hiPSC^{W101C}-CMs, this was absent in response to the XAF-1407-induced blockade of the $I_{K,ACh}$ current. Thus, the slow-pacemaking phenotype with hyperpolarized MDPs in RA-treated hiPSC^{W101C}-CMs under baseline conditions is important due to the constitutively active GIRK channels.

In voltage clamp measurements (Figure 2), the acetylcholine analog CCh activated a robust $I_{K,ACh}$ current in RA-treated hiPSC^{corr}-CMs and hiPSC^{W101C}-CMs, but CCh had a tendency to lower $I_{K,ACh}$ density in hiPSC^{W101C}-CMs. Next, the effect of 10 μM CCh on spontaneous APs in both hiPSC-CMs groups was tested. In the RA-treated hiPSC^{corr}-CMs, CCh caused a significant slowing of spontaneous activity and MDP hyperpolarization (Figure 4B, Supplemental Table S2), consistent with previous findings in RA-treated control lines of hiPSC-CMs findings [32,33]. On average, CCh decreased the frequency of spontaneous activity by approx. 47.6% from 2.1 ±0.2 Hz to 1.1 ± 0.1 Hz (n = 8, $p < 0.001$ (paired t-test)), while the MDP increased by about 3.1% from −68.8 ± 0.8 mV to −71.0 ± 0.6 mV (n = 8, $p < 0.01$ (paired t-test)) in hiPSC^{corr}-CMs (Supplemental Table S2). The AP upstroke velocity (v_{max}) and AP amplitude (APA) increased significantly due to CCh application, which both well reflect the increased availability of the cardiac sodium channels [40,41] due to the MDP hyperpolarization. CCh did not affect AP durations (APDs) in hiPSC^{corr}-CMs (Supplemental Table S2). In the RA-treated hiPSC^{W101C}-CMs, CCh induced a small, but significant, slowing of spontaneous activity and MDP hyperpolarization (Figure 4B, Supplemental Table S1). The frequency decreased by approx. 23.5% and the MDP increased by 1.0 ± 0.4%. Other AP parameters were unaffected, likely because the CCh-induced changes in frequency and MDP are rather limited. The application of XAF-1407 in the continued presence of CCh completely abolished the CCh-induced effects in both the hiPSC^{corr}-CMs and hiPSC^{W101C}-CMs (Figure 4C, Supplemental Tables S1 and S2), indicating that they were mainly due to the activation of $I_{K,ACh}$.

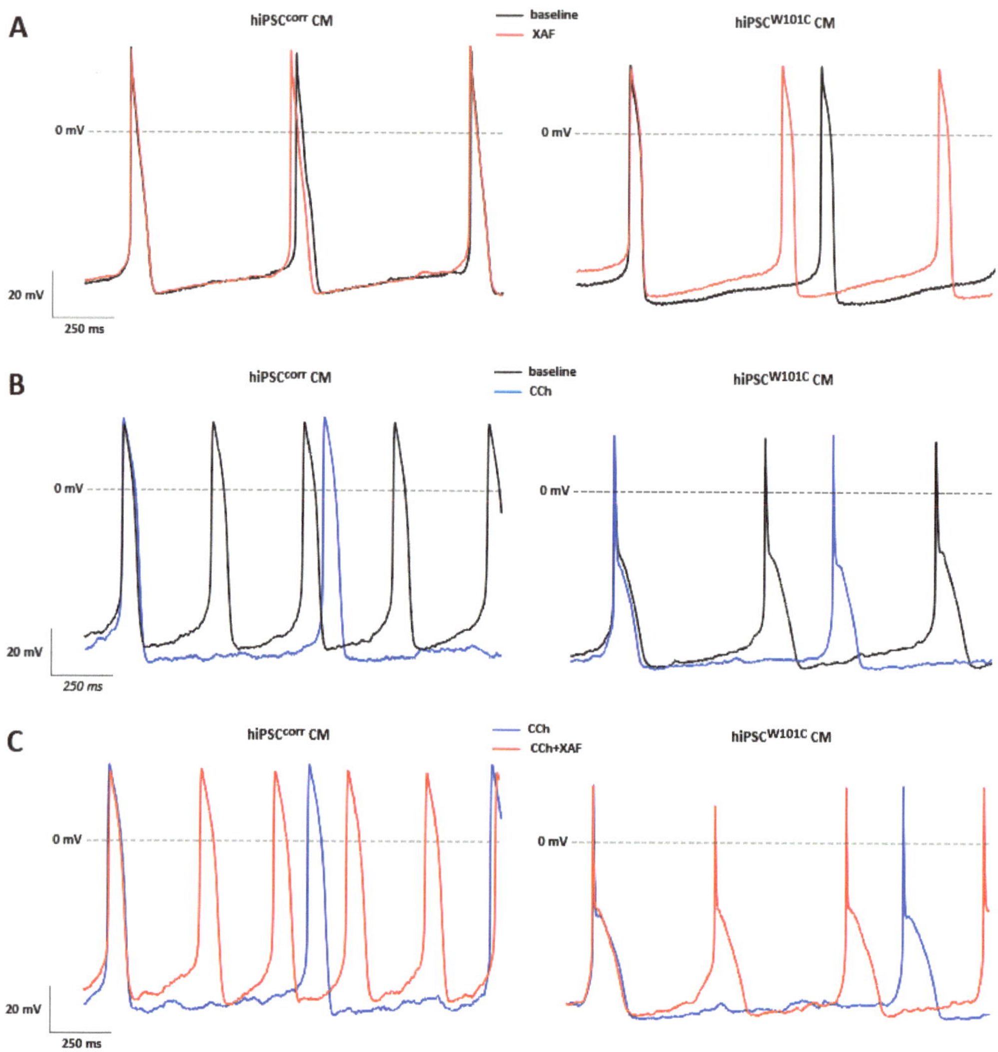

Figure 4. Effects of AP modulating substances in RA-treated hiPSC^{corr}-CMs and hiPSC^{W101C}-CMs. (**A**) Typical APs of a hiPSC^{corr}-CM and hiPSC^{W101C}-CM in the absence and presence of XAF. (**B**) Typical APs of a hiPSC^{corr}-CM and hiPSC^{W101C}-CM in the absence and presence of CCh. (**C**) Typical APs of a hiPSC^{corr}-CM and hiPSC^{W101C}-CM in the presence of CCh and after additional application of XAF.

2.5. hiPSC^{W101C}-Derived Ventricular-like Cardiomyocytes Do Not Show an Electrophysiological Phenotype in Comparison with Controls (hiPSC^{corr}-CMs)

Although $I_{K,ACh}$ is not functionally present in control ventricular-like hiPSC-CMs [32,42], it could not be excluded that this is also the case in untreated, ventricular-like hiPSC-CMs with the W101C *KCNJ5* variant. Repeating the experiments of Figure 2, but now in ventricular-like hiPSC-CMs, demonstrated that $I_{K,ACh}$ was not present in both ventricular-like hiPSC^{W101C}-CMs and hiPSC^{corr}-CMs (Figure 5A–C). Consequently, the AP parameters did not differ under baseline conditions, except for a lower V_{max} in hiPSC^{W101C}-CMs (Supplemental Tables S3 and S4). Neither XAF, nor CCh and XAF in the presence of

CCh altered the AP properties in ventricular-like hiPSC^{W101C}-CMs and hiPSC^{corr}-CMs (Figure 5D–F, Supplemental Tables S3 and S4). Thus, the W101C *KCNJ5* variant did not induce an altered phenotype in ventricular-like hiPSC-CMs. In addition, because the electrophysiological properties were largely similar in ventricular-like hiPSC^{W101C}-CMs and hiPSC^{corr}-CMs, additional side effects on other ion channels can be excluded.

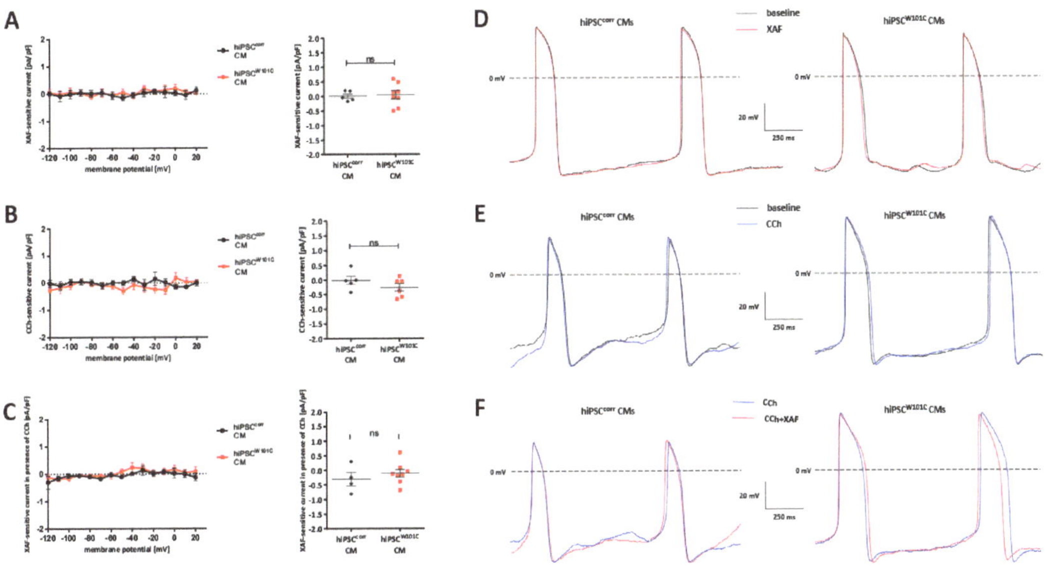

Figure 5. Electrophysiological properties in isogenic untreated, ventricular-like hiPSC^{corr}-CMs and hiPSC^{W101C}-CMs. (**A**) I–V relationships of XAF-sensitive currents under baseline conditions and dot plots of the XAF-sensitive currents at −120 mV (**right panel**). ns = not significant. (**B**) CCh-sensitive currents and dot plots of CCh-sensitive currents at −120 mV (**right panel**). ns = not significant. (**C**) I–V relationships of XAF-sensitive currents in the presence of CCh and dot plots of XAF-sensitive currents in the presence of CCh at −120 mV (**right panel**). ns = not significant. (**D**) Typical APs of hiPSC^{corr}-CM and hiPSC^{W101C}-CM in the absence and presence of XAF. (**E**) Typical APs of a hiPSC^{corr}-CM and hiPSC^{W101C}-CM in the absence and presence of CCh. (**F**) Typical APs of a hiPSC^{corr}-CM and hiPSC^{W101C}-CM in the presence of CCh and after additional application of XAF-1407.

3. Discussion

As set out in the introduction, the W101C *KCNJ5* variant is associated with bradycardia [9]. To study the exact electrophysiological mechanism underlying the slower pacemaking activity in a human cardiomyocytes model, a patient-derived hiPSC line of a patient carrying the familial pathogenic variant and having corrected the mutation were generated, using CRISPR/Cas9-based gene editing, to create an isogenic control hiPSC line. The hiPSC lines were differentiated into ventricular-like hiPSC-CMs, as well as into atrial-like hiPSC-CMs that robustly expressed GIRK channels using RA-treatment during differentiation (atrial-like CMs). In this study, the focus was on the phenotype of hiPSC-CMs with the W101C *KCNJ5*, and the potential pharmacological rescue by XAF-1407, rather than studying the mechanisms and/or modulation of the constitutively active GIRK due to the W101C *KCNJ5* mutation. However, the studied mutation is located in the first transmembrane domain of GIRK4 [9], which is outside the GIRK4 channels regions such as the N and C terminal known for activation and modulatory binding sites [43,44] or outside the pore region of the GIRK4 channel which is important for selectivity changes [45]. This suggests that the basic modulation of the W101C *KCNJ5* mutant channel will not be affected. Although further studies are required, at least Kir3.1 + Kir3.4W101C complexes in oocytes are still able to respond to M2 receptor stimulation [9] and selectivity is not affected as demonstrated in the present study (Figure 2) and in the work of [9]. Constitutive agonist-independent $I_{K,ACh}$ is also observed in response to atrial fibrillation (AF) [46], and these basal active channels has been extensively studies in the last two decades, as reviewed in detail by Voigt and colleagues [44]. Among others, the novel PKC isoform, PKCε, which is enhanced by AF [47], may cause an increased open channel probability of cardiac GIRK channels by phosphorylation of the channel. The cytosolic C-terminal end of the GIRK4 channel, and especially Ser418, is critical for basal activity and the PKCε-mediated augmentation, as recently demonstrated by Gada et al. [48], thus excluding such a mechanism for the constitutively active GIRK due to the W101C *KCNJ5* mutation. In addition, Logothetis and colleagues created various other mutations in GIRK4 [49,50] and they demonstrated that S143T GIRK4, which is within the pore region of the channel, gives rise to highly active homomeric channels. Although this mutation was used to investigate the individual properties of each subunit of the GIRK4 channel in detail [51,52], the mechanisms of the constitutive active current due to S143T GIRK4 alone were not studied. Kuss et al., [9], however, suggests that the constitutively active GIRK due to the W101C *KCNJ5* have an alteration of the spermidine binding site for W101 *KCNJ5* as cause for the increased $I_{K,ACh}$.

The main findings of the present study are as follows: (1) in RA-treated hiPSC-CMs, the W101C *KCNJ5* variant resulted in a constitutively active $I_{K,ACh}$ current under baseline conditions; (2) CCh application activated $I_{K,ACh}$ in both RA-treated hiPSC^{corr}-CMs and hiPSC^{W101C}-CMs; (3) the total $I_{K,ACh}$ density was similar in RA-treated hiPSC^{corr}-CMs and hiPSC^{W101C}-CMs; (4) RA-treated hiPSC^{W101C}-CMs had a lower spontaneous (pacemaking) activity due to a more hyperpolarized MDP at baseline conditions; (5) $I_{K,ACh}$ inhibition rescued the spontaneous beating rate in hiPSC^{W101C}-CMs; (6) the W101C *KCNJ5* variant did not result in electrical abnormalities in untreated ventricular-like hiPSC^{W101C}-CMs. Thus, the present study demonstrates that the familial W101C *KCNJ5* variant results in a slowed pacemaking activity caused by constitutively active GIRK channels. Importantly, the inhibiting of the constitutively active $I_{K,ACh}$ rescues the bradycardic phenotype which is of importance for clinical applications.

Of note, in RA-treated hiPSC-CMs, no significant changes were found in AP repolarization as indicated by the lack of APD changes in response to CCh, although there is a substantial $I_{K,ACh}$ present at positive potentials due to relatively weak inward rectification (Figure 2). The absence of AP repolarization changes is in line with the work of Veerman et al. [22] and Li et al. [23], who found that CCh did not lead to major alterations in APD in RA-treated hiPSC-CMs, except when $I_{K,ACh}$ is largely increased by a *GNB5* variant [32,33]. In contrast, muscarinic receptor activation in human atrial tissue [53,54] and freshly isolated human isolated cardiomyocytes [55] results in an AP shortening. The exact reason for this discrepancy is unknown, but it may be related to the still somewhat immature electrophysiological properties of hiPSC-CMs [56] and potential differences in $I_{K,ACh}$ density between hiPSC-CMs (Figure 2) and human atrial cells [55]. Early- and late-stage development differences in $I_{K,ACh}$ are known due to differences in Kir3.1 and Kir3.4 expression, but our $I_{K,ACh}$ I–V relationship better matches the late-stage than the early-stage development, due to the weaker inward rectification. Further studies using identical experimental protocols are required to solve this issue in detail. Nevertheless, muscarinic receptor activation in freshly isolated rabbit SAN cells does not affect the APD [15], indicating that our used cells' response is closer to SAN cells than human atrial myocytes. This is supported by the slowing of spontaneously activity in our hiPSC-CMs due to muscarinic receptor activation, which is a hallmark for cardiac pacemaker cells [57].

CCh neither induced an $I_{K,ACh}$ nor affected the pacemaker frequency or MDP in our ventricular-like hiPSC-CMs. These findings are in line with previous studies [32,33,42]. In addition, we found no effects on APD in response to CCh, although it is known that muscarinic cholinergic agonists result in a significant decrease in L-type Ca^{2+} currents $I_{Ca,L}$ in ventricular-like hiPSC-CMs [54]. Consequently, an AP shortening may be assumed; however, it was not present here. The lack of major APD changes is in agreement with other studies [32,33,42,54] and is likely due to the muscarinic-agonists-induced decrease in I_{Kr} [54] and I_{Ks} [58], which limits major changes in the net repolarizing current [54].

In the present study, a limited number of cells was used in some phases of the study, especially when the effects of the $I_{K,ACh}$ blockade and CCh were tested. However, we have performed paired experiments, thus considerably raising the power of statistics at these small numbers of experiments. In addition, the principal findings of the voltage clamp experiments were confirmed by our AP measurements. Furthermore, we found that the E_{rev} of $I_{K,ACh}$ is slightly more positive than the calculated E_K, in agreement with previous findings in RA-treated hiPSC-CMs [42] and human embryonic stem cell-derived cardiomyocytes [30]. The exact reason is speculative but may be related to a slightly reduced K^+ selectivity of GRIK channels in hiPSC-CMs. We exclude the immature phenotype of hiPSC-CMs as the reason. Although the ratio between Kir3.1 and 3.4 differs between early- and late-stage development, with consequent differences in rectification, the reversal potential is not different between early- and late-stage development [59]. Nevertheless, because the E_{rev} is similar in RA-treated hiPSC^{corr}-CMs and hiPSC^{W101C}-CMs (Figure 2D), we exclude mutation-induced loss of ion selectivity, as is shown for some other *KCNJ5* mutations [45]. In addition, RA-treated hiPSC-CMs were used as a surrogate for SAN cells, similar to those employed by Veerman et al. [22]. Very recently, however, differentiation protocols have been developed to create SAN-like hiPSC-CMs [33], but this protocol is not yet commonly used in pacemaker research and was not available at the start of our study. Nevertheless, RA-treated hiPSC-CMs exhibit *HCN4* and spontaneous activity that is sensitive for the pacemaker current (I_f) blocker, ivabradine [33], and they display $I_{K,ACh}$ suggesting that they are suitable to study pacemaker-related changes [32]. In fact, spontaneous activity changes in response of CCh and I_f blockade are largely similar in SAN-like and RA-treated hiPSC-CMs [33], suggesting that our results are not hampered by our used differentiation protocol.

In conclusion, our study shows that the W101C *KCNJ5* mutant—when analyzed in a patient-derived disease model—results in slowed pacemaking by a constitutively active $I_{K,ACh}$ and that $I_{K,ACh}$ inhibition may be useful for clinical applications.

4. Materials and Methods

4.1. Human iPSC (hiPSC) Generation

The study was approved by the Ethics Committee of the Westfälische Wilhelms University (file number 2010-048-f-s, 2020-302-f-S, 2022-103-f-S). Written informed consent was obtained from the patient.

For the generation of the *KCNJ5*-hiPSC line, PBMCs with the W101C *KCNJ5* mutation were isolated from the patient's EDTA blood via density gradient centrifugation with Ficoll™ (Sigma-Aldrich, Taufkirchen, Germany). Therefore, the blood plasma was removed by centrifuging the EDTA blood (1000× *g*, 10 min) and the cell pellet was washed with a PBS-buffer mix (DPBS, 2 mM ETDA; 1% BSA) and frozen at −80 °C with Cryo-SFM media until reprogramming.

The Cytotune™-iPS 2.0 Sendai Reprogramming Kit (Thermo Fisher Scientific, Carlsbad, CA, USA) was used for the hiPSC generation process. Cells were seeded for transduction with three Sendai-virus-based vectors (KLF4, SOX2, OCT3/4, c-MYC) as stated in manufacture's advice. After low-density seeding, clones were picked and cultivated until checking for virus clearance by immunocytochemistry staining (anti-Sendai Rabbit polyclonal antibody; MBL International, Woburn, MA, USA).

4.2. Human iPSC Culture

hiPSCs with W101C *KCNJ5* variant were cultured in FTDA medium (DMEM/F12 (Gibco, Life Technologies, Paisley, UK) with 2 mM L-Glutamine (ThermoFisher), 1× ITS (Corning, Bedford, MA, USA), 0.1% HSA (Lucerna-Chem, Lucerne, Switzerland), CD lipid concentrate (Gibco, Life Technologies, Paisley, UK), 50 nM Dorsomorphin (Santa Cruz Biotechnology, Heidelberg, Germany), 2.5 ng/mL Activin A (STEMCELL Technologies,

Vancouver, BC, Canada), 0.5 ng/mL TGFß1 (ThermoFisher), 30 ng/mL FGF2 (PeproTech, Cranbury, NJ, USA)) on Matrigel (Corning)- or Geltrex (Thermo Fisher Scientific)-coated plates [60,61]. Cells were examined daily under a light microscope for differentiated sites, density, and morphology.

At 90–100% confluence of the monolayer, hiPSC were passaged with Accutase (Sigma-Aldrich, Taufkirchen, Germany) and 10 µM Y-27632 (Abcam, Amsterdam, Netherlands). For maintenance, 500,000 to 600,000 cells per 6-well plate were laid out in new FTDA medium with ROCK-inhibitor.

4.3. hiPSC Editing Using CRISPR/Cas9 (hiPSC^{corr} Generation)

For the in silico design of the crRNA, the Custom Alt-R® CRISP/Cas9 guide RNA design tool was used and the HDR-repair template was designed with the Custom Alt-R® CRISPR-Cas9 guide RNA Design-Tool (https://eu.idtdna.com/site/order/designtool/index/CRISPR_CUSTOM; accessed on 10 August 2019) from Integrated DNA Technologies (IDT, Leuven, Belgium). Additionally, we used CRISPOR (http://crispor.tefor.net/, version 4.95). CRISPOR is a web-based tool which helps to design, evaluate, clone and select guide sequences for the CRISPR/Cas9 system [62,63]. crRNA with a low off-target and high on-target score and specificity towards the mutant allele were used for CRISPR/Cas9-mediated HDR repair (Supplemental Table S5).

First, a duplex with the designed crRNA and tracrRNA was formed and mixed with a ratio of 1:1.2 ALT-R Cas9 enzyme (IDT) for RNP-complex formation according to the manufacturer's advice. For RNP delivery to the KCNJ5 W101C variant-hiPSC (hiPSC^{W101C}) the P4 Primary Cell 4D-Nucleofector™ Kit (Lonza, Cologne, Germany) was used. Therefore, 800,000 hiPSC were harvested with Accutase (Sigma-Aldrich) and 10 µM Y-27632 (Abcam), transferred to a sterile centrifugation tube, centrifuged at 200× g (2 min), and the cell pellet resolved in 100 µL of nucleofector solution (Kit P4 Primary Cell 4D-Nucleofector™). Then, 1 µL of ssODN and 1 µL of Alt-R® Cas9 Electroporation Enhancer (IDT) was added to the cells. In total, 4 µL of the previously formed RNP complex was added and the cell suspension was transferred to the nucleofection cuvette. On the 4D nucleofector system (Lonza), the CA-137 program was used. After nucleofection was complete, cells were incubated at 37 °C for 5 min in an incubator, 500 µL of the mTESR plus (STEMCELL Technologies) with Clone R (STEMCELL Technologies) was added, and another 500 µL of medium was laid out on a 12-well plate. After 48 h, cells were detached using Accutase (Sigma-Aldrich) + 10 µM Y-27632 (Abcam) and separated via low-density seeding on a 6-well plate to allow clonal growth. When sufficient cell growth had occurred, the cells were split. Here, one part was laid out on a 24-well plate and the other part was used for DNA isolation and mutation analysis.

4.4. Sequencing Analysis

DNA isolation of the young CRISPR/Cas9 clones was carried out based on the protocol of Li et al. [64]. For this purpose, 10 µL of the cell suspension was taken during splitting, centrifuged at 200× g for 2 min, and the supernatant was discarded. The cell pellet was resuspended in 5 µL 1× PCR buffer (Thermo Fischer Scientific) containing 1 µL protease K solution [20 µg/µL] (Qiagen, Hilden, Germany) and incubated for 5 h at 56 °C. To inactivate the protease K, the preparation was heated for 15 min at 95 °C in a thermal cycler. The DNA concentration obtained was determined and diluted to approx. 30 ng/µL with AE buffer for subsequent sequencing.

The sequencing reaction was performed using the BigDye® Terminator v3.1. cycle sequencing kit (Applied Biosystems, Life Technologies, Waltham, MA, USA) and specific primer pairs for KCNJ5 (Supplemental Table S5).

4.5. Karyotype Analysis

Karyotypes of hiPSC lines were tested with the NanoString nCounter Elements technology with a customized NanoString Probe Set based on nCounter® Human Karyotype

Panel following the manufacturer's advice. In short, 200 ng gDNA was fragmented by AluI digestion. After hybridization with the Capture and Reporter Probe Set at 65 °C, the samples were analyzed on the nCounter® Sprint Profiler (NanoString Technologies, Seattle, WA, USA) and evaluated based on the manufacturer's protocol for copy number variation (CNV).

4.6. Cardiac Differentiation of hiPSC

On the day of differentiation, the confluent monolayer was detached with Accutase + 10 μM Y-27632 and the remaining cell suspension was centrifuged (200× *g*, 2 min). The cell pellet was resuspended in differentiation medium (KO-DMEM (Gibco), 1× Penicillin/Streptomycin/Glutamine (Thermo Fisher Scientific), 5 μg/mL ITS (Corning), 10 μM Y-27632 (Abcam), 200 ng/mL FGF2 (PeproTech), 1 nM CHIR-99021 (STEMCELL Technologies), and 0.75 to 1.0 nM BMP-4 (tested before) (STEMCELL Technologies) [61], and the cell number determined. For KCNJ5-hiPSC lines, 550,000 cells were transferred to a new tube, filled with differentiation medium and mixed with the optimal amount of BMP-4 (tested before) (STEMCELL Technologies), and laid out on matrigel-coated 24-well plates. After 24 h, the differentiation medium was replaced with TS-ASC (KO-DMEM (Gibco), 5.5 mg/L Transferrin (Sigma-Aldrich), 6.75 μg/L Selenium (Sigma-Aldrich), 1× Penicillin/Streptomycin/Glutamine (Thermo Fisher Scientific), and 250 μM ascorbate (Sigma-Aldrich)) [61]. On Day 2 and 3, TS-ASC was supplemented with 0.5 mM C59 (R&D Systems, Minneapolis, MN, USA) for inhibition of the Wnt signaling pathway. Subsequently, the cells were supplied with TS-ASC until the onset of cell contraction.

For atrial cardiac-like cells, 0.5 μM all-trans-retinoic acid (RA; Sigma-Aldrich) was applied during Day 3 and 4 of differentiation to promote the acquisition of atrial-like fate and expression of the GIRK channels [30–32]. To exclude the effect of the RA solvent, an equal amount of DMSO was added to the ventricular-like hiPSC-CM during differentiation.

For electrophysiological experiments, hiPSC-CM cultures were enzymatically dissociated into single cells and plated at a low density on glass slides coated with 0.1% gelatine/Matrigel in KO-THAI media (KO-DMEM (Gibco), 1× Penicillin/Streptomycin/Glutamine (Thermo Fisher Scientific), 0.2% human serum albumin (Lucerna-Chem), 250 μM ascorbate (Sigma-Aldrich), 5 μg/mL ITS (Corning), and 0.004% (*v*/*v*) Thioglycerol (Sigma-Aldrich)). All experiments were performed on cells from at least 3 independent differentiation replicates.

4.7. Immunofluorescence Staining and Imaging

For the detection of OCT3/4, NANOG, the cells were stained using the Human Pluripotent Stem Cell Marker Antibody Panel Plus from R&D Systems. For the immunofluorescence detection of SOX2, an antibody from Abcam (ab137385) was used.

4.8. RT-qPCR

Total RNA was prepared using the Quick-RNA Miniprep- Kit (Zymo Research Corp., Irvine, CA, USA) according to the manufacturer's instructions. Reverse transcription was performed using the QuantiTect Reverse Transcription Kit from Qiagen according to the manufacturer's recommendation. The cDNA obtained was stored at −20 °C until use.

For the RT-qPCR, the QuantiTect SYBR Green PCR Kit (Qiagen) or the Luna Universal qPCR Master Mix (NEB, Ipswich, MA, USA) was used according to the manufacturer's protocol and primers (Supplemental Table S5).

4.9. Cellular Electrophysiology in hiPSC-CMs

4.9.1. Data Acquisition

Spontaneously beating hiPSC-CMs showing regular, synchronous contractions were selected for patch-clamp recordings. Membrane currents and APs were recorded with the amphotericin-perforated patch-clamp technique using an Axopatch 200B amplifier (Molecular Devices, Sunnyvale, CA, USA). Signals were low-pass filtered with a cut-off of 5 kHz and digitized at 5 kHz. Voltage control and data acquisition were realized with custom-made software Scope (version 04.04.27; kindly provided by J. Zegers) and analysis was performed with the custom-made software, MacDaq (version 8.0; kindly provided by A. van Ginneken). The experiments were performed at 36 ± 0.2 °C and the potentials were corrected for the calculated liquid junction potential [65]. Cells were superfused with modified Tyrode's solution containing (in mM): NaCl 140, KCl 5.4, $CaCl_2$ 1.8, $MgCl_2$ 1.0, glucose 5.5, and HEPES 5.0; pH was set to 7.4 with NaOH. Patch pipettes were pulled from borosilicate glass (Harvard apparatus, UK) and had resistances of 2.0–3.0 MΩ after filling with the pipette solution containing (in mM): K-gluconate 125, KCl 20, NaCl 5, amphotericin-B 0.44, and HEPES 10; pH was set to 7.2 with KOH. Cell membrane capacitance (C_m) was estimated by dividing the time constant of the decay of the capacitive transient in response to 5 mV hyperpolarizing voltage clamp steps from −40 mV by the series resistance. For membrane current measurements, series resistance was compensated by ≥70%.

4.9.2. Voltage Clamp Experiments

Membrane currents were measured in voltage clamp mode during a descending RAMP protocol from 20 to −120 mV from a holding potential of −40 mV [30,66]. We measured membrane currents sensitive to 100 nM XAF-1407 and 10 μM CCh. Current densities were calculated at a fixed potential during the RAMP, i.e., −120 mV, −110 mV, −100 mV, etc., (Figure 2B) by dividing the current amplitudes by C_m.

4.9.3. Current Clamp Experiments

APs were recorded in current clamp mode from spontaneously beating single hiPSC-CMs. APs were recorded under baseline conditions and after 4–5 min of application of the various drugs, including 10 μM CCh to activate $I_{K,Ach}$ [30,32] and/or 100 nM XAF-1407 to block $I_{K,ACH}$ [37,38]. APs were characterized by maximal diastolic potential (MDP), maximum upstroke velocity (V_{max}), maximum AP amplitude (APA), AP plateau amplitude (APA_{plat} -defined as the potential difference between MDP and potential at 20 ms after the upstroke), and the duration at 20, 50, and 90% repolarization (APD_{20}, APD_{50}, and APD_{90}, respectively). The diastolic depolarization rate (DDR) was measured over the 50 ms time interval starting at MDP + 1 mV. MDP + 1 mV was used rather than MDP because the time at which the MDP + 1 mV was reached could be determined more reliably than the time at which the MDP was reached [67]. The parameter values obtained from 10 consecutive APs were averaged.

4.10. Statistics

Data are presented as the mean ± SEM. Statistical analysis was carried out with GraphPad Prism 6 software (GraphPad Software, Inc., Boston, MA, USA). Normality and equal variance assumptions were tested with the Kolmogorov–Smirnov, D'Agostino and Pearson omnibus, or the Shapiro–Wilk test, respectively. For normally distributed data, the (un)paired *t*-test was used and for non-normally distributed data, we used the Mann–Whitney–Rank sum test or the Wilcoxon test. $p < 0.05$ was considered statistically significant.

Supplementary Materials: The following supporting information can be downloaded at: https://www.mdpi.com/article/10.3390/ijms242015290/s1.

Author Contributions: Conceptualization, A.K., S.D. and A.O.V.; methodology, A.K., A.O.V., T.Š. and G.M.; software, A.K., A.O.V. and G.M.; formal analysis, A.K. and A.O.V.; investigation, A.K. and A.O.V.; writing—original draft: A.K. and A.O.V.; writing—review and editing: S.D., T.Š. and E.S.-B.; supervision: E.S.-B.; funding acquisition: E.S.-B. All authors have read and agreed to the published version of the manuscript.

Funding: This research received external funding by the Leducq Foundation (FANTASY Network: FIGHTING AGAINST SINUS NODE DYSFUNCTION AND ASSOCIATED ARRHYTHMIAS).

Institutional Review Board Statement: This study was conducted in accordance with the Declaration of Helsinki and approved by the Ethics Committee of the Westfälische Wilhelms University (2010-048-f-s, 2020-302-f-S, 2022-103-f-S).

Informed Consent Statement: Written informed consent was obtained from the patient.

Data Availability Statement: Data available on request due to privacy or ethical restrictions.

Acknowledgments: We would like to thank Bärbel Ulmer for karyotype probe design and Elisabeth Krämer, Nanostring Core Facility (both UKE, Hamburg). Andreas van Impel, who assisted us in confocal microscopy (University Münster), and Jassin Hamidi and Ramona Krampe for support in everyday cell culture. We thank Maike Kreutzenbeck and Rebecca Dieterich (Center for Physiology and Pathophysiology, Medical Faculty and University Hospital Cologne) for technical support in hiPSC generation and Sendai-Virus clearance of W101C *KCNJ5* hiPSC.

Conflicts of Interest: G.M. is an employee and stockholder at DiNAQOR AG. The other authors declare no conflict of interest.

References

1. Hawks, M.K.; Paul, M.L.B.; Malu, O.O. Sinus Node Dysfunction. *Am. Fam. Physician* **2021**, *104*, 179–185.
2. Monfredi, O.; Boyett, M.R. Sick sinus syndrome and atrial fibrillation in older persons-A view from the sinoatrial nodal myocyte. *J. Mol. Cell Cardiol.* **2015**, *83*, 88–100. [CrossRef]
3. De Ponti, R.; Marazzato, J.; Bagliani, G.; Leonelli, F.M.; Padeletti, L. Sick Sinus Syndrome. *Card. Electrophysiol. Clin.* **2018**, *10*, 183–195. [CrossRef]
4. Manoj, P.; Kim, J.A.; Kim, S.; Li, T.; Sewani, M.; Chelu, M.G.; Li, N. Sinus node dysfunction: Current understanding and future directions. *Am. J. Physiol. Heart Circ. Physiol.* **2023**, *324*, H259–H278. [CrossRef]
5. Sathnur, N.; Ebin, E.; Benditt, D.G. Sinus Node Dysfunction. *Cardiol. Clin.* **2023**, *41*, 349–367. [CrossRef]
6. Porta-Sanchez, A.; Priori, S.G. Genetic Abnormalities of the Sinoatrial Node and Atrioventricular Conduction. *Cardiol. Clin.* **2023**, *41*, 333–347. [CrossRef]
7. Wallace, M.J.; El Refaey, M.; Mesirca, P.; Hund, T.J.; Mangoni, M.E.; Mohler, P.J. Genetic Complexity of Sinoatrial Node Dysfunction. *Front. Genet.* **2021**, *12*, 654925. [CrossRef]
8. Baruscotti, M.; Bottelli, G.; Milanesi, R.; DiFrancesco, J.C.; DiFrancesco, D. HCN-related channelopathies. *Pflugers Arch.* **2010**, *460*, 405–415. [CrossRef]
9. Kuss, J.; Stallmeyer, B.; Goldstein, M.; Rinne, S.; Pees, C.; Zumhagen, S.; Seebohm, G.; Decher, N.; Pott, L.; Kienitz, M.C.; et al. Familial Sinus Node Disease Caused by a Gain of GIRK (G-Protein Activated Inwardly Rectifying K^+ Channel) Channel Function. *Circ. Genom. Precis. Med.* **2019**, *12*, e002238. [CrossRef]
10. Krapivinsky, G.; Gordon, E.A.; Wickman, K.; Velimirovic, B.; Krapivinsky, L.; Clapham, D.E. The G-protein-gated atrial K+ channel IKACh is a heteromultimer of two inwardly rectifying K^+-channel proteins. *Nature* **1995**, *374*, 135–141. [CrossRef]
11. Ferrer, J.; Nichols, C.G.; Makhina, E.N.; Salkoff, L.; Bernstein, J.; Gerhard, D.; Wasson, J.; Ramanadham, S.; Permutt, A. Pancreatic islet cells express a family of inwardly rectifying K^+ channel subunits which interact to form G-protein-activated channels. *J. Biol. Chem.* **1995**, *270*, 26086–26091. [CrossRef]
12. Choi, M.; Scholl, U.I.; Yue, P.; Bjorklund, P.; Zhao, B.; Nelson-Williams, C.; Ji, W.; Cho, Y.; Patel, A.; Men, C.J.; et al. K^+ channel mutations in adrenal aldosterone-producing adenomas and hereditary hypertension. *Science* **2011**, *331*, 768–772. [CrossRef]
13. Wickman, K.; Krapivinsky, G.; Corey, S.; Kennedy, M.; Nemec, J.; Medina, I.; Clapham, D.E. Structure, G protein activation, and functional relevance of the cardiac G protein-gated K^+ channel, IKACh. *Ann. N. Y. Acad. Sci.* **1999**, *868*, 386–398. [CrossRef]
14. Ravens, U.; Poulet, C.; Wettwer, E.; Knaut, M. Atrial selectivity of antiarrhythmic drugs. *J. Physiol.* **2013**, *591*, 4087–4097. [CrossRef]
15. Verkerk, A.O.; Geuzebroek, G.S.; Veldkamp, M.W.; Wilders, R. Effects of acetylcholine and noradrenalin on action potentials of isolated rabbit sinoatrial and atrial myocytes. *Front. Physiol.* **2012**, *3*, 174. [CrossRef]
16. DiFrancesco, D.; Ducouret, P.; Robinson, R.B. Muscarinic modulation of cardiac rate at low acetylcholine concentrations. *Science* **1989**, *243*, 669–671. [CrossRef]
17. Choi, H.S.; Wang, D.Y.; Noble, D.; Lee, C.O. Effect of isoprenaline, carbachol, and Cs^+ on Na^+ activity and pacemaker potential in rabbit SA node cells. *Am. J. Physiol.* **1999**, *276*, H205–H214. [CrossRef]
18. Zaza, A.; Robinson, R.B.; DiFrancesco, D. Basal responses of the L-type Ca^{2+} and hyperpolarization-activated currents to autonomic agonists in the rabbit sino-atrial node. *J. Physiol.* **1996**, *491 Pt 2*, 347–355. [CrossRef]
19. Campos-Rios, A.; Rueda-Ruzafa, L.; Lamas, J.A. The Relevance of GIRK Channels in Heart Function. *Membranes* **2022**, *12*, 1119. [CrossRef]
20. Wang, J.; Irnaten, M.; Neff, R.A.; Venkatesan, P.; Evans, C.; Loewy, A.D.; Mettenleiter, T.C.; Mendelowitz, D. Synaptic and neurotransmitter activation of cardiac vagal neurons in the nucleus ambiguus. *Ann. N. Y. Acad. Sci.* **2001**, *940*, 237–246. [CrossRef]
21. Marschall, C.; Moscu-Gregor, A.; Klein, H.G. Variant panorama in 1,385 index patients and sensitivity of expanded next-generation sequencing panels in arrhythmogenic disorders. *Cardiovasc. Diagn. Ther.* **2019**, *9*, S292–S298. [CrossRef] [PubMed]

22. Kokunai, Y.; Nakata, T.; Furuta, M.; Sakata, S.; Kimura, H.; Aiba, T.; Yoshinaga, M.; Osaki, Y.; Nakamori, M.; Itoh, H.; et al. A Kir3.4 mutation causes Andersen-Tawil syndrome by an inhibitory effect on Kir2.1. *Neurology* **2014**, *82*, 1058–1064. [CrossRef] [PubMed]
23. Yamada, N.; Asano, Y.; Fujita, M.; Yamazaki, S.; Inanobe, A.; Matsuura, N.; Kobayashi, H.; Ohno, S.; Ebana, Y.; Tsukamoto, O.; et al. Mutant *KCNJ3* and *KCNJ5* Potassium Channels as Novel Molecular Targets in Bradyarrhythmias and Atrial Fibrillation. *Circulation* **2019**, *139*, 2157–2169. [CrossRef] [PubMed]
24. Proost, D.; Saenen, J.; Vandeweyer, G.; Rotthier, A.; Alaerts, M.; Van Craenenbroeck, E.M.; Van Crombruggen, J.; Mortier, G.; Wuyts, W.; Vrints, C.; et al. Targeted Next-Generation Sequencing of 51 Genes Involved in Primary Electrical Disease. *J. Mol. Diagn.* **2017**, *19*, 445–459. [CrossRef]
25. Akdis, D.; Saguner, A.M.; Medeiros-Domingo, A.; Schaller, A.; Balmer, C.; Steffel, J.; Brunckhorst, C.; Duru, F. Multiple clinical profiles of families with the short QT syndrome. *Europace* **2018**, *20*, f113–f121. [CrossRef]
26. Asatryan, B.; Schaller, A.; Seiler, J.; Servatius, H.; Noti, F.; Baldinger, S.H.; Tanner, H.; Roten, L.; Dillier, R.; Lam, A.; et al. Usefulness of Genetic Testing in Sudden Cardiac Arrest Survivors With or Without Previous Clinical Evidence of Heart Disease. *Am. J. Cardiol.* **2019**, *123*, 2031–2038. [CrossRef]
27. van Lint, F.H.M.; Mook, O.R.F.; Alders, M.; Bikker, H.; Lekanne Dit Deprez, R.H.; Christiaans, I. Large next-generation sequencing gene panels in genetic heart disease: Yield of pathogenic variants and variants of unknown significance. *Neth. Heart J.* **2019**, *27*, 304–309. [CrossRef]
28. Meyer, K.M.; Malhotra, N.; Kwak, J.S.; El Refaey, M. Relevance of *KCNJ5* in Pathologies of Heart Disease. *Int. J. Mol. Sci.* **2023**, *24*, 10849. [CrossRef]
29. Duan, S.; Du, J. Sinus node dysfunction and atrial fibrillation-Relationships, clinical phenotypes, new mechanisms, and treatment approaches. *Ageing Res. Rev.* **2023**, *86*, 101890. [CrossRef]
30. Devalla, H.D.; Schwach, V.; Ford, J.W.; Milnes, J.T.; El-Haou, S.; Jackson, C.; Gkatzis, K.; Elliott, D.A.; Chuva de Sousa Lopes, S.M.; Mummery, C.L.; et al. Atrial-like cardiomyocytes from human pluripotent stem cells are a robust preclinical model for assessing atrial-selective pharmacology. *EMBO Mol. Med.* **2015**, *7*, 394–410. [CrossRef]
31. Marczenke, M.; Piccini, I.; Mengarelli, I.; Fell, J.; Ropke, A.; Seebohm, G.; Verkerk, A.O.; Greber, B. Cardiac Subtype-Specific Modeling of K(v)1.5 Ion Channel Deficiency Using Human Pluripotent Stem Cells. *Front. Physiol.* **2017**, *8*, 469. [CrossRef] [PubMed]
32. Veerman, C.C.; Mengarelli, I.; Koopman, C.D.; Wilders, R.; van Amersfoorth, S.C.; Bakker, D.; Wolswinkel, R.; Hababa, M.; de Boer, T.P.; Guan, K.; et al. Genetic variation in *GNB5* causes bradycardia by augmenting the cholinergic response via increased acetylcholine-activated potassium current (I(K,ACh)). *Dis. Model. Mech.* **2019**, *12*, dmm037994. [CrossRef] [PubMed]
33. Li, J.; Wiesinger, A.; Fokkert, L.; Boukens, B.J.; Verkerk, A.O.; Christoffels, V.M.; Boink, G.J.J.; Devalla, H.D. Molecular and electrophysiological evaluation of human cardiomyocyte subtypes to facilitate generation of composite cardiac models. *J. Tissue Eng.* **2022**, *13*, 20417314221127908. [CrossRef] [PubMed]
34. Lemme, M.; Ulmer, B.M.; Lemoine, M.D.; Zech, A.T.L.; Flenner, F.; Ravens, U.; Reichenspurner, H.; Rol-Garcia, M.; Smith, G.; Hansen, A.; et al. Atrial-like Engineered Heart Tissue: An In Vitro Model of the Human Atrium. *Stem Cell Rep.* **2018**, *11*, 1378–1390. [CrossRef]
35. Cyganek, L.; Tiburcy, M.; Sekeres, K.; Gerstenberg, K.; Bohnenberger, H.; Lenz, C.; Henze, S.; Stauske, M.; Salinas, G.; Zimmermann, W.H.; et al. Deep phenotyping of human induced pluripotent stem cell-derived atrial and ventricular cardiomyocytes. *JCI Insight* **2018**, *3*. [CrossRef]
36. Goldfracht, I.; Protze, S.; Shiti, A.; Setter, N.; Gruber, A.; Shaheen, N.; Nartiss, Y.; Keller, G.; Gepstein, L. Generating ring-shaped engineered heart tissues from ventricular and atrial human pluripotent stem cell-derived cardiomyocytes. *Nat. Commun.* **2020**, *11*, 75. [CrossRef]
37. Fenner, M.F.; Carstensen, H.; Dalgas Nissen, S.; Melis Hesselkilde, E.; Scott Lunddahl, C.; Adler Hess Jensen, M.; Loft-Andersen, A.V.; Sattler, S.M.; Platonov, P.; El-Haou, S.; et al. Effect of selective I(K,ACh) inhibition by XAF-1407 in an equine model of tachypacing-induced persistent atrial fibrillation. *Br. J. Pharmacol.* **2020**, *177*, 3778–3794. [CrossRef]
38. Sobota, V.; Gatta, G.; van Hunnik, A.; van Tuijn, I.; Kuiper, M.; Milnes, J.; Jespersen, T.; Schotten, U.; Verheule, S. The Acetylcholine-Activated Potassium Current Inhibitor XAF-1407 Terminates Persistent Atrial Fibrillation in Goats. *Front. Pharmacol.* **2020**, *11*, 608410. [CrossRef]
39. Linz, B.; Thostrup, A.H.; Saljic, A.; Rombouts, K.; Hertel, J.N.; Hohl, M.; Milnes, J.; Tfelt-Hansen, J.; Linz, D.; Jespersen, T. Pharmacological inhibition of acetylcholine-regulated potassium current (I (K,ACh)) prevents atrial arrhythmogenic changes in a rat model of repetitive obstructive respiratory events. *Heart Rhythm O2* **2022**, *3*, 97–104. [CrossRef]
40. Berecki, G.; Wilders, R.; de Jonge, B.; van Ginneken, A.C.; Verkerk, A.O. Re-evaluation of the action potential upstroke velocity as a measure of the Na^+ current in cardiac myocytes at physiological conditions. *PLoS ONE* **2010**, *5*, e15772. [CrossRef]
41. Krishnan, S.C.; Antzelevitch, C. Sodium channel block produces opposite electrophysiological effects in canine ventricular epicardium and endocardium. *Circ. Res.* **1991**, *69*, 277–291. [CrossRef] [PubMed]
42. Horvath, A.; Lemoine, M.D.; Loser, A.; Mannhardt, I.; Flenner, F.; Uzun, A.U.; Neuber, C.; Breckwoldt, K.; Hansen, A.; Girdauskas, E.; et al. Low Resting Membrane Potential and Low Inward Rectifier Potassium Currents Are Not Inherent Features of hiPSC-Derived Cardiomyocytes. *Stem Cell Rep.* **2018**, *10*, 822–833. [CrossRef] [PubMed]

43. Luscher, C.; Slesinger, P.A. Emerging roles for G protein-gated inwardly rectifying potassium (GIRK) channels in health and disease. *Nat. Rev. Neurosci.* **2010**, *11*, 301–315. [CrossRef] [PubMed]
44. Voigt, N.; Abu-Taha, I.; Heijman, J.; Dobrev, D. Constitutive activity of the acetylcholine-activated potassium current IK,ACh in cardiomyocytes. *Adv. Pharmacol.* **2014**, *70*, 393–409. [CrossRef]
45. Charmandari, E.; Sertedaki, A.; Kino, T.; Merakou, C.; Hoffman, D.A.; Hatch, M.M.; Hurt, D.E.; Lin, L.; Xekouki, P.; Stratakis, C.A.; et al. A novel point mutation in the *KCNJ5* gene causing primary hyperaldosteronism and early-onset autosomal dominant hypertension. *J. Clin. Endocrinol. Metab.* **2012**, *97*, E1532–E1539. [CrossRef]
46. Cha, T.J.; Ehrlich, J.R.; Chartier, D.; Qi, X.Y.; Xiao, L.; Nattel, S. Kir3-based inward rectifier potassium current: Potential role in atrial tachycardia remodeling effects on atrial repolarization and arrhythmias. *Circulation* **2006**, *113*, 1730–1737. [CrossRef]
47. Makary, S.; Voigt, N.; Maguy, A.; Wakili, R.; Nishida, K.; Harada, M.; Dobrev, D.; Nattel, S. Differential protein kinase C isoform regulation and increased constitutive activity of acetylcholine-regulated potassium channels in atrial remodeling. *Circ. Res.* **2011**, *109*, 1031–1043. [CrossRef]
48. Gada, K.D.; Chang, M.; Chandrashekar, A.; Plant, L.D.; Noujaim, S.F.; Logothetis, D.E. Mechanism of PKCepsilon regulation of cardiac GIRK channel gating. *Proc. Natl. Acad. Sci. USA* **2023**, *120*, e2212325120. [CrossRef]
49. Chan, K.W.; Sui, J.L.; Vivaudou, M.; Logothetis, D.E. Control of channel activity through a unique amino acid residue of a G protein-gated inwardly rectifying K^+ channel subunit. *Proc. Natl. Acad. Sci. USA* **1996**, *93*, 14193–14198. [CrossRef]
50. Vivaudou, M.; Chan, K.W.; Sui, J.L.; Jan, L.Y.; Reuveny, E.; Logothetis, D.E. Probing the G-protein regulation of GIRK1 and GIRK4, the two subunits of the KACh channel, using functional homomeric mutants. *J. Biol. Chem.* **1997**, *272*, 31553–31560. [CrossRef]
51. He, C.; Zhang, H.; Mirshahi, T.; Logothetis, D.E. Identification of a potassium channel site that interacts with G protein betagamma subunits to mediate agonist-induced signaling. *J. Biol. Chem.* **1999**, *274*, 12517–12524. [CrossRef]
52. He, C.; Yan, X.; Zhang, H.; Mirshahi, T.; Jin, T.; Huang, A.; Logothetis, D.E. Identification of critical residues controlling G protein-gated inwardly rectifying K^+ channel activity through interactions with the beta gamma subunits of G proteins. *J. Biol. Chem.* **2002**, *277*, 6088–6096. [CrossRef] [PubMed]
53. Petersen, J.; Castro, L.; Bengaard, A.K.P.; Pecha, S.; Ismaili, D.; Schulz, C.; Sahni, J.; Steenpass, A.; Meier, C.; Reichenspurner, H.; et al. Muscarinic Receptor Activation Reduces Force and Arrhythmias in Human Atria Independent of IK,ACh. *J. Cardiovasc. Pharmacol.* **2022**, *79*, 678–686. [CrossRef] [PubMed]
54. Verkerk, A.O.; Doszpod, I.J.; Mengarelli, I.; Magyar, T.; Polyak, A.; Paszti, B.; Efimov, I.R.; Wilders, R.; Koncz, I. Acetylcholine Reduces L-Type Calcium Current without Major Changes in Repolarization of Canine and Human Purkinje and Ventricular Tissue. *Biomedicines* **2022**, *10*, 2987. [CrossRef]
55. Koumi, S.; Arentzen, C.E.; Backer, C.L.; Wasserstrom, J.A. Alterations in muscarinic K^+ channel response to acetylcholine and to G protein-mediated activation in atrial myocytes isolated from failing human hearts. *Circulation* **1994**, *90*, 2213–2224. [CrossRef] [PubMed]
56. Seibertz, F.; Sutanto, H.; Dulk, R.; Pronto, J.R.D.; Springer, R.; Rapedius, M.; Liutkute, A.; Ritter, M.; Jung, P.; Stelzer, L.; et al. Electrophysiological and calcium-handling development during long-term culture of human-induced pluripotent stem cell-derived cardiomyocytes. *Basic. Res. Cardiol.* **2023**, *118*, 14. [CrossRef]
57. MacDonald, E.A.; Rose, R.A.; Quinn, T.A. Neurohumoral Control of Sinoatrial Node Activity and Heart Rate: Insight From Experimental Models and Findings From Humans. *Front. Physiol.* **2020**, *11*, 170. [CrossRef]
58. Freeman, L.C.; Kass, R.S. Cholinergic inhibition of slow delayed-rectifier K^+ current in guinea pig sino-atrial node is not mediated by muscarinic receptors. *Mol. Pharmacol.* **1995**, *47*, 1248–1254.
59. Fleischmann, B.K.; Duan, Y.; Fan, Y.; Schoneberg, T.; Ehlich, A.; Lenka, N.; Viatchenko-Karpinski, S.; Pott, L.; Hescheler, J.; Fakler, B. Differential subunit composition of the G protein-activated inward-rectifier potassium channel during cardiac development. *J. Clin. Investig.* **2004**, *114*, 994–1001. [CrossRef]
60. Frank, S.; Zhang, M.; Scholer, H.R.; Greber, B. Small molecule-assisted, line-independent maintenance of human pluripotent stem cells in defined conditions. *PLoS ONE* **2012**, *7*, e41958. [CrossRef]
61. Zhang, M.; Schulte, J.S.; Heinick, A.; Piccini, I.; Rao, J.; Quaranta, R.; Zeuschner, D.; Malan, D.; Kim, K.P.; Ropke, A.; et al. Universal cardiac induction of human pluripotent stem cells in two and three-dimensional formats: Implications for in vitro maturation. *Stem Cells* **2015**, *33*, 1456–1469. [CrossRef] [PubMed]
62. Haeussler, M.; Schonig, K.; Eckert, H.; Eschstruth, A.; Mianne, J.; Renaud, J.B.; Schneider-Maunoury, S.; Shkumatava, A.; Teboul, L.; Kent, J.; et al. Evaluation of off-target and on-target scoring algorithms and integration into the guide RNA selection tool CRISPOR. *Genome Biol.* **2016**, *17*, 148. [CrossRef]
63. Concordet, J.P.; Haeussler, M. CRISPOR: Intuitive guide selection for CRISPR/Cas9 genome editing experiments and screens. *Nucleic Acids Res.* **2018**, *46*, W242–W245. [CrossRef] [PubMed]
64. Li, H.L.; Gee, P.; Ishida, K.; Hotta, A. Efficient genomic correction methods in human iPS cells using CRISPR-Cas9 system. *Methods* **2016**, *101*, 27–35. [CrossRef] [PubMed]
65. Barry, P.H.; Lynch, J.W. Liquid junction potentials and small cell effects in patch-clamp analysis. *J. Membr. Biol.* **1991**, *121*, 101–117. [CrossRef]

66. Lomax, A.E.; Rose, R.A.; Giles, W.R. Electrophysiological evidence for a gradient of G protein-gated K^+ current in adult mouse atria. *Br. J. Pharmacol.* **2003**, *140*, 576–584. [CrossRef]
67. Verkerk, A.O.; Wilders, R.; Zegers, J.G.; van Borren, M.M.; Ravesloot, J.H.; Verheijck, E.E. Ca^{2+}-activated Cl^- current in rabbit sinoatrial node cells. *J. Physiol.* **2002**, *540*, 105–117. [CrossRef]

International Journal of
Molecular Sciences

MDPI

Article

Selective Inhibition of Pulmonary Vein Excitability by Constitutively Active GIRK Channels Blockade in Rats

Ian Findlay [1], Côme Pasqualin [2], Angèle Yu [2], Véronique Maupoil [2] and Pierre Bredeloux [2,*]

[1] Laboratoire de Pharmacologie, Faculté de Pharmacie, Université de Tours, 37200 Tours, France; ian.findlay@univ-tours.fr

[2] EA4245, Transplantation, Immunologie et Inflammation, Université de Tours, 37200 Tours, France; come.pasqualin@univ-tours.fr (C.P.); angele.yu@univ-tours.fr (A.Y.); veronique.maupoil@univ-tours.fr (V.M.)

* Correspondence: pierre.bredeloux@univ-tours.fr; Tel.: +33-247-367-205

Abstract: Pulmonary veins (PV) are the main source of ectopy, triggering atrial fibrillation. This study investigated the roles of G protein-coupled inwardly rectifying potassium (GIRK) channels in the PV and the left atrium (LA) of the rat. Simultaneous intracellular microelectrode recording from the LA and the PV of the rat found that in the presence or absence of acetylcholine, the GIRK channel blocker tertiapin-Q induced AP duration elongation in the LA and the loss of over-shooting AP in the PV, suggesting the presence of constitutively active GIRK channels in these tissues. Patch-clamp recordings from isolated myocytes showed that tertiapin-Q inhibited a basal inwardly rectified background current in PV cells with little effect in LA cells. Experiments with ROMK1 and KCa1.1 channel blockers ruled out the possibility of an off-target effect. Western blot showed that GIRK4 subunit expression was greater in PV cardiomyocytes, which may explain the differences observed between PV and LA in response to tertiapin-Q. In conclusion, GIRK channels blockade abolishes AP only in the PV, providing a molecular target to induce electrical disconnection of the PV from the LA.

Keywords: rat; cardiomyocytes; pulmonary veins; left atria; GIRK channels; tertiapin-Q; constitutive activity; electrophysiology

Citation: Findlay, I.; Pasqualin, C.; Yu, A.; Maupoil, V.; Bredeloux, P. Selective Inhibition of Pulmonary Vein Excitability by Constitutively Active GIRK Channels Blockade in Rats. *Int. J. Mol. Sci.* **2023**, *24*, 13629. https://doi.org/10.3390/ijms241713629

Academic Editors: Yosuke Okamoto and Kunichika Tsumoto

Received: 21 July 2023
Revised: 25 August 2023
Accepted: 1 September 2023
Published: 4 September 2023

1. Introduction

Ectopic activities arising from the cardiac sleeves of the human pulmonary vein (PV) are responsible for triggering atrial fibrillation [1]. Over the past few years, numerous studies have been conducted to characterize differences in the physiology of atrial and PV cardiomyocytes in an attempt to understand this phenomena (For review, [2]). In a rat PV-Left atria (LA) preparation, we showed that selective activation of α1-adrenergic receptors resulted in the abolition of electrical conduction within the PV, whereas electrical activity was maintained in the LA of the rat [3]. This effect was reversed by adenosine, which activates the G protein-coupled inwardly rectifying potassium (GIRK) channels [4,5], suggesting a role of these channels in PV electrical conduction.

Cardiac GIRK channels are composed of four subunits which are associated to form hetero-(Kir3.1/3.4) or homotetramers (Kir3.4/3.4) involved in the response of the atria to the vagal/parasympathetic nervous system activation [6]. Remodeling of atrial electrophysiology during the development of cardiac pathologies such as tachycardia and fibrillation have been found to include the development of constitutively active GIRK channels in man and a number of animal models [7–10]. The mechanisms leading to the conversion from GIRK activation by the $\beta\gamma$ subunits of G-protein coupled receptors into spontaneous opening are still unclear [11–14].

In the atria, the peptide toxin tertiapin-Q (TQ), which selectively blocks the GIRK/Kir3.n channels family [15–17], reverses action potential (AP) shortening induced by acetylcholine (ACh). TQ induces also by itself atrial AP elongation in pathological [8,10,18] as well as in young and healthy animal models [19–21]. In consequence, GIRK channels could also provide a constitutively active background K^+ conductance in the atria under physiological conditions.

This study was undertaken to compare the contribution of GIRK channels on AP in PV and LA cardiomyocytes. We found that TQ increased action potential duration (APD) in the LA, whereas a depolarization of the membrane inducing AP abolition was observed in the PV. Taken as a whole, these results suggest that GIRK channels may contribute more significantly to the regulation of constitutively active background K^+ conductance in PV than in LA cardiomyocytes, offering a new pharmacological approach to inhibit the conduction of PV ectopic activity to the LA, triggering atrial fibrillation.

2. Results

2.1. Differential Effects of Tertiapin-Q upon the PV and the LA in Presence of Acetylcholine

The contribution of GIRK channels was investigated by studying the effect of TQ in presence of ACh on AP, simultaneously recorded at the frequency of 5 Hz in the PV and the LA of the same preparation (n = 5).

In the PV, 10^{-7} M ACh induced a slight but not statistically significant hyperpolarization (p = 0.071) with a significant (p = 0.025) increase in AP peak (Figure 1a,b,e). Unexpectedly, in the continued presence of Ach, the addition of 150 nM TQ led to a significant depolarization of PV from -76.6 ± 1.1 mV to -57.8 ± 1.2 mV ($p < 0.001$) and to the suppression of AP, which are replaced by small electrotonic waves (Figure 1c,e). In the LA, neither 10^{-7} M ACh alone nor ACh and 150 nM TQ had any significant effect upon DMP or AP peaks (Figure 1a–d). ACh alone reduced LA APD_{90} from 43.8 ± 1.4 ms to 35.9 ± 1.0 ms (p = 0.002). The addition of TQ then increased LA APD_{90} to 57.2 ± 3.0 ms ($p < 0.001$). This latter value was also significantly greater than the APD_{90}, which had been recorded under basal conditions (p = 0.004).

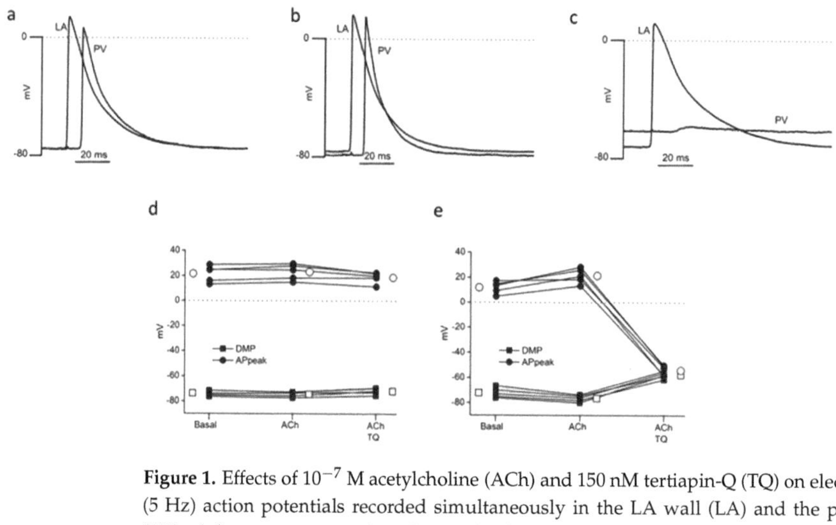

Figure 1. Effects of 10^{-7} M acetylcholine (ACh) and 150 nM tertiapin-Q (TQ) on electrically evoked (5 Hz) action potentials recorded simultaneously in the LA wall (LA) and the pulmonary vein (PV) of the same preparation. Pairs of individual traces (**a**–**c**) are from a continuous recording obtained under basal conditions (**a**), then during ACh superfusion (**b**), and finally in the presence of ACh with the addition of tertiapin-Q (**c**). (**d**,**e**) represent the results obtained from individual microelectrode penetrations in the LA and PV of five preparations, respectively. Squares represent diastolic membrane potential (DMP); circles represent action potential peak values (AP peak). Open symbols represent mean values. SEM are not visible on the figure because they are smaller than the symbols.

These results indicate that 150 nM TQ had effects over and above the simple reversal of the effects of 10^{-7} M ACh. Traces in Figure 1c clearly show that, in the same preparation, TQ abolished AP in the PV, whereas they were still recorded in the LA.

2.2. Direct Effects of Tertiapin-Q upon the PV and the LA

We then examined the effects of TQ upon AP recorded simultaneously in the PV and LA of the same preparations in the absence of ACh (Figure 2, n = 10).

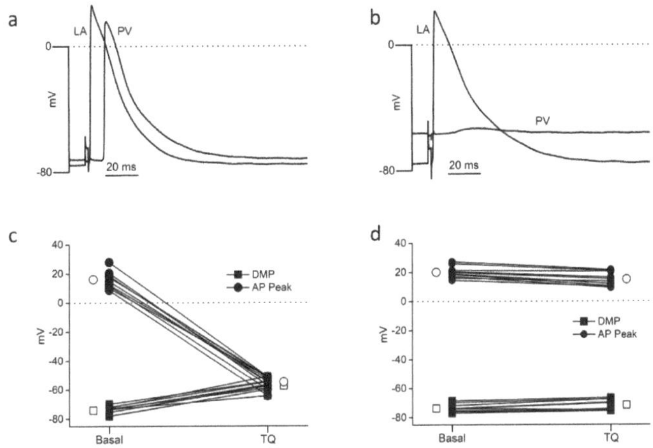

Figure 2. Effects of 150 nM tertiapin-Q (TQ) on electrically evoked (5 Hz) action potentials recorded simultaneously by intracellular microelectrodes placed in the LA wall (LA) and the pulmonary vein (PV) of the same preparation. Pairs of individual traces (**a**,**b**) are from a continuous recording obtained from one preparation under basal conditions (**a**), then during the superfusion of tertiapin-Q (**b**). (**c**,**d**) represent the results obtained from individual microelectrode penetrations in the PV and LA of ten preparations, respectively. Squares represent diastolic membrane potential (DMP); circles represent action potential peak values (AP peak). Open symbols represent mean values. SEM are not visible on the figure because they are smaller than the symbols.

In these conditions, 150 nM TQ evoked a significant depolarization of PV from -74.1 ± 0.8 mV to -57.4 ± 1.7 mV ($p < 0.001$) associated with the replacement of AP by small electrotonic waves (Figure 2a–c). In the LA of the same preparations, TQ had no significant effect upon either the DMP ($p = 0.173$) or AP peak ($p = 0.077$) (Figure 2a,b,d), but it significantly increased LA APD_{90} from 45.7 ± 1.5 ms to 58.1 ± 1.3 ms ($p < 0.001$).

To investigate if TQ had direct effect upon the whole rat heart in our experimental conditions, its effect on sino-atrial node AP was studied. The application of 150 nM TQ alone had no effect upon spontaneous AP frequency (4.5 ± 0.01 Hz in basal condition versus 4.5 ± 0.1 Hz in TQ; $p = 0.876$), phase 4 duration (93.6 ± 6 ms in basal condition versus 90.4 ± 6 ms in TQ; $p = 0.72$), and APD_{90} (72.9 ± 2.9 ms versus 75.5 ± 3.6 ms in TQ; $p = 0.636$) recorded in 5 different preparations. On the other hand, TQ was able to significantly reverse the effects of 10^{-6} M of the muscarinic receptor agonist carbachol upon these parameters (Table S1).

2.3. Reversal of Tertiapin-Q Effects upon the PV and the LA

2.3.1. Reversal of Tertiapin-Q Effects by Acetylcholine

In the continuous presence of 150 nM TQ, we showed that 10^{-7} M ACh had no significant effect upon PV DMP or AP peak (Figure 3a,c; Table 1). Increasing the concentration of ACh to 10^{-6} M led to a significant hyperpolarization of PV (Table 1) and the restoration of over-shooting AP in 4 of the 7 preparations (Figure 3a,c). However, both PV DMP and AP peak were significantly less than those recorded in 10^{-6} M ACh under control conditions

(Table 1). In the LA of the same preparations, while the different concentrations of ACh had no significant effect upon either DMP or AP peak (Figure 3b,d; Table 1), a concentration-dependent decrease of LA APD_{90} was observed (Figure 3b; Table 2), though LA APD_{90} was significantly greater than equivalent data obtained from control experiments in ACh alone (Table 2).

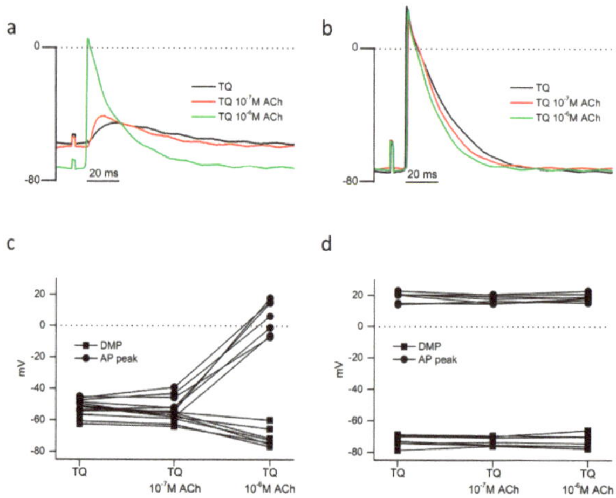

Figure 3. Concentration-dependent effects of acetylcholine (ACh) upon electrically evoked (5 Hz) action potentials in the PV and LA in the continuous presence of 150 nM tertiapin-Q (TQ). Traces are from a continuous recording obtained from one preparation in the PV (**a**) and LA (**b**). Data points linked by lines represent the results obtained from individual microelectrode penetrations in the PV (**c**) and LA (**d**) of 7 different preparations. Squares represent diastolic membrane potential (DMP); circles represent action potential peak values (AP peak). The mean and SEM values of the data are shown in Table 1.

Table 1. Effects of acetylcholine (ACh) upon the PV and LA of the rat in the presence (n = 7) and the absence (n = 10) of 150 nM tertiapin-Q (TQ). Each preparation was stimulated electrically at 5 Hz. The data shown for tertiapin-Q represent the means ± SEM of the individual data points shown in Figure 3. *p* values in italics are given for statistical comparisons between data for diastolic membrane potential (DMP, in mV) and action potential peak values (AP peak, in mV) recorded under TQ and TQ + 10^{-7} M ACh, TQ + 10^{-7} M Ach, and TQ + 10^{-6} M ACh and TQ + 10^{-6} M ACh and 10^{-6} M ACh alone. The data for the effects of 10^{-6} M ACh alone were obtained from a separate series of experiments.

	TQ		**TQ+ 10^{-7} M ACh**		**TQ+ 10^{-6} M ACh**		**10^{-6} M ACh**
PV							
DMP	−55.3 ± 1.8	*p* = 0.145	−58.7 ± 1.3	*p* < 0.001	−70.9 ± 2.3	*p* = 0.012	−77.2 ± 1.0
AP peak	−49.2 ± 1.3	*p* = 0.988	−49.2 ± 2.5	*p* < 0.001	5.6 ± 4.0	*p* < 0.001	23.4 ± 1.3
LA							
DMP	−72.7 ± 1.3	*p* = 0.895	−72.9 ± 1.1	*p* = 0.959	−72.8 ± 1.6	*p* = 0.042	−76.3 ± 0.7
AP peak	19.0 ± 1.2	*p* = 0.417	17.7 ± 1.0	*p* = 0.401	18.9 ± 0.9	*p* = 0.471	17.7 ± 1.1

Table 2. Concentration-effect of acetylcholine (ACh) under control conditions or in the presence of 150 nM tertiapin-Q upon action potential duration (APD_{90}, in ms) of the left atria electrically stimulated at 5 Hz. *p* values in italics are given for statistical comparisons of the basal, 10^{-7} M and 10^{-6} M ACh values are recorded under their respective experimental condition (control or tertiapin-Q). *p* values in bold are given for statistical comparisons between data obtained under control and tertiapin-Q condition in the absence or presence of ACh. APD_{90} values were obtained from experiments shown in Figure 3.

	Basal		10^{-7} M ACh		10^{-6} M ACh
Control (n = 10)	38.4 ± 0.9	*p* < 0.001	29.9 ± 1.2	*p* < 0.001	19.3 ± 1.3
	***p* < 0.001**		***p* < 0.001**		***p* < 0.001**
Tertiapin-Q (n = 9)	52.8 ± 1.6	*p* = 0.002	45.0 ± 1.0	*p* = 0.001	39.6 ± 1.0

2.3.2. Reversal of Tertiapin-Q Effects through Kir6.2 Channel Activation

We used the pinacidil analog P1075 to activate Kir6.2 (KATP) channels that are unaffected by TQ [17]. In PV, in the presence of 150 nM TQ, P1075 evoked a strong hyperpolarization from −56.8 ± 1.7 to −78.3 ± 1.0 mV ($p < 0.001$), and the recovery of over-shooting AP from the electrotonic waves at −54.0 ± 1.9 to 19.3 ± 0.4 mV ($p < 0.001$) in each of the 7 different preparations (Figure 4a,c). In the same preparations, P1075 had no effect upon either DMP (−74.7 ± 1.3 versus −76.3 ± 1.2 mV: $p = 0.383$) or AP peak (16.5 ± 1.2 versus 19.5 ± 0.8 mV: $p = 0.063$) in the LA (Figure 4b,d). P1075 significantly reduced LA APD under control conditions and in the presence of 150 nM TQ (Figure 4b; Table 3). In contrast to the results obtained with ACh (Table 2), there was only a slight difference between LA APD recorded in P1075 in the presence and the absence of TQ (Table 3).

Taken as a whole, these results show that TQ had a direct effect upon both the PV and LA of the rat, suggesting the existence of a constitutively active GIRK channels population in these tissues.

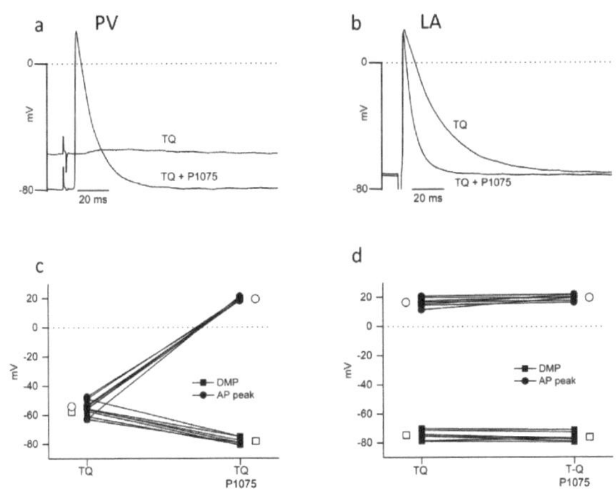

Figure 4. Recovery of over-shooting action potentials in the pulmonary vein (PV) by activation of Kir6.2 (KATP) channels with the pinacidil analog P1075 in the presence of tertiapin-Q (TQ). Electrically evoked (5 Hz) action potentials simultaneously recorded by intracellular microelectrodes placed in the LA wall (LA) and the PV. The pairs of individual traces are from a continuous recording in the PV (**a**) and the LA (**b**) in the presence of first 150 nM tertiapin-Q, then following the addition of 30 μM P1075. (**c**,**d**). Data points linked by lines represent the results obtained from individual microelectrode penetrations in the PV (**c**) and LA (**d**) of seven different preparations, respectively. Squares represent diastolic membrane potential (DMP); circles represent action potential peak values (AP peak). Open symbols represent mean values. SEM are not visible on the figure because they are smaller than the symbols.

Table 3. Effects 30 μM P1075 under control conditions or in the presence of 150 nM tertiapin-Q upon action potential duration (APD_{90}, values in ms) of the left atria electrically stimulated at 5 Hz. *p* values in italics are given for statistical comparisons between data obtained with P1075 and the basal values recorded under their respective experimental condition (control or tertiapin-Q). *p* values in bold are given for statistical comparisons between data obtained under control and tertiapin-Q condition in the absence or presence P0175. APD_{90} values were obtained from experiments shown in Figure 4.

	Basal		P1075
Control (n = 9)	42.5 ± 2.1	*p* < 0.001	11.5 ± 1.2
	***p* < 0.001**		***p* = 0.020**
Tertiapin-Q (n = 7)	56.4 ± 2.1	*p* < 0.001	16.3 ± 1.5

2.4. Selectivity of Tertiapin-Q against GIRK Channels in PV and LA Cardiomyocytes

The initial study describing the synthesis of TQ demonstrated that it had a high affinity for GIRK channels, whereas IRK1 (Kir2.1) channels were essentially insensitive to this molecule [22]. However, TQ was also shown to be effective against the Maxi-K^+ (KCa1.1) channels [23] and ROMK1 (Kir1.1) [15], the latter being expressed in rat PV cardiomyocytes [24]. The Maxi-K^+ channel blocker iberiotoxin [25] and the Kir1.1 channel blocker VU591 [26] were therefore tested upon the quiescent membrane potential of rat PV cardiomyocytes [27] and compared with a subsequent exposure to TQ (Figure 5a). Neither iberiotoxin (Figure 5b: $p = 0.974$) nor VU591 (Figure 5c: $p = 0.698$) had any significant effect. In both cases the subsequent application of TQ evoked a significant depolarization ($p < 0.001$ and $p = 0.022$, respectively). We conclude that, in the rat PV, the effects of TQ were not due to an action of the toxin on its other known molecular targets.

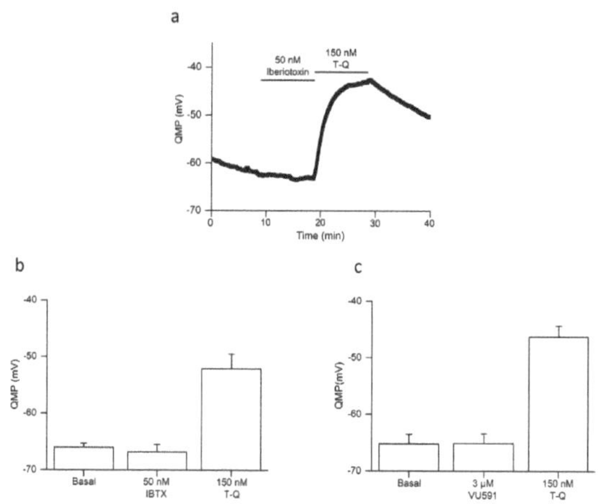

Figure 5. Comparison of the effects of iberiotoxin (IBTX), VU591, and tertiapin-Q (TQ) upon the quiescent membrane potential of cardiac muscle in the pulmonary vein (PV) of the rat. (**a**) Representative experiment comparing the effects of first iberiotoxin and then tertiapin-Q on the quiescent membrane potential (QMP). Data values for QMP were obtained by analyzing the otherwise continuous recording every second. The protocol consisted of that developed by Doisne et al. [27], where penetration of cardiac muscle in the PV was performed during 5 Hz electrical stimulation showing over-shooting action potentials. Electrical stimulation then ceased, and the preparation was allowed 10–15 min to establish an approximately steady-state QMP. In this example 10 min of superfusion of iberiotoxin was followed by 10 min superfusion of tertiapin-Q, as indicated by the bars above the recording. Columns and bars graphs represent mean and SEM values of QMP from 5 experiments comparing the effects of iberiotoxin and tertiapin-Q (**b**) and from 4 experiments comparing the effect of VU591 and tertiapin-Q (**c**).

2.5. Constitutively Active GIRK Channels in PV and LA Cardiomyocytes

Finally the existence of a constitutively active ACh-activated K^+ current (IK_{Ach}) was studied in PV and LA using the whole-cell patch clamp technique on enzymatically isolated cardiomyocytes. The voltage clamp protocol consisted of a pre-pulse to −40 mV from the holding potential of −70 mV to inactivate the Na^+ current. Then a voltage ramp from −120 to +60 mV at a rate of 18 mV/s which inactivated the Ca^{2+} current. I/V ramps were recorded in the absence of ACh, first under control conditions and then in the presence of 300 nM TQ. The currents blocked by TQ obtained by the subtraction $I_{control}$–I_{TQ} in PV (red trace) and LA (black trace) in absence of ACh are illustrated Figure 6. In the PV cardiomyocytes, the TQ sensitive basal current showed an I/V curve with the characteristics of an IK_{ACh} [7]. It has a large inward component for membrane potentials below the potassium equilibrium potential ($p = 0.029$ versus LA at −100 mV). At potentials positive to E_K, there is a clear outgoing current ($p = 0.023$ versus LA at −40 mV) which shows inward rectification. On the other hand, there was little or no TQ sensitive current in the cardiomyocytes of the LA

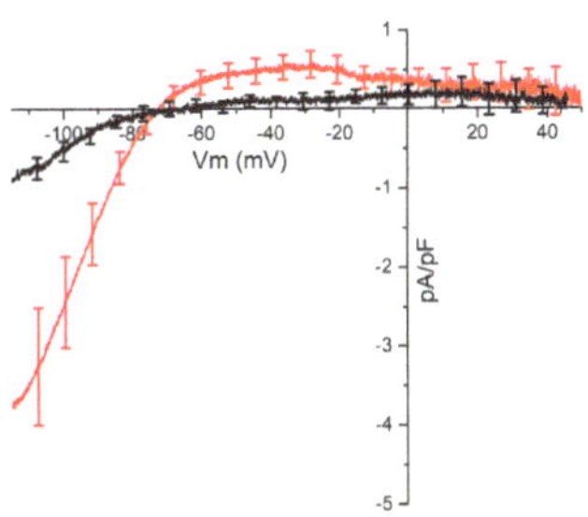

Figure 6. Tertiapin-Q (300 nM) sensitive currents in isolated left atria (black trace) and pulmonary vein cardiomyocytes (red trace) recorded under control condition in the absence of acetylcholine. Data are expressed as the means of voltage ramp obtained from 8 LA and 10 PV cardiomyocytes. For clarity, vertical bars representing SEM are not included for all data points.

In an attempt to explain this difference between PV and LA cardiomyocytes, the expression of the GIRK1 (Kir3.1) and GIRK4 (Kir3.4) subunits of the cardiac GIRK channel were examined with the western blot technique (Figure 7). Both proteins responsible for the IK_{ACh} current are expressed in both PV and LA cardiomyocytes (Figure 7a). However, while the GIRK1 protein expression did not differ between the 2 tissues ($p = 0.273$ vs. LA, Figure 7b), the GIRK4 protein expression is 1.73 times greater in PV cardiomyocytes ($p = 0.014$ vs. LA, Figure 7c). This suggests the expression of GIRK channels composed of GIRK4 homo-tetramers in the PV as well as GIRK1/4 hetero-tetramers.

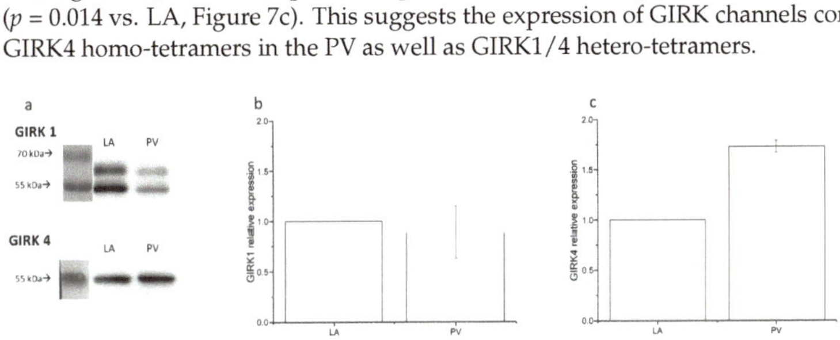

Figure 7. GIRK1 and GIRK4 subunits expression in the left atria (LA) and pulmonary vein (PV) cardiomyocytes. (**a**) Representative western blots of GIRK1 and GIRK4 in LA and PV cardiomyocytes. Bar graphs represent GIRK1 (**b**) and GIRK4 expression (**c**) in the PV relative to that in the LA. Data expressed as mean values ± SEM (n = 3, 2 rats/sample).

3. Discussion

The main finding of this study is that blocking constitutively active GIRK channels selectively abolishes excitability in the rat PV.

We found that TQ was able to reverse the effect of ACh upon LA APD and PV diastolic membrane potential beyond the values recorded in basal conditions, suggesting that it might have effects of its own in LA, but more particularly in PV cardiomyocytes. This was confirmed when we observed that TQ in the absence of ACh blocked an IK_{ACh} current and provoked a depolarization, leading to the reduction of AP to small electrotonic waves in PV cardiomyocytes, whereas it only induced an elongation of APD_{90} in LA, and had no effect on spontaneous AP of the sino-atrial node.

The fact that TQ was not having a generalized effect upon the heart is sustained by different studies. An APD elongation in the absence of ACh has been reported in dog [8,19] and guinea pig [20] LA and PV cardiomyocytes, whereas no effect was observed in ventricular APD [20] and the standard ECG parameters [16] in isolated guinea pig hearts. ACh and tertiapin were also devoid of effect in canine ventricular and purkinje cells [28]. In rat ventricles, TQ had no effect at the physiological stimulation frequency of 5 Hz but rather, prolonged APD and provoked a slight depolarization when the hearts were paced at 1 Hz [29]. Although functional, GIRK channels do not appear to have a significant impact on ventricular physiology or arrythmogenesis in mice [30]. In PV cardiomyocytes, TQ was reported to induce spontaneous AP in both the quiescent guinea pig PV [31] and the PV of rats with abdominal aorto-venocaval shunt [32]. These data, combined with the fact that we showed that TQ caused depolarization in quiescent rat PV without inducing spontaneous AP, support a selective loss of excitability in the cardiac muscle of the PV.

We interpret our results as arising from the inhibition of constitutively active GIRK channels by TQ, since they were obtained in the absence of Ach, partially reversed by the application of 10^{-6} M Ach, and fully reversed by the activation of Kir6.2 ATP-sensitive channels that are unaffected by TQ [17]. This suggests that they results from the selective blockade of GIRK channels and not from a non-selective blockade of multiple K^+ channels, especially as TQ also has no significant affinity for Kir2.1 channels [17,22,33,34]. This was confirmed in control experiments with iberiotoxin and VU591, which ruled out the possibility that these results were due to the action of TQ on its other known targets i.e., Kir1.1 and KCa1.1. Finally, TQ in the absence of ACh reduced an IK_{ACh}-like current in PV cardiomyocytes.

The differences observed in response to TQ between LA and PV cardiac muscle could be explained at least in part by the fact that the expression of GIRK4 is greater in PV than in LA cardiomyocytes, suggesting that GIRK4 homo-tetramers formation is more likely in PV than in LA cardiomyocytes. The absence of a differential reaction to TQ in the dog might then result from the equivalent distribution of GIRK1 and GIRK4 in the LA and PV [19,35]. While GIRK1/4 hetero-tetramers are also unaffected by intracellular Na^+, the spontaneous activity of GIRK4 homo-tetramer channels increases with the intracellular Na^+ concentration [12,36]. However, the link between the enhanced basal Na^+ permeability of rat PV myocytes [37] and the effects of TQ upon PV membrane potential requires further investigation.

Another mechanism with similar electrophysiological effects as TQ upon the rat PV has been found to be α1-adrenergic receptor activation. Doisne et al. [27] first demonstrated a clear physiological distinction between the cardiac muscle of the LA and the PV of the rat. They showed that a combination of α1- and β1-adrenergic receptor stimulation evoked automatic electrical activity in the PV while having only a minor effect upon the membrane potential in the LA. They demonstrated that the biphasic response of the PV membrane potential to norepinephrine resulted from hyperpolarization associated with β1-adrenergic receptor stimulation and depolarization associated with α1-adrenergic receptor activation. Then, Bredeloux et al. [3] demonstrated that the depolarization induced by α-adrenergic receptor stimulation reduced PV AP peak to electrotonic waves, leading to the loss of electrical conduction from the LA along the PV. The GIRK channel agonist adenosine [5] induced PV membrane hyperpolarization and reversed these phenomena. A number of studies have demonstrated inhibition of GIRK channels with the α-adrenergic receptor agonist phenylephrine by means of PIP2 depletion in atrial cardiomyocytes [38–41]. In the same way, Deng et al. [42] showed that the chloride current blocker DCPIB reduced the activity of a number of the members of the Kir channel family, including GIRK1/4 hetero- and GIRK4 homo-tetramers, by interfering with PIP2 binding sites. Thus inhibition of constitutively active GIRK channels might explain the depolarization induced by α1-adrenergic receptors in rat PV cardiomyocytes [3,27] and the depolarization of the PV induced by DCPIB (Figure S3). TQ would evoke the same effects by direct block of the GIRK channel pore [34,43].

Application of longitudinal stretch to the rat PV also causes depolarization and a significant reduction of AP amplitude [44]. This phenomenon could be reversed by the chloride current blockers 4AC, DIDS, and DCPIB and has been linked to the activation of an I_{swell} chloride current in isolated myocytes from the rat PV. Since stretch was not applied to the atrium in those experiments, we do not know if this was specific to the PV, as it is the case for α-adrenergic stimulation [3,27] and here with TQ. We questioned whether PV depolarization induced by longitudinal stretch [44] and α-adrenergic stimulation [3,27] shared a common mechanism. Although ACh could reverse depolarization induced by stretch [45], neither DIDS nor DCPIB reversed depolarization induced by the α-adrenergic agonist cirazoline (Figures S1 and S2). Thus, we cannot explain the depolarization of the rat PV induced by α-adrenergic stimulation as resulting from the activation of an anionic current, while there are arguments in favor of GIRK channel inhibition.

In conclusion, all these data suggest that it is possible to induce electrical disconnection of the PV from the LA by one or more approaches such as α-adrenergic receptors stimulation [3,45], GIRK channel block with TQ, and activation of the I_{swell} anion conductance [44]. The common factor is the PV depolarization that leads to the reduction of AP peak and concurrent reduction of AP dV/dt_{max}. This suggests a graded voltage-dependent inactivation of INa^+ rather than direct block of conduction between LA and PV. The recovery of over-shooting AP in the PV by activation of GIRK (Figure 3) [3,45] and IK_{ATP} or block of I_{swell} [44] is associated with PV hyperpolarization that leads to the increase in AP amplitude and dV/dt_{max}, suggesting recovery from inactivation of INa^+.

Nevertheless, whatever the process, little effect is found upon the membrane potential of the LA wall. Since a significant difference in the quiescent membrane potentials of the PV and LA has been found [27], it may be proposed, therefore, that it is the regulation of the cardiac muscle membrane potential that underlies the responses of the PV and the lack of response of the LA wall. The regulation of membrane potential in the LA and the PV requires further investigation to identify molecular targets allowing the future development of more specific pharmacological treatments of AF. These could include constitutively active GIRK channels, whose blockade would prevent the conduction of ectopic electrical activity from the PV to the LA, as well as its remodelling and progression of the disease from paroxysmal to persistent form of AF. This could offer new therapeutic prospects for patients with paroxysmal AF who would not benefit from surgical pulmonary veins isolation (PVI) due to certain contraindications. Moreover, despite the high success rate of surgical PVI, arrhythmia recurrence related to PV reconnections are commonly observed [46,47] and require additional interventions. Thus, constitutively active GIRK blockade could also be an interesting adjunctive therapy by preventing functional electrical PV reconnection and arrhythmia recurrence, thus avoiding the need for new invasive procedures.

4. Materials and Methods

The protocols used in this study had been approved by the local Animal Care and Use Committee (Comité d'Ethique en Expérimentation Animale Val de Loire, Tours, France. Permit number 2022032119225830).

Male Wistar rats (CER Janvier, Le Genest-St Isle, France) were anaesthetized by intraperitoneal injection of pentobarbital (60 mg kg^{-1}). After intravenous injection of heparin

(500 IU kg^{-1}), the heart and lung block was rapidly removed and placed in a dissecting dish that contained cold (+4 °C) cardioplegic solution (110 mM NaCl, 10 mM $NaHCO_3$, 16 mM KCl, 16 mM $MgCl_2$, 1.2 mM $CaCl_2$, and 10 mM glucose).

4.1. Intracellular Microelectrode Recordings

The preparations used in this study consisted of either the LA and the superior PV, as previously described [3], or the right atrium and the superior vena cava. These were pinned out endocardial side up to the floor of an organ bath through which flowed Krebs-Henseleit solution (119 mM NaCl, 25 mM $NaHCO_3$, 4.7 mM KCl, 1.18 mM KH_2PO_4, 1.17 mM $MgSO_4$, 1.36 mM $CaCl_2$, and 5.5 mM glucose; pH 7.4 equilibrated with 95% O_2 and 5% CO_2) at 5 mL/min and maintained between 35–36 °C.

Electrically evoked action potentials were recorded with either one or two glass capillary microelectrodes filled with 3 M KCl (20–30 MΩ) connected to a Duo 773 Electrometer amplifier (World Precision Instruments, Aston, UK). Voltage was filtered at 10 kHz low pass, displayed on an oscilloscope, and transferred at 40 kHz A/D via a PowerLab 4/25 interface (ADInstruments, Chalgrove, UK) to a PC computer running Chart 5.0 software. Electrical stimuli consisted of 2 ms duration square wave pulses generated by a Master-9 programmable stimulator (AMPI, Jerusalem, Israel) through a WPI A360 stimulus isolation unit and delivered to the left atrial wall of the preparation via two fine shielded Ag/AgCl wires. Depending on the preparations, evoked action potentials were recorded in either of the superior PV, as previously described [3].

Analysis of action potential parameters was performed with the Peak Parameters Module of Chart 5 software. The threshold for the action potential was set at >2 mV from the diastolic membrane potential.

4.2. Whole Cell Patch Clamp Experiments

Cardiomyocytes were enzymatically isolated as previously described [37,48].

Background potassium currents were recorded from PV and LA cardiomyocytes plated onto glass coverslips and maintained at room temperature (22–25 °C). During the patch clamp experiments, the cells were locally superfused by gravity with a Tyrode solution containing (in mM): NaCl 140; KCl 5.37; $CaCl_2$ 1.36; $MgCl_2$ 1; NaH_2PO_4 0.33; HEPES 10; Glucose 10 (pH 7.4 with NaOH). The recording pipettes were pulled from borosilicate glass capillaries (Clark Electromedical Instruments, Edenbridge, UK) to a tip resistance of 2–4 MΩ. The pipette solution contained (in mM): DL-Aspartic acid K^+ salt 115; KCl 25; $MgCl_2$ 1; ATP-Mg 5; GTP Tris 0.1; NaH_2PO_4 5; EGTA 10; HEPES 10 (pH 7.3 with KOH). An Axopatch 200 A amplifier connected to a PC computer, running Clampex (pClamp 9 software, Axon Instruments, Union City, CA, USA) through a Digidata 1200 A interface was used to control voltage and record currents from a holding potential (HP) of −70 mV. Data were acquired at a sampling frequency of 10 kHz and 2 kHz low-pass filtered with an 8-pole Bessel filter. Membrane capacitance (Cm) was measured by integration of the capacitive currents in response to a series of 10 hyperpolarizing pulses applied from the HP (amplitude and duration: 10 mV, 10 ms) and then averaged. The pipette and cell capacitances were compensated by 80%.

4.3. Preparation of Tissue Homogenates and Western Blotting

Tissue lysates were prepared from flash-frozen two pooled rat PV and LA. Samples were homogenized in lysis buffer (in mM: 30 histidine, 250 sucrose, 1× protease inhibitor cocktail cOmplete ULTRA (Roche Diagnostics, Singapore), 0.6 PMSF and 1 DTT) on ice. Insoluble material was removed by centrifugation at 11,000× *g* for 15 min, and the protein concentration of the supernatants was quantified according to the Bradford protein assay.

Proteins (30 µg) were then separated by SDS-PAGE on 9% tris-glycine gels and transferred to PVDF membrane (0.45 µm; Millipore, Burlington, MA, USA) using a wet transfer unit. After blocking of the membrane with 7.5% BSA in 0.1% TBS-Tween for 1 h at room temperature, the proteins of interest were labeled by the primary antibodies anti-GIRK1 and anti-GIRK4 (rabbit polyclonals, Alomone Labs, Jerusalem, Israel) overnight at 4 °C. The membranes were then incubated 1 h at room temperature with anti-rabbit HRP conjugated (Jackson ImmunoResearch, Ely, Cambridgeshire, UK) secondary antibody. The immunoblots were developed with Luminata Forte substrate (Millipore) and G: Box Chemi XR5 chemiluminescence imaging system (Syngene, Bengaluru, India). The protein-signal densities were normalized to the corresponding β-actin-signal densities. The analysis of the western blotting images was performed using ImageJ software 1.53f.

4.4. Statistical Analysis

Statistical analysis was performed by one way ANOVA, Student's *t*-test, or Mann-Whitney test. Mood's median test was used for western blot experiments. *p* values are given except for values less than 0.001 which are assigned a value < 0.001.

4.5. Chemicals

Chemicals were of reagent grade and obtained from either Merck KG (Darmstadt, Germany) or Sigma-Aldrich (Saint Quentin Fallavier, France). TQ and P1075 were obtained from Tocris Bioscience (Bristol, UK).

Supplementary Materials: The following supporting information can be downloaded at: https://www.mdpi.com/article/10.3390/ijms241713629/s1.

Author Contributions: Conceptualization, I.F., V.M. and P.B.; methodology, I.F.; validation, I.F. and P.B.; formal analysis, I.F., C.P. and P.B.; investigation, I.F., P.B., C.P., A.Y.; resources, V.M.; data curation, I.F.; writing—original draft preparation, I.F and P.B.; writing—review and editing, I.F., P.B., C.P., A.Y. and V.M.; visualization, P.B. and I.F.; supervision, V.M.; project administration, V.M. All authors have read and agreed to the published version of the manuscript.

Funding: This research received no external funding.

Institutional Review Board Statement: The animal study protocol was approved by the local Animal Care and Use Committee (Comité d'Ethique en Expérimentation Animale Val de Loire, Tours, France, permit number 2022032119225830).

Data Availability Statement: All data generated or analysed during this study are available from the corresponding author on reasonable request.

Acknowledgments: We thank Anissa Terbah and Clément Bion for their technical support during patch-clamp experiments.

Conflicts of Interest: The authors declare no conflict of interest.

References

1. Haïssaguerre, M.; Jaïs, P.; Shah, D.C.; Takahashi, A.; Hocini, M.; Quiniou, G.; Garrigue, S.; Le Mouroux, A.; Le Métayer, P.; Clémenty, J. Spontaneous Initiation of Atrial Fibrillation by Ectopic Beats Originating in the Pulmonary Veins. *N. Engl. J. Med.* **1998**, *339*, 659–666. [CrossRef] [PubMed]
2. Bredeloux, P.; Pasqualin, C.; Bordy, R.; Maupoil, V.; Findlay, I. Automatic Activity Arising in Cardiac Muscle Sleeves of the Pulmonary Vein. *Biomolecules* **2021**, *12*, 23. [CrossRef]
3. Bredeloux, P.; Findlay, I.; Pasqualin, C.; Hocini, M.; Bernus, O.; Maupoil, V. Selective Inhibition of Electrical Conduction within the Pulmonary Veins by A1-Adrenergic Receptors Activation in the Rat. *Sci. Rep.* **2020**, *10*, 5390. [CrossRef] [PubMed]
4. Ragazzi, E.; Wu, S.N.; Shryock, J.; Belardinelli, L. Electrophysiological and Receptor Binding Studies to Assess Activation of the Cardiac Adenosine Receptor by Adenine Nucleotides. *Circ. Res.* **1991**, *68*, 1035–1044. [CrossRef] [PubMed]
5. Wang, X.; Liang, B.; Skibsbye, L.; Olesen, S.-P.; Grunnet, M.; Jespersen, T. GIRK Channel Activation via Adenosine or Muscarinic Receptors Has Similar Effects on Rat Atrial Electrophysiology. *J. Cardiovasc. Pharmacol.* **2013**, *62*, 192–198. [CrossRef]
6. Hibino, H.; Inanobe, A.; Furutani, K.; Murakami, S.; Findlay, I.; Kurachi, Y. Inwardly Rectifying Potassium Channels: Their Structure, Function, and Physiological Roles. *Physiol. Rev.* **2010**, *90*, 291–366. [CrossRef]
7. Dobrev, D.; Friedrich, A.; Voigt, N.; Jost, N.; Wettwer, E.; Christ, T.; Knaut, M.; Ravens, U. The G Protein-Gated Potassium Current I(K,ACh) Is Constitutively Active in Patients with Chronic Atrial Fibrillation. *Circulation* **2005**, *112*, 3697–3706. [CrossRef]
8. Cha, T.-J.; Ehrlich, J.R.; Chartier, D.; Qi, X.-Y.; Xiao, L.; Nattel, S. Kir3-Based Inward Rectifier Potassium Current: Potential Role in Atrial Tachycardia Remodeling Effects on Atrial Repolarization and Arrhythmias. *Circulation* **2006**, *113*, 1730–1737. [CrossRef]

9. Voigt, N.; Friedrich, A.; Bock, M.; Wettwer, E.; Christ, T.; Knaut, M.; Strasser, R.H.; Ravens, U.; Dobrev, D. Differential Phosphorylation-Dependent Regulation of Constitutively Active and Muscarinic Receptor-Activated IK,ACh Channels in Patients with Chronic Atrial Fibrillation. *Cardiovasc. Res.* **2007**, *74*, 426–437. [CrossRef]
10. Voigt, N.; Maguy, A.; Yeh, Y.-H.; Qi, X.; Ravens, U.; Dobrev, D.; Nattel, S. Changes in I K, ACh Single-Channel Activity with Atrial Tachycardia Remodelling in Canine Atrial Cardiomyocytes. *Cardiovasc. Res.* **2008**, *77*, 35–43. [CrossRef]
11. Rishal, I.; Porozov, Y.; Yakubovich, D.; Varon, D.; Dascal, N. Gbetagamma-Dependent and Gbetagamma-Independent Basal Activity of G Protein-Activated K+ Channels. *J. Biol. Chem.* **2005**, *280*, 16685–16694. [CrossRef] [PubMed]
12. Mintert, E.; Bösche, L.I.; Rinne, A.; Timpert, M.; Kienitz, M.-C.; Pott, L.; Bender, K. Generation of a Constitutive Na+-Dependent Inward-Rectifier Current in Rat Adult Atrial Myocytes by Overexpression of Kir3.4. *J. Physiol.* **2007**, *585*, 3–13. [CrossRef] [PubMed]
13. Suzuki, K.; Matsumoto, A.; Nishida, H.; Reien, Y.; Maruyama, H.; Nakaya, H. Termination of Aconitine-Induced Atrial Fibrillation by the KACh-Channel Blocker Tertiapin: Underlying Electrophysiological Mechanism. *J. Pharmacol. Sci.* **2014**, *125*, 406–414. [CrossRef]
14. Kienitz, M.-C.; Mintert-Jancke, E.; Hertel, F.; Pott, L. Differential Effects of Genetically-Encoded $G\beta\gamma$ Scavengers on Receptor-Activated and Basal Kir3.1/Kir3.4 Channel Current in Rat Atrial Myocytes. *Cell Signal.* **2014**, *26*, 1182–1192. [CrossRef]
15. Jin, W.; Klem, A.M.; Lewis, J.H.; Lu, Z. Mechanisms of Inward-Rectifier K+ Channel Inhibition by Tertiapin-Q. *Biochemistry* **1999**, *38*, 14294–14301. [CrossRef]
16. Drici, M.D.; Diochot, S.; Terrenoire, C.; Romey, G.; Lazdunski, M. The Bee Venom Peptide Tertiapin Underlines the Role of I(KACh) in Acetylcholine-Induced Atrioventricular Blocks. *Br. J. Pharmacol.* **2000**, *131*, 569–577. [CrossRef] [PubMed]
17. Kitamura, H.; Yokoyama, M.; Akita, H.; Matsushita, K.; Kurachi, Y.; Yamada, M. Tertiapin Potently and Selectively Blocks Muscarinic K^+ Channels in Rabbit Cardiac Myocytes. *J. Pharmacol. Exp. Ther.* **2000**, *293*, 196–205.
18. Koo, S.-H.; Wakili, R.; Heo, J.-H.; Chartier, D.; Kim, H.-S.; Kim, S.-J.; Lee, J.-W.; Qi, X.Y.; Nattel, S.; Cha, T.-J. Role of Constitutively Active Acetylcholine-Mediated Potassium Current in Atrial Contractile Dysfunction Caused by Atrial Tachycardia Remodelling. *Europace* **2010**, *12*, 1490–1497. [CrossRef]
19. Ehrlich, J.R.; Cha, T.-J.; Zhang, L.; Chartier, D.; Villeneuve, L.; Hébert, T.E.; Nattel, S. Characterization of a Hyperpolarization-Activated Time-Dependent Potassium Current in Canine Cardiomyocytes from Pulmonary Vein Myocardial Sleeves and Left Atrium. *J. Physiol.* **2004**, *557*, 583–597. [CrossRef]
20. Matsuda, T.; Takeda, K.; Ito, M.; Yamagishi, R.; Tamura, M.; Nakamura, H.; Tsuruoka, N.; Saito, T.; Masumiya, H.; Suzuki, T.; et al. Atria Selective Prolongation by NIP-142, an Antiarrhythmic Agent, of Refractory Period and Action Potential Duration in Guinea Pig Myocardium. *J. Pharmacol. Sci.* **2005**, *98*, 33–40. [CrossRef]
21. Yeh, Y.-H.; Ehrlich, J.R.; Qi, X.; Hébert, T.E.; Chartier, D.; Nattel, S. Adrenergic Control of a Constitutively Active Acetylcholine-Regulated Potassium Current in Canine Atrial Cardiomyocytes. *Cardiovasc. Res.* **2007**, *74*, 406–415. [CrossRef] [PubMed]
22. Jin, W.; Lu, Z. Synthesis of a Stable Form of Tertiapin: A High-Affinity Inhibitor for Inward-Rectifier K+ Channels. *Biochemistry* **1999**, *38*, 14286–14293. [CrossRef] [PubMed]
23. Kanjhan, R. Tertiapin-Q Blocks Recombinant and Native Large Conductance K+ Channels in a Use-Dependent Manner. *J. Pharmacol. Exp. Ther.* **2005**, *314*, 1353–1361. [CrossRef] [PubMed]
24. Michelakis, E.D.; Weir, E.K.; Wu, X.; Nsair, A.; Waite, R.; Hashimoto, K.; Puttagunta, L.; Knaus, H.G.; Archer, S.L. Potassium Channels Regulate Tone in Rat Pulmonary Veins. *Am. J. Physiol. Lung Cell. Mol. Physiol.* **2001**, *280*, L1138–L1147. [CrossRef]
25. Candia, S.; Garcia, M.L.; Latorre, R. Mode of Action of Iberiotoxin, a Potent Blocker of the Large Conductance Ca^{2+}-Activated K+ Channel. *Biophys. J.* **1992**, *63*, 583–590. [CrossRef]
26. Bhave, G.; Chauder, B.A.; Liu, W.; Dawson, E.S.; Kadakia, R.; Nguyen, T.T.; Lewis, L.M.; Meiler, J.; Weaver, C.D.; Satlin, L.M.; et al. Development of a Selective Small-Molecule Inhibitor of Kir1.1, the Renal Outer Medullary Potassium Channel. *Mol. Pharmacol.* **2011**, *79*, 42–50. [CrossRef]
27. Doisne, N.; Maupoil, V.; Cosnay, P.; Findlay, I. Catecholaminergic Automatic Activity in the Rat Pulmonary Vein: Electrophysiological Differences between Cardiac Muscle in the Left Atrium and Pulmonary Vein. *Am. J. Physiol. Heart Circ. Physiol.* **2009**, *297*, H102–H108. [CrossRef]
28. Calloe, K.; Goodrow, R.; Olesen, S.-P.; Antzelevitch, C.; Cordeiro, J.M. Tissue-Specific Effects of Acetylcholine in the Canine Heart. *Am. J. Physiol. Heart Circ. Physiol.* **2013**, *305*, H66–H75. [CrossRef]
29. Liang, B.; Nissen, J.D.; Laursen, M.; Wang, X.; Skibsbye, L.; Hearing, M.C.; Andersen, M.N.; Rasmussen, H.B.; Wickman, K.; Grunnet, M.; et al. G-Protein-Coupled Inward Rectifier Potassium Current Contributes to Ventricular Repolarization. *Cardiovasc. Res.* **2014**, *101*, 175–184. [CrossRef]
30. Anderson, A.; Kulkarni, K.; Marron Fernandez de Velasco, E.; Carlblom, N.; Xia, Z.; Nakano, A.; Martemyanov, K.A.; Tolkacheva, E.G.; Wickman, K. Expression and Relevance of the G Protein-Gated K+ Channel in the Mouse Ventricle. *Sci. Rep.* **2018**, *8*, 1192. [CrossRef]
31. Tsuneoka, Y.; Irie, M.; Tanaka, Y.; Sugimoto, T.; Kobayashi, Y.; Kusakabe, T.; Kato, K.; Hamaguchi, S.; Namekata, I.; Tanaka, H. Permissive Role of Reduced Inwardly-Rectifying Potassium Current Density in the Automaticity of the Guinea Pig Pulmonary Vein Myocardium. *J. Pharmacol. Sci.* **2017**, *133*, 195–202. [CrossRef] [PubMed]

32. Hamaguchi, S.; Tsuneoka, Y.; Tanaka, A.; Irie, M.; Tsuruta, M.; Nakayama, T.; Namekata, I.; Nada, M.; Aimoto, M.; Takahara, A.; et al. Manifestation of Automaticity in the Pulmonary-Vein Myocardium of Rats with Abdominal Aorto-Venocaval Shunt. *J. Pharmacol. Sci.* **2015**, *128*, 212–215. [CrossRef] [PubMed]
33. Ramu, Y.; Klem, A.M.; Lu, Z. Short Variable Sequence Acquired in Evolution Enables Selective Inhibition of Various Inward-Rectifier K+ Channels. *Biochemistry* **2004**, *43*, 10701–10709. [CrossRef] [PubMed]
34. Hilder, T.A.; Chung, S.-H. Conductance Properties of the Inwardly Rectifying Channel, Kir3.2: Molecular and Brownian Dynamics Study. *Biochim. Biophys. Acta* **2013**, *1828*, 471–478. [CrossRef]
35. Arora, R.; Ng, J.; Ulphani, J.; Mylonas, I.; Subacius, H.; Shade, G.; Gordon, D.; Morris, A.; He, X.; Lu, Y.; et al. Unique Autonomic Profile of the Pulmonary Veins and Posterior Left Atrium. *J. Am. Coll. Cardiol.* **2007**, *49*, 1340–1348. [CrossRef]
36. Touhara, K.K.; Wang, W.; MacKinnon, R. The GIRK1 Subunit Potentiates G Protein Activation of Cardiac GIRK1/4 Hetero-Tetramers. *eLife* **2016**, *5*, e15750. [CrossRef]
37. Malécot, C.O.; Bredeloux, P.; Findlay, I.; Maupoil, V. A TTX-Sensitive Resting Na+ Permeability Contributes to the Catecholaminergic Automatic Activity in Rat Pulmonary Vein. *J. Cardiovasc. Electrophysiol.* **2015**, *26*, 311–319. [CrossRef]
38. Cho, H.; Nam, G.B.; Lee, S.H.; Earm, Y.E.; Ho, W.K. Phosphatidylinositol 4,5-Bisphosphate Is Acting as a Signal Molecule in Alpha(1)-Adrenergic Pathway via the Modulation of Acetylcholine-Activated K(+) Channels in Mouse Atrial Myocytes. *J. Biol. Chem.* **2001**, *276*, 159–164. [CrossRef]
39. Meyer, T.; Wellner-Kienitz, M.C.; Biewald, A.; Bender, K.; Eickel, A.; Pott, L. Depletion of Phosphatidylinositol 4,5-Bisphosphate by Activation of Phospholipase C-Coupled Receptors Causes Slow Inhibition but Not Desensitization of G Protein-Gated Inward Rectifier K+ Current in Atrial Myocytes. *J. Biol. Chem.* **2001**, *276*, 5650–5658. [CrossRef]
40. Bender, K.; Wellner-Kienitz, M.-C.; Pott, L. Transfection of a Phosphatidyl-4-Phosphate 5-Kinase Gene into Rat Atrial Myocytes Removes Inhibition of GIRK Current by Endothelin and α-Adrenergic Agonists. *FEBS Lett.* **2002**, *529*, 356–360. [CrossRef]
41. Cho, H.; Lee, D.; Lee, S.H.; Ho, W.-K. Receptor-Induced Depletion of Phosphatidylinositol 4,5-Bisphosphate Inhibits Inwardly Rectifying K+ Channels in a Receptor-Specific Manner. *Proc. Natl. Acad. Sci. USA* **2005**, *102*, 4643–4648. [CrossRef]
42. Deng, W.; Mahajan, R.; Baumgarten, C.M.; Logothetis, D.E. The ICl,Swell Inhibitor DCPIB Blocks Kir Channels That Possess Weak Affinity for PIP2. *Pflugers Arch.* **2016**, *468*, 817–824. [CrossRef] [PubMed]
43. Patel, D.; Kuyucak, S.; Doupnik, C.A. Structural Determinants Mediating Tertiapin Block of Neuronal Kir3.2 Channels. *Biochemistry* **2020**, *59*, 836–850. [CrossRef]
44. Egorov, Y.V.; Lang, D.; Tyan, L.; Turner, D.; Lim, E.; Piro, Z.D.; Hernandez, J.J.; Lodin, R.; Wang, R.; Schmuck, E.G.; et al. Caveolae-Mediated Activation of Mechanosensitive Chloride Channels in Pulmonary Veins Triggers Atrial Arrhythmogenesis. *J. Am. Heart Assoc.* **2019**, *8*, e012748. [CrossRef] [PubMed]
45. Egorov, Y.V.; Rosenshtraukh, L.V.; Glukhov, A.V. Arrhythmogenic Interaction Between Sympathetic Tone and Mechanical Stretch in Rat Pulmonary Vein Myocardium. *Front. Physiol.* **2020**, *11*, 237. [CrossRef]
46. Santangeli, P.; Lin, D. Catheter Ablation of Paroxysmal Atrial Fibrillation: Have We Achieved Cure with Pulmonary Vein Isolation? *Methodist Debakey Cardiovasc. J.* **2015**, *11*, 71–75. [CrossRef] [PubMed]
47. Bond, R.; Olshansky, B.; Kirchhof, P. Recent Advances in Rhythm Control for Atrial Fibrillation. *F1000Research* **2017**, *6*, 1796. [CrossRef]
48. Pasqualin, C.; Yu, A.; Malécot, C.O.; Gannier, F.; Cognard, C.; Godin-Ribuot, D.; Morand, J.; Bredeloux, P.; Maupoil, V. Structural Heterogeneity of the Rat Pulmonary Vein Myocardium: Consequences on Intracellular Calcium Dynamics and Arrhythmogenic Potential. *Sci. Rep.* **2018**, *8*, 3244. [CrossRef]

International Journal of
Molecular Sciences

Review

The Role of TRPM4 in Cardiac Electrophysiology and Arrhythmogenesis

Yaopeng Hu *, Jiehui Cang, Keizo Hiraishi, Takayuki Fujita and Ryuji Inoue *

Department of Physiology, Fukuoka University School of Medicine, Fukuoka 814-0180, Japan
* Correspondence: huyaopeng@fukuoka-u.ac.jp (Y.H.); inouery@fukuoka-u.ac.jp (R.I.)

Abstract: The transient receptor potential melastatin 4 (TRPM4) channel is a non-selective cation channel that activates in response to increased intracellular Ca^{2+} levels but does not allow Ca^{2+} to pass through directly. It plays a crucial role in regulating diverse cellular functions associated with intracellular Ca^{2+} homeostasis/dynamics. TRPM4 is widely expressed in the heart and is involved in various physiological and pathological processes therein. Specifically, it has a significant impact on the electrical activity of cardiomyocytes by depolarizing the membrane, presumably via Na^+ loading. The TRPM4 channel likely contributes to the development of cardiac arrhythmias associated with specific genetic backgrounds and cardiac remodeling. This short review aims to overview what is known so far about the TRPM4 channel in cardiac electrophysiology and arrhythmogenesis, highlighting its potential as a novel therapeutic target to effectively prevent and treat cardiac arrhythmias.

Keywords: TRPM4 channel; Ca^{2+} homeostasis; cardiac electrophysiology; arrhythmogenesis

Citation: Hu, Y.; Cang, J.; Hiraishi, K.; Fujita, T.; Inoue, R. The Role of TRPM4 in Cardiac Electrophysiology and Arrhythmogenesis. *Int. J. Mol. Sci.* **2023**, *24*, 11798. https://doi.org/10.3390/ijms241411798

Academic Editor: Joachim Neumann

Received: 30 June 2023
Revised: 20 July 2023
Accepted: 20 July 2023
Published: 22 July 2023

1. Introduction

The transient receptor potential melastatin 4 (TRPM4) channel is a Ca^{2+}-activated non-selective cationic (NSca) channel, with a unitary conductance of approximately 20pS [1]. The TRPM4 channel protein is ubiquitously expressed in many kinds of cells, where it participates in intracellular Ca^{2+} homeostasis and modify excitability by influencing the membrane potential. Among the widespread expressed tissues, TRPM4 protein is also abundant in the heart [2]. Several lines of evidence suggest that TRPM4 channels may be involved in arrhythmogenicity in the heart, with both inherited and acquired traits. For example, a gain-of-function mutation (GOF) on its distal N-terminal domain has been reported to produce degenerative changes in the cardiac Purkinje system, and has been identified in a few pedigrees of Jewish and French families that manifest progressive conduction blocks and associated sudden death [3]. In spontaneously hypertensive rats (SHRs), long-term pressure overload produces hypertrophic changes in the heart accompanied by upregulation of TRPM4 channel proteins and their excessive activities [4]. These changes are further associated with the prolongation of QT interval in electrocardiogram, which is a risk factor for lethal arrhythmias. In murine hearts exposed to acute anoxic insults, early afterdepolarization (EAD)-like oscillations in the repolarization phase of action potential (AP) occur, which are selectively inhibited by a TRPM4 channel blocker, 9-phenathrol [5]. All these observations are consistent with the idea that the TRPM4 channel may play non-trivial roles in arrhythmogenesis.

This review succinctly summarizes the characteristic properties of the TRPM4 channel and discusses its implications in cardiac arrhythmogenicity.

2. Biophysical Properties of TRPM4

The TRPM4 channel has a large cytosolic domain. Until 2017, the structure of the TRPM4 channel remained unknown. However, since then, several TRPM4 structures have been resolved in different ligand-bound states using single-particle cryo-electron microscopy (EM). All of these structures were determined for the closed state of the TRPM4 channel, while its open-state conformation is still elusive [6–9].

Like many other TRPM subfamily members, TRPM4 has an N-terminal TRPM homology region (MHR) domain, a transmembrane domain (TMD) that includes six transmembrane helices, a C-terminal coiled-coil domain, a TRP helix, and a C-terminal domain (CTD). The ion-conducting pore domain is formed by the transmembrane S5 and S6 helices and surrounded by the S1–S4 domain. These two crucial domains are connected through the S4–S5 linker, forming a domain-swapped conformation that may play an important role in the gating of the TRPM4 channel [6]. The S1–S4 domain of the TRPM4 channel contains binding sites for Ca^{2+} and other ligands [8], as well as a voltage sensor-like domain (VSLD) reminiscent of the counterpart of classical voltage-gated ion channels [10]. However, the VSLD of TRP channels shows a small number of gating charges; in TRPM4, it is only −0.7e [11], which may be mediated by polar residues in and around the putative fourth transmembrane domain. Nonetheless, this charge appears important for the channel's voltage dependence, though more compelling evidence such as from mutagenesis studies is necessary.

While not primary itself, the membrane depolarization strongly modulates the TRPM4 channel's activity; hence, the TRPM4 channel shows a prominent dependence on the physiological range of membrane potentials, once activated by an increase in the intracellular Ca^{2+} level ($[Ca^{2+}]_i$). This is similar to other types of Ca^{2+}-activated channels, such as BK_{Ca} [12]. The voltage dependence of the TRPM4 channel is an important feature in cardiac tissues, where it affects the membrane potential over a wide range. Incorporation of this property into simulation models shows that the TRPM4 channel is most prominently activated at the late repolarization phase of action potentials (APs) and when the resting $[Ca^{2+}]_i$ is high (in the high submicromolar and micromolar ranges), and is substantially active even near the maximum diastolic potential (MDP) [13]. Therefore, TRPM4 may contribute primarily to the prolongation of APs and, to a lesser extent, to the generation of abnormal diastolic depolarizations that can occur after complete termination of APs. While TRPM4 may not be a major contributor to pace-making potentials, it can still play a role in modulating cardiac excitability and arrhythmogenesis.

The Ca^{2+} binding sites of the TRPM4 channel are thought to exist within a hydrophilic pocket (Glu^{828}, Gln^{831}, Asn^{865}) of the cytosolic parts of S2 and S3 transmembrane helices. Beneath these binding sites is a very narrow space that allows cytosolic Ca^{2+} to pass through, between the S2-S3 linker and the TRP domain. One of the only two positively charged amino acid residues of the S4 domain (Arg^{905}) is located just above the S2–S3 linker [6]. This positioning may facilitate coordination and priming effects for voltage-dependent opening. It has also been reported that phosphatidylinositol bisphosphate (PIP_2) binding as well as calmodulin interaction are involved in maintaining and enhancing the Ca^{2+} sensitivity of the TRPM4 channel [14]. Therefore, understanding and evaluating the Ca^{2+} sensitivity of the TRPM4 channel at physiological or submicromolar Ca^{2+} levels is very important in elucidating its exact roles in arrhythmogenicity [13].

3. Physiological Roles of TRPM4 in the Heart

(1) Contribution of TRPM4 to sinoatrial (SA) nodal and other cardiac automaticity

In the SA node (SAN), intracellular Ca^{2+} oscillations generated by the Ca^{2+} clock play a critical role in regulating cardiac automaticity. During the diastolic depolarization phase, the voltage-dependent Cav1.3 L-type Ca^{2+} channels (LTCC) activate, then Ca^{2+} influx into the SAN cell triggers Ca^{2+} release from the sarcoplasmic reticulum (SR) via the ryanodine receptors (RyR) [15]. A subsequent rise in $[Ca^{2+}]_i$ activates the Na^+-Ca^{2+} exchange (NCX), which in turn extrudes Ca^{2+} from the cell in exchange for Na^+. This generates an inward Na^+ current that contributes to the depolarization of the membrane potential and increases the rate of pace-making diastolic depolarization (DD) [16].

TRPM4 channels possibly contribute to the regulation of cardiac automaticity by modulating both the Ca^{2+} clock and membrane clock in the SAN [17]. It has been suggested that the TRPM4 channel may contribute to the inward Na^+ current during the DD phase [18]. The TRPM4 channel may also have additional effects on the cardiac automaticity, such as modulating the inward driving force for Ca^{2+} and $[Ca^{2+}]_i$ [19]. It has been reported that inhibition of TRPM4 channels by 9-phenanthrol reduces the heart rate in mice, rats and rabbits, suggesting that the TRPM4 channel may act as an accelerator of DD when the heart rate is decreased, to avoid bradycardia [18,20].

Overall, the TRPM4 channel plays a critical role in regulating cardiac automaticity by modulating the DD slope and $[Ca^{2+}]_i$ in the SAN, and its precise contribution to the inward Na^+ current and the DD phase is still an ongoing focus of investigation.

(2) Role of TRPM4 channel in the atrial myocardium

The TRPM4 channel has been shown to be involved in atrial electrophysiology [21]. It is expressed in mice, rats, and human atrial cardiomyocytes [21–23]. The electrophysiological function of the TRPM4 channel in atrial APs was evaluated in isolated atrial cardiomyocytes by using TRPM4 knockout (*Trpm4 KO*) mice and a selective inhibitor, 9-phenanthrol. Inhibition of the TRPM4 channel shortened the AP duration of isolated atrial cardiomyocytes compared to those in wild-type (WT), but not knockout animals. The duration of atrial APs was also shorter in *Trpm4 KO* compared to WT animals [21].

The TRPM4 channel is reported to be responsive to shear stress induced by IP_3 receptor-mediated Ca^{2+} releases in rat atrial cardiomyocytes [23]. TRPM4 has also been implicated in aldosterone-induced atrial arrhythmias [24]. In the same study, disorganization of connexin-43 (Cx43) in atria was observed in *Trpm4 KO* mice, more than in WT mice. This phenomenon may be involved in the occurrence of atrial electrical disturbances. Additionally, TRPM4 channel may contribute to the growth of human and mice atrial fibroblasts; both expression and functional currents of TRPM4 has been shown to increase under cultured conditions [25]. Presumably, the TRPM4 channel may be engaged in some way commit to the process of atrial fibrosis [25,26].

In summary, the physiological role of TRPM4 in the atrial myocardium appears to be complex and context-dependent, with a variety of consequences observed under different conditions. It may thus be reasonable to speculate a nontrivial role(s) of TRPM4 in the pathogenesis of atrial fibrillation (AF), through alterations of both atrial electrophysiology and remodeling.

(3) Role of TRPM4 channel in the ventricular myocardium and Purkinje conduction system

The physiological role of the TRPM4 channel in the ventricular myocardium remains controversial. Discrepancies in the contribution of the TRPM4 channel between the atrium and ventricle has been reported, following comparisons of both the transcript level and functional current. Although highly detected in the atrium, the TRPM4 channel is much less expressed in the ventricular myocardium [27]. The definitive contribution of the TRPM4 channel to the AP duration of canine ventricular cardiomyocytes has been confirmed in the latest study by using a potent and highly selective inhibitor, 4-chloro-2-[[2-(2-chlorophenoxy) acetyl] amino] benzoic acid (CBA) [28]. In contrast, the *Trpm4* mRNA detection and functional Ca^{2+}-activated nonselective cation current were greatly enhanced in the ventricular cardiomyocytes of SHRs compared to those in the Wistar-Kyoto (WKY) rat [4]. The same study observed QT prolongation in SHRs, which is accompanied by an increased TRPM4 channel activity. Furthermore, the TRPM4 channel was involved in the positive inotropic effect of β-adrenergic stimulation in the ventricular myocardium [29].

The network of terminal Purkinje fibers (PFs) carries electrical impulses to the ventricular myocardium, playing a central role in the excitation-contraction cycle of the ventricle. PFs demonstrate a unique electrophysiology with complicated intracellular Ca^{2+} cycling [30]. The TRPM4 channel is most abundantly expressed in PFs, compared to the other human heart tissues [31], suggesting its significant contribution to the electrical properties of PFs [32]. The propagation failure of PFs caused by TRPM4 channel overexpression has been observed in silico [33]. More intriguingly, optical mapping of ectopic activation induced by mechanical stimulation in PFs demonstrated a clear link with the activation of

TRPM4 channels [34]. The involvement of the TRPM4 channel in the electrophysiology of PFs suggests its potential role in cardiac conduction and ventricular arrhythmias.

4. Involvement of TRPM4 in Cardiac Arrhythmias under Pathophysiological Conditions

(1) Pathophysiology of the TRPM4 channel under hypertrophic and remodeling conditions

Both *Trpm4* mRNA expression and TRPM4-like currents were found to be increased in hypertrophied ventricular cardiomyocytes from SHRs, compared to normotensive WKY rats [4]. These changes were also accompanied by abnormal prolongation of QT intervals. Thus, the authors speculated that, in hypertrophied hearts, overactivation of TRPM4 channels may act as a proarrhythmic substrate for early and delayed depolarizations. Subsequent studies employing genetically engineered TRPM4-transgenic mice and numerical simulations confirmed this possibility [13,35–37]. Intriguingly, using the same cardiomyocyte-specific *Trpm4 KO* mice [29], TRPM4 channels were shown to act differently as either negative or positive regulators for angiotensin II-induced cardiac hypertrophy [29] and pressure overload-induced cardiac hypertrophy [38]. At present, the reason for this discrepancy is unclear and awaits further investigation. It is likely that compromised Ca^{2+} homeostasis and Ca^{2+} overload, two prominent features of remodeling and stressed hearts, facilitate TRPM4 overactivation and concurrent arrhythmias (Figure 1). The TRPM4 channel specifically contributes to Ca^{2+} overload-induced background current that enhances ectopic excitability, and its nonspecific blocker, meclofenamate, significantly suppresses catecholaminergic polymorphic ventricular tachycardia (CPVT)-associated arrhythmias at doses appropriate for TRPM4 channel inhibition [39].

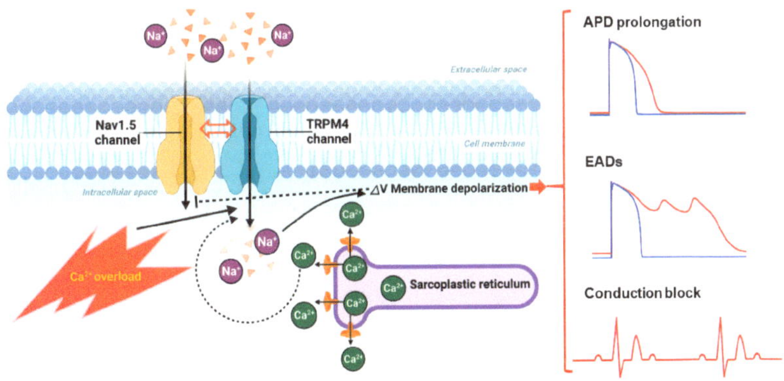

Figure 1. Ca^{2+} overloading conditions lead to TRPM4 overactivation inducing pleiotropic electrical abnormalities in cardiomyocytes. Under pathological conditions such as ischemia and heart failure, Ca^{2+} overload occurs with spontaneous Ca^{2+} release events [40], leading to the overactivation of TRPM4 channels. This may result in increased Na^+ influx through TRPM4 channels particularly during the late phase of AP or even after AP termination, which in turn causes AP prolongation as well as a depolarizing shift of the resting membrane potential. When this is combined with genetic mutations and/or tissue remodeling, both of which could facilitate TRPM4 overactivation, early after-depolarizations (EADs) can be triggered, or reduction of the conduction velocity or conduction block can ensue, due to substantial inactivation of Nav1.5 channels. Moreover, there is evidence for a functional coupling between TRPM4 and Nav1.5 channels [41], which may further complicate the consequences of TRPM4 overactivation. (More details are provided in the text).

Increased expression and activation of the TRPM4 channel could lead to higher susceptibility to hypertrophy- and stress-induced ventricular arrhythmias in association with aberrant Ca^{2+} homeostasis [42]. TRPM7, albeit much less abundant in the atrium than TRPM4, is significantly upregulated in human atrial myocytes isolated from the patients with atrial fibrillation [43]. Mice subjected to subcutaneous aldosterone infusion and high-salt diet underwent proarrhythmic changes in the atrium, such as enhanced triggered arrhythmias (EAD, DAD) and shortened AP, both of which were eliminated by ablation of the *Trpm4* gene. In contrast, complex morphologic changes occurred due to the *Trpm4* gene ablation, specifically, dilation of the left atrium and thickening of the septum and ventricular posterior wall. Intriguingly, only the latter two of these changes were abolished by aldosterone and high-salt treatment. The basal heart rate was insensitive to aldosterone and high salt in normal mice, but was responsive to it in the *Trpm4 KO* mice [24]. Although the mechanisms underlying remain unclear, the above findings strongly favor the vital participation of TRPM4 channels in electrical and remodeling.

(2) Role of TRPM4 in ischemia-reperfusion-related arrhythmias

Reperfusion of an ischemic heart could cause severe additional damage to the myocardium and induce lethal arrhythmias [44]. Excessive production of reactive oxygen species (ROS) and intracellular Ca^{2+} overload during ischemia-reperfusion play crucial roles in the genesis of both tissue injury and arrhythmias [45]. It has been reported that in the Langendorff ischemia-reperfusion model of rat hearts pretreated with the TRPM4 channel blocker 9-phenanthrol significantly improved contractile function and limited the infarcted area [46]. Furthermore, TRPM4 channel inhibition and silencing greatly suppressed ROS-induced injury on H9C2 cardiomyocytes [47]. The destructive contribution of TRPM4 to ROS-induced cardiac injury is potentially linked to the reduction of mitochondrial membrane potential and intracellular ATP level [48]. During ischemia-reperfusion, the electrical disturbances from PFs, where the TRPM4 channel is most abundantly and functionally expressed [32], account for the occurrence of ventricular arrhythmias [49]. The first report on the anti-arrhythmic effect of the TRPM4 channel-selective blocker 9-phenanthrol was from the treatment of arrhythmias induced by hypoxia and re-oxygenation in murine ventricles: perfusion with 9-phenanthrol demonstrated a spectacular dose-dependent abolishment of EADs [5]. Interestingly, an in vivo study of a mouse model, where acute ischemia was induced by ligation of the left anterior descending (LAD) coronary artery for 30 min, showed that *Trpm4 KO* mice are much less likely to develop ischemia-induced arrhythmias. In contrast, no significant difference in arrhythmic responses between *Trpm4 KO* mice and WT mice was observed after effective reperfusion [39].

5. TRPM4 Channelopathy in Inherited Cardiac Arrhythmias

Inherited cardiac arrhythmias are a group of genetic disorders that affect the electrical activity of the heart, to cause abnormal heart rhythms and potentially life-threatening arrhythmias. These disorders could be caused by mutations in the genes encoding ion channels that are responsible for controlling the flow of ions across the cell membrane and maintaining the normal rhythm of the heart [50]. Different kinds of inherited cardiac arrhythmias, including conduction blocks, Brugada syndrome (BrS), and long QT syndrome (LQTS), have been linked to dozens of *Trpm4* gene mutations, as evidenced by genetic linkage analyses and subsequent cohort studies (Table 1) [2,51,52]. It was often found that the same genotype variant of TRPM4 produces multiple phenotypes of arrhythmias while its multiple genotype mutations cause an arrhythmia with similar clinical features. The exact mechanism(s) underlying this redundancy is not yet fully understood.

Table 1. TRPM4 channelopathy associated with cardiac arrhythmias.

Amino Acid Alteration	Effect	Arrhythmia Phenotype	Underlying Mechanism	Reference
p.E7K	GOF	PFHBI	Impaired SUMOylation and increased surface expression/altered PIP_2 interaction	[31,53–55]
p.C20S	LOF	SUD	Decreased protein expression	[56]
p.A101T	LOF	CHB	Increased degradation rate/decreased protein expression	[57]
p.A101T/P1204L	LOF	IVF	Increased degradation rate/decreased protein expression	[57]
p.Q131H	Not determined	RBBB	Not determined	[58]
p.R144W	Not determined	BrS	Not determined	[52]
p.T160M	Not determined	LQTS	Involved with KCNQ1 mutation	[59]
p.R164W	GOF	RBBB	Impaired SUMOylation and increased surface expression	[3]
p.D198G	No effect	AVB	Not determined	[55]
p.T286T	LOF	PFHBI	Failure to produce functional protein	[60]
p.Q293R	Not determined	AVB	Not determined	[58]
p.I376T	GOF	PFHBI	Increased surface expression	[61]
p.A380V	LOF	SUD	Decreased surface expression/rapid desensitization	[56]
p.A432T	GOF/LOF	AVB/RBBB/BrS	Decreased surface expression or impaired SUMOylation and increased surface expression/delayed deactivation	[3,52,55,58]
p.A432T/G582S	LOF	AVB	Decreased surface expression	[55]
p.V441M	LOF	LQTS	Not determined	[51]
p.R499W	LOF	LQTS	Not determined	[51]
p.R499P	Not determined	LQTS	Not determined	[51]
p.W525X	LOF	BrS	absence of functional protein	[62]
p.G555R	Not determined	BrS	Not determined	[52]
p.G582S	GOF	AVB/RBBB/BrS	increased surface expression	[52,55,58]
p.L595V	LOF	SUD	Decreased surface expression	[56]
p.T677I	No effect	AVB	Not determined	[55]
p.G737R	Not determined	BrS	Not determined	[52]
p.F773I	Not determined	BrS	Not determined	[52]
p.P779R	LOF	BrS	Decreased protein expression	[52]
p.Y790H	Not determined	AVB	Not determined	[58]
p.G844D	GOF	LQTS/RBBB/BrS	Impaired SUMOylation and increased surface expression	[3,51,52]
p.Q854R	GOF	RBBB/BrS	Reduced degradation rate/increased protein expression	[52,57,58]
p.T873I	No effect	BrS	increased surface expression	[52]
p.K914R	GOF	AVB	Delayed deactivation	[58,63]
p.K914X	LOF	BrS	Decreased protein expression/decreased surface expression	[52]
p.V921I	No effect	AVB	Not determined	[55]
p.P970S	Not determined	RBBB	Not determined	[58]
p.S1044C	LOF	CHB	Increased degradation rate/decreased protein expression	[57]
p.L1075P	No effect	BrS	Increased surface expression	[52]
p.I1082S	LOF	SUD	Not determined	[56]
p.P1204L	Not determined	BrS	Not determined	[58]

PFHB1: progressive familial heart block type I; SUD: sudden unexpected death; CHB: complete heart block; IVF: idiopathic ventricular fibrillation; RBBB: right bundle branch block; BrS: Brugada syndrome; LQTS: long QT syndrome; AVB: atrioventricular block.

(1) TRPM4 variants in cardiac conduction block

The TRPM4-p.E7K mutant was the first to be identified as a point mutation of the TRPM4 channel in a few pedigrees of patient families that manifest progressive familial conduction block type I (PFHB I) [31]. Due to impaired deregulation of the small ubiquitin modifier conjunction (SUMOylation), the TRPM4-p.E7K mutant exhibited a GOF effect, associated with greater current density and increased protein expression at the plasma membrane [31]. It has also been reported that enhanced PIP_2 affinity and altered channel kinetics can facilitate its activation, especially around the resting membrane potential, presumably contributive to brady-arrhythmogenicity [53,54]. Several other GOF mutations of the TRM4 channel (p.R164W, p.A432T, and p.G844D) have been discovered in patient families suffering from heart block, which also involves with the same mechanism of reduced SUMOylation [3]. In contrast, electrophysiological kinetic analysis of the TRPM4-p.A432T mutant suggested that a four-fold slower deactivation, rather than excessive cell surface expression, accounted for the augmented membrane current [64]. Using a combination of electrophysiology, rapid treatments of intracellular Ca^{2+} with UV-flash photolysis, and molecular docking analysis, this research group further investigated another GOF mutant, TRPM4-p.K914R, identified from two patients with heart disease in the same family [63]. TRPM4-p.K914R demonstrated slower activation and deactivation kinetics leading to increased membrane currents. Furthermore, the 914th lysine residue was crucial as the nanoscopic interface between the S4-S5 linker, the MHR, and TRP-domain, which determines TRPM4's functional behavior [63].

The majority of TRPM4 variations related to familial conduction block are, thus far, GOF phenotypes. However, several loss-of-function (LOF) mutations of TRPM4 are also identified in inherited cardiac conduction defects [57,60]. The mechanism underlying such complex genotype-to-phenotype relationships is unclear. For instance, a LOF mutation, TRPM4-p.T286T, was found in patients with ventricular noncompaction and cardiac conduction disease, which eventually requires implantation of a permanent pacemaker [60]. This study revealed that inhibition of TRPM4 channel activity by its selective blocker 9-phenanthrol in human induced pluripotent stem cell–derived cardiomyocytes (hiPSC-CMs) resulted in decreased mRNA levels of HEY2, TBX5, and NKX2-5, transcription factors that play important roles in postnatal conduction system maturation [60,65]. This result is consistent with that of a gene invalidation study: *Trpm4 KO* mice displayed Luciani–Wenckebach atrioventricular blocks [66]. Taken together, the above lines of evidence support the view that downregulated function of the TRPM4 channel in immature hearts could influence the development of the conduction system and myocardial structures.

(2) TRPM4 variants in BrS

BrS is an inherited cardiac disease associated with a significant risk of lethal arrhythmias that lead to sudden death [67]. About 25% of BrS patients have the LOF mutations in the SCN5A gene that encode the α-subunit of the cardiac voltage-dependent sodium channel (Nav1.5) [68]. Among the other genes, TRPM4 variants explain 2.7% to 6% of total BrS cases [52,62]. However, the exact pathogenic mechanism by which both LOF and GOF mutations of TRPM4 can lead to BrS remain elusive [52].

By systematically analyzing the genotype-to-phenotype relationship of TRPM4 mutation for BrS, new evidence was found of LOF mutations of TRPM4, especially in a heterozygous status, that may not be sufficient to generate BrS [62]. However, an important physical interaction between TRPM4 and Nav1.5 was revealed: a significant decrease in the Nav1.5 current was observed in cardiomyocytes from *Trpm4 KO* mice [41]. Further studies are needed to understand how such interactions modulate cardiac electrical activity and the significance of the interactions in the pathogenesis of BrS.

(3) TRPM4 variants in LQTS

LQTS is a congenital and arrhythmogenic ion channel disorder characterized by QT prolongation in the electrocardiogram (ECG) [69]. It is potentially life-threatening because of delayed ventricular repolarization. Mutations in three major LQTS genes (KCNQ1, KCNH2, and SCN5A) were shown to impair the AP repolarization and contribute to approximately 75% of the disorder [70]. TRPM4 variants were also found to be involved in approximately 2% of LQTS cases [51], and two of these variants, p.R499W and p.V441M, were shown to display smaller TRPM4-mediated currents compared to those in wild type [51]. There is however a caveat against simply interpreting this result as the direct cause for abnormal delayed AP repolarization. It is suggested that the role of TRPM4 variants on QT interval might be multifactorial [51], as they may modulate other membrane currents as well. It is also possible that the impact of TRPM4 mutations may only become apparent in concert with the dysfunction of other AP formation channels such as KCNQ1, KCNJ2, and CACNA1C. These channels are known to alter the QT interval via a more complex mechanism than directly affecting the AP duration. For instance, a heterozygous

KCNQ1/TRPM4 dual mutation has been found in a LQTS patient, in whom verapamil treatment successfully reduced defibrillator discharge frequency by shortening the QT interval [71]. However, the underlying mechanism in the contribution of TRPM4 variants to LQTS still remains unclear.

6. Conclusions and Perspective

TRPM4 has recently emerged as a new factor in cardiac arrhythmogenesis of both acquired and inherited types. More than thirty single mutations in the *TRPM4* gene are now connected to risky arrhythmic manifestations, and functional upregulation/dysregulation of TRPM4 channels with ECG abnormalities is reported to be involved in pathological cardiac remodeling (in heart failure and other hypertrophic cardiomyopathies). To fully understand such multiplicity and complexity and to develop effective therapeutic interventions against them, mechanistically focused studies, that delve deeper than studies of genotype-phenotype associations, are needed. In this regard, there are several challenging but promising directions. For example, transgenic animal models can be more avidly combined with optogenetic approaches to deepen our tissue-level knowledge, in vivo or ex vivo [72,73]. The single nuclei RNA sequencing can be utilized to acquire more direct information about the geometrical heterogeneity of TRPM4 and the other arrhythmic substrates across the whole heart tissue [74], in which, appropriate heterocellular cardiac organoid models could be fabricated to disentangle the tissue-level complexity in more depth [75]. Continuous monitoring of genetically at-risk individuals, using wearable detection devices, may facilitate the uncovering of otherwise unknown triggers for TRPM4-associated arrhythmias [76]. Atomic-level structural analysis will also increase detailed knowledge of processes involved in the altered gating of the TRPM4 channel which predisposes individuals to arrhythmias [77]. All these lines of quantitative information could be integrated into multi-scale, multi-hierarchical numerical models to simulate the 2- or 3-dimensional propagation of excitation waves and their disturbances (reentries, blocks) in a more precise fashion, ultimately providing a useful tool for risk evaluation and in silico design of treatments.

The pathophysiological impact of TRPM4 dysfunction is further complicated by its tight link with $[Ca^{2+}]_i$ dynamics and possible modification of the other types of cardiac ion channels. Although the primary consequence of TRPM4 channel activation is presumed to prolong AP duration or QT intervals, it is also known that genetic deletion of this channel induces pleiotropic changes in the ECG [66]. In this respect, the latest finding that TRPM4 interacts with Nav1.5 [41] might offer an additional clue to deciphering why a single mutation of the *TRPM4* gene results in multiple phenotypes and vice versa, as observed in LQTS and BrS.

In preclinical studies using transgenic mice, a number of selective TRPM4 blockers have been shown to suppress arrhythmias [5,13,39,78], suggesting that targeting this channel may serve as a new therapeutic strategy. In most investigations, however, only low concentrations of 9-phenanthrol inhibited TRPM4 channels relatively selectively. Intriguingly, recent investigations identified two promising aryloxyacyl-anthranilic acid compounds, NBA and CBA, as novel TRPM4 antagonists. In particular, CBA exhibited a greater potency than 9-phenanthrol in inhibiting human TRPM4; however, had no effect on its mouse homologue [79]. These unique features may serve as a good starting point to design a new generation of TRPM4 blockers with better pharmacological profiles.

In future studies, advancements in gene editing technologies like CRISPR-Cas9 may enable direct modification of the TRPM4 gene in cells, tissues or organs. Gene editing could be utilized to correct deleterious mutations associated with TRPM4-related arrhythmias or to prevent the overactivation of TRPM4 channels in cardiac tissues undergoing pathological remodeling, which would otherwise increase the arrhythmogenicity.

Obviously, further detailed investigations about TRPM4 physiology and pathophysiology are needed to substantiate the utility of the above-mentioned therapeutic means and elucidate their true clinical benefits for arrhythmia treatments.

Author Contributions: Conceptualization, Y.H. and R.I.; Software, Y.H., J.C. and K.H.; Validation, Y.H. and R.I.; Writing—Original Draft Preparation, Y.H. and R.I.; Writing—Review & Editing, Y.H., T.F. and R.I.; Visualization, Y.H. and J.C.; Project Administration, Y.H. and R.I.; Funding Acquisition, Y.H. and R.I. All authors have read and agreed to the published version of the manuscript.

Funding: This work was supported by JSPS grants-in-aid to R.I. (No. 20K07269) and Y.H. (No. 20K16126), and by funding from Fukuoka University to Y.H. (No. GW2312).

Institutional Review Board Statement: Not applicable.

Informed Consent Statement: Not applicable.

Data Availability Statement: The research data described in this paper are available on request.

Acknowledgments: Part of Figure 1 was created with BioRender.com.

Conflicts of Interest: The authors declare no conflict of interest.

References

1. Launay, P.; Fleig, A.; Perraud, A.-L.; Scharenberg, A.M.; Penner, R.; Kinet, J.-P. TRPM4 is a Ca^{2+}-activated nonselective cation channel mediating cell membrane depolarization. *Cell* **2002**, *109*, 397–407. [PubMed]
2. Guinamard, R.; Bouvagnet, P.; Hof, T.; Liu, H.; Simard, C.; Sallé, L. TRPM4 in cardiac electrical activity. *Cardiovasc. Res.* **2015**, *108*, 21–30.
3. Liu, H.; El Zein, L.; Kruse, M.; Guinamard, R.; Beckmann, A.; Bozio, A.; Kurtbay, G.; Mégarbané, A.; Ohmert, I.; Blaysat, G. Gain-of-function mutations in TRPM4 cause autosomal dominant isolated cardiac conduction disease. *Circ. Cardiovasc. Genet.* **2010**, *3*, 374–385. [PubMed]
4. Guinamard, R.; Demion, M.; Magaud, C.; Potreau, D.; Bois, P. Functional expression of the TRPM4 cationic current in ventricular cardiomyocytes from spontaneously hypertensive rats. *Hypertension* **2006**, *48*, 587–594. [CrossRef]
5. Simard, C.; Sallé, L.; Rouet, R.; Guinamard, R. Transient receptor potential melastatin 4 inhibitor 9-phenanthrol abolishes arrhythmias induced by hypoxia and re-oxygenation in mouse ventricle. *Br. J. Pharmacol.* **2012**, *165*, 2354–2364.
6. Autzen, H.E.; Myasnikov, A.G.; Campbell, M.G.; Asarnow, D.; Julius, D.; Cheng, Y. Structure of the human TRPM4 ion channel in a lipid nanodisc. *Science* **2018**, *359*, 228–232.
7. Duan, J.; Li, Z.; Li, J.; Santa-Cruz, A.; Sanchez-Martinez, S.; Zhang, J.; Clapham, D.E. Structure of full-length human TRPM4. *Proc. Natl. Acad. Sci. USA* **2018**, *115*, 2377–2382. [PubMed]
8. Guo, J.; She, J.; Zeng, W.; Chen, Q.; Bai, X.-C.; Jiang, Y. Structures of the calcium-activated, non-selective cation channel TRPM4. *Nature* **2017**, *552*, 205–209.
9. Winkler, P.A.; Huang, Y.; Sun, W.; Du, J.; Lü, W. Electron cryo-microscopy structure of a human TRPM4 channel. *Nature* **2017**, *552*, 200–204.
10. Zhao, Y.; McVeigh, B.M.; Moiseenkova-Bell, V.Y. Structural pharmacology of TRP channels. *J. Mol. Biol.* **2021**, *433*, 166914.
11. Nilius, B.; Prenen, J.; Droogmans, G.; Voets, T.; Vennekens, R.; Freichel, M.; Wissenbach, U.; Flockerzi, V. Voltage dependence of the Ca^{2+}-activated cation channel TRPM4. *J. Biol. Chem.* **2003**, *278*, 30813–30820.
12. Lee, U.S.; Cui, J. BK channel activation: Structural and functional insights. *Trends Neurosci.* **2010**, *33*, 415–423. [PubMed]
13. Hu, Y.; Duan, Y.; Takeuchi, A.; Hai-Kurahara, L.; Ichikawa, J.; Hiraishi, K.; Numata, T.; Ohara, H.; Iribe, G.; Nakaya, M. Uncovering the arrhythmogenic potential of TRPM4 activation in atrial-derived HL-1 cells using novel recording and numerical approaches. *Cardiovasc. Res.* **2017**, *113*, 1243–1255. [PubMed]
14. Nilius, B.; Prenen, J.; Tang, J.; Wang, C.; Owsianik, G.; Janssens, A.; Voets, T.; Zhu, M.X. Regulation of the Ca^{2+} sensitivity of the nonselective cation channel TRPM4. *J. Biol. Chem.* **2005**, *280*, 6423–6433. [PubMed]
15. Torrente, A.G.; Mesirca, P.; Neco, P.; Rizzetto, R.; Dubel, S.; Barrere, C.; Sinegger-Brauns, M.; Striessnig, J.; Richard, S.; Nargeot, J. L-type Cav1. 3 channels regulate ryanodine receptor-dependent Ca^{2+} release during sino-atrial node pacemaker activity. *Cardiovasc. Res.* **2016**, *109*, 451–461. [PubMed]
16. Capel, R.A.; Terrar, D.A. The importance of Ca^{2+}-dependent mechanisms for the initiation of the heartbeat. *Front. Physiol.* **2015**, *6*, 80.
17. Lakatta, E.G.; Maltsev, V.A.; Vinogradova, T.M. A coupled SYSTEM of intracellular Ca^{2+} clocks and surface membrane voltage clocks controls the timekeeping mechanism of the heart's pacemaker. *Circ. Res.* **2010**, *106*, 659–673.
18. Hof, T.; Simard, C.; Rouet, R.; Sallé, L.; Guinamard, R. Implication of the TRPM4 nonselective cation channel in mammalian sinus rhythm. *Heart Rhythm* **2013**, *10*, 1683–1689. [CrossRef]
19. Abriel, H.; Syam, N.; Sottas, V.; Amarouch, M.Y.; Rougier, J.-S. TRPM4 channels in the cardiovascular system: Physiology, pathophysiology, and pharmacology. *Biochem. Pharmacol.* **2012**, *84*, 873–881.
20. Demion, M.; Bois, P.; Launay, P.; Guinamard, R. TRPM4, a Ca^{2+}-activated nonselective cation channel in mouse sino-atrial node cells. *Cardiovasc. Res.* **2007**, *73*, 531–538. [CrossRef]
21. Simard, C.; Hof, T.; Keddache, Z.; Launay, P.; Guinamard, R. The TRPM4 non-selective cation channel contributes to the mammalian atrial action potential. *J. Mol. Cell. Cardiol.* **2013**, *59*, 11–19. [PubMed]

22. Guinamard, R.; Chatelier, A.; Demion, M.; Potreau, D.; Patri, S.; Rahmati, M.; Bois, P. Functional characterization of a Ca^{2+}-activated non-selective cation channel in human atrial cardiomyocytes. *J. Physiol.* **2004**, *558*, 75–83. [PubMed]
23. Son, M.J.; Kim, J.C.; Kim, S.W.; Chidipi, B.; Muniyandi, J.; Singh, T.D.; So, I.; Subedi, K.P.; Woo, S.H. Shear stress activates monovalent cation channel transient receptor potential melastatin subfamily 4 in rat atrial myocytes via type 2 inositol 1, 4, 5-trisphosphate receptors and Ca^{2+} release. *J. Physiol.* **2016**, *594*, 2985–3004.
24. Simard, C.; Ferchaud, V.; Sallé, L.; Milliez, P.; Manrique, A.; Alexandre, J.; Guinamard, R. TRPM4 participates in aldosterone-salt-induced electrical atrial remodeling in mice. *Cells* **2021**, *10*, 636. [CrossRef]
25. Simard, C.; Magaud, C.; Adjlane, R.; Dupas, Q.; Sallé, L.; Manrique, A.; Bois, P.; Faivre, J.-F.; Guinamard, R. TRPM4 non-selective cation channel in human atrial fibroblast growth. *Pflügers Arch. -Eur. J. Physiol.* **2020**, *472*, 1719–1732.
26. Dienes, C.; Kovács, Z.M.; Hézső, T.; Almássy, J.; Magyar, J.; Bányász, T.; Nánási, P.P.; Horváth, B.; Szentandrássy, N. Pharmacological Modulation and (Patho) Physiological Roles of TRPM4 Channel—Part 2: TRPM4 in Health and Disease. *Pharmaceuticals* **2022**, *15*, 40.
27. Guinamard, R.; Demion, M.; Chatelier, A.; Bois, P. Calcium-activated nonselective cation channels in mammalian cardiomyocytes. *Trends Cardiovasc. Med.* **2006**, *16*, 245–250. [CrossRef]
28. Dienes, C.; Hézső, T.; Kiss, D.Z.; Baranyai, D.; Kovács, Z.M.; Szabó, L.; Magyar, J.; Bányász, T.; Nánási, P.P.; Horváth, B. Electrophysiological Effects of the Transient Receptor Potential Melastatin 4 Channel Inhibitor (4-Chloro-2-(2-chlorophenoxy) acetamido) Benzoic Acid (CBA) in Canine Left Ventricular Cardiomyocytes. *Int. J. Mol. Sci.* **2021**, *22*, 9499. [CrossRef]
29. Kecskés, M.; Jacobs, G.; Kerselaers, S.; Syam, N.; Menigoz, A.; Vangheluwe, P.; Freichel, M.; Flockerzi, V.; Voets, T.; Vennekens, R. The Ca $^{2+}$-activated cation channel TRPM4 is a negative regulator of angiotensin II-induced cardiac hypertrophy. *Basic Res. Cardiol.* **2015**, *110*, 43. [CrossRef]
30. Haissaguerre, M.; Vigmond, E.; Stuyvers, B.; Hocini, M.; Bernus, O. Ventricular arrhythmias and the His–Purkinje system. *Nat. Rev. Cardiol.* **2016**, *13*, 155–166.
31. Kruse, M.; Schulze-Bahr, E.; Corfield, V.; Beckmann, A.; Stallmeyer, B.; Kurtbay, G.; Ohmert, I.; Schulze-Bahr, E.; Brink, P.; Pongs, O. Impaired endocytosis of the ion channel TRPM4 is associated with human progressive familial heart block type I. *J. Clin. Investig.* **2009**, *119*, 2737–2744. [CrossRef]
32. Hof, T.; Sallé, L.; Coulbault, L.; Richer, R.; Alexandre, J.; Rouet, R.; Manrique, A.; Guinamard, R. TRPM4 non-selective cation channels influence action potentials in rabbit Purkinje fibres. *J. Physiol.* **2016**, *594*, 295–306. [CrossRef]
33. Gaur, N.; Hof, T.; Haissaguerre, M.; Vigmond, E.J. Propagation failure by TRPM4 overexpression. *Biophys. J.* **2019**, *116*, 469–476. [CrossRef]
34. Hurley, M.; Kaur, S.; Walton, R.; Power, A.; Haïssaguerre, M.; Bernus, O.; Ward, M.-L.; White, E. Endocardial role in arrhythmias induced by acute ventricular stretch and the involvement of Purkinje fibres, in isolated rat hearts. *Curr. Res. Physiol.* **2023**, *6*, 100098. [CrossRef] [PubMed]
35. Hedon, C.; Lambert, K.; Chakouri, N.; Thireau, J.; Aimond, F.; Cassan, C.; Bideaux, P.; Richard, S.; Faucherre, A.; Le Guennec, J.-Y. New role of TRPM4 channel in the cardiac excitation-contraction coupling in response to physiological and pathological hypertrophy in mouse. *Prog. Biophys. Mol. Biol.* **2021**, *159*, 105–117. [CrossRef] [PubMed]
36. Hu, Y.; Kaschitza, D.R.; Essers, M.; Arullampalam, P.; Fujita, T.; Abriel, H.; Inoue, R. Pathological activation of CaMKII induces arrhythmogenicity through TRPM4 overactivation. *Pflügers Arch.-Eur. J. Physiol.* **2021**, *473*, 507–519. [CrossRef] [PubMed]
37. Pironet, A.; Syam, N.; Vandewiele, F.; Van den Haute, C.; Kerselaers, S.; Pinto, S.; Vande Velde, G.; Gijsbers, R.; Vennekens, R. AAV9-mediated overexpression of TRPM4 increases the incidence of stress-induced ventricular arrhythmias in mice. *Front. Physiol.* **2019**, *10*, 802. [CrossRef]
38. Guo, Y.; Yu, Z.-Y.; Wu, J.; Gong, H.; Kesteven, S.; Iismaa, S.E.; Chan, A.Y.; Holman, S.; Pinto, S.; Pironet, A. The Ca^{2+}-activated cation channel TRPM4 is a positive regulator of pressure overload-induced cardiac hypertrophy. *eLife* **2021**, *10*, e66582. [CrossRef]
39. Vandewiele, F.; Pironet, A.; Jacobs, G.; Kecskés, M.; Wegener, J.; Kerselaers, S.; Hendrikx, L.; Verelst, J.; Philippaert, K.; Oosterlinck, W. TRPM4 inhibition by meclofenamate suppresses Ca^{2+}-dependent triggered arrhythmias. *Eur. Heart J.* **2022**, *43*, 4195–4207. [CrossRef] [PubMed]
40. Györke, S.; Carnes, C. Dysregulated sarcoplasmic reticulum calcium release: Potential pharmacological target in cardiac disease. *Pharmacol. Ther.* **2008**, *119*, 340–354. [CrossRef]
41. Ozhathil, L.C.; Rougier, J.-S.; Arullampalam, P.; Essers, M.C.; Ross-Kaschitza, D.; Abriel, H. Deletion of Trpm4 alters the function of the Nav1. 5 channel in murine cardiac myocytes. *Int. J. Mol. Sci.* **2021**, *22*, 3401. [CrossRef]
42. Robson, A. A nonsteroidal anti-inflammatory drug suppresses Ca^{2+}-dependent triggered arrhythmias. *Nat. Cardiovasc. Res.* **2022**, *1*, 697. [CrossRef]
43. Zhang, Y.-H.; Sun, H.-Y.; Chen, K.-H.; Du, X.-L.; Liu, B.; Cheng, L.-C.; Li, X.; Jin, M.-W.; Li, G.-R. Evidence for functional expression of TRPM7 channels in human atrial myocytes. *Basic Res. Cardiol.* **2012**, *107*, 282. [CrossRef] [PubMed]
44. Neri, M.; Riezzo, I.; Pascale, N.; Pomara, C.; Turillazzi, E. Ischemia/reperfusion injury following acute myocardial infarction: A critical issue for clinicians and forensic pathologists. *Mediat. Inflamm.* **2017**, *2017*, 7018393. [CrossRef]
45. Perrelli, M.-G.; Pagliaro, P.; Penna, C. Ischemia/reperfusion injury and cardioprotective mechanisms: Role of mitochondria and reactive oxygen species. *World J. Cardiol.* **2011**, *3*, 186. [CrossRef] [PubMed]
46. Wang, J.; Takahashi, K.; Piao, H.; Qu, P.; Naruse, K. 9-Phenanthrol, a TRPM4 inhibitor, protects isolated rat hearts from ischemia–reperfusion injury. *PLoS ONE* **2013**, *8*, e70587. [CrossRef]

47. Piao, H.; Takahashi, K.; Yamaguchi, Y.; Wang, C.; Liu, K.; Naruse, K. Transient receptor potential melastatin-4 is involved in hypoxia-reoxygenation injury in the cardiomyocytes. *PLoS ONE* **2015**, *10*, e0121703. [CrossRef]
48. Wang, C.; Chen, J.; Wang, M.; Naruse, K.; Takahashi, K. Role of the TRPM4 channel in mitochondrial function, calcium release, and ROS generation in oxidative stress. *Biochem. Biophys. Res. Commun.* **2021**, *566*, 190–196. [CrossRef]
49. Ideker, R.E.; Kong, W.; Pogwizd, S. Purkinje fibers and arrhythmias. *Pacing Clin. Electrophysiol. PACE* **2009**, *32*, 283. [CrossRef]
50. Schwartz, P.J.; Ackerman, M.J.; Antzelevitch, C.; Bezzina, C.R.; Borggrefe, M.; Cuneo, B.F.; Wilde, A.A. Inherited cardiac arrhythmias. *Nat. Rev. Dis. Primers* **2020**, *6*, 58. [CrossRef] [PubMed]
51. Hof, T.; Liu, H.; Sallé, L.; Schott, J.-J.; Ducreux, C.; Millat, G.; Chevalier, P.; Probst, V.; Guinamard, R.; Bouvagnet, P. TRPM4 non-selective cation channel variants in long QT syndrome. *BMC Med. Genet.* **2017**, *18*, 31. [CrossRef] [PubMed]
52. Liu, H.; Chatel, S.; Simard, C.; Syam, N.; Salle, L.; Probst, V.; Morel, J.; Millat, G.; Lopez, M.; Abriel, H. Molecular genetics and functional anomalies in a series of 248 Brugada cases with 11 mutations in the TRPM4 channel. *PLoS ONE* **2013**, *8*, e54131. [CrossRef] [PubMed]
53. Hu, Y.; Li, Q.; Kurahara, L.-H.; Shioi, N.; Hiraishi, K.; Fujita, T.; Zhu, X.; Inoue, R. An Arrhythmic mutation E7K facilitates TRPM4 channel activation via enhanced PIP2 interaction. *Cells* **2021**, *10*, 983. [CrossRef]
54. Hu, Y.; Li, Q.; Shen, Y.; Fujita, T.; Zhu, X.; Inoue, R. Theoretical Investigation of the Mechanism by which A Gain-of-Function Mutation of the TRPM4 Channel Causes Conduction Block. *Int. J. Mol. Sci.* **2021**, *22*, 8513. [CrossRef]
55. Syam, N.; Chatel, S.; Ozhathil, L.C.; Sottas, V.; Rougier, J.S.; Baruteau, A.; Baron, E.; Amarouch, M.Y.; Daumy, X.; Probst, V. Variants of transient receptor potential melastatin member 4 in childhood atrioventricular block. *J. Am. Heart Assoc.* **2016**, *5*, e001625. [CrossRef]
56. Subbotina, E.; Williams, N.; Sampson, B.A.; Tang, Y.; Coetzee, W.A. Functional characterization of TRPM4 variants identified in sudden unexpected natural death. *Forensic Sci. Int.* **2018**, *293*, 37–46. [CrossRef]
57. Bianchi, B.; Ozhathil, L.C.; Medeiros-Domingo, A.; Gollob, M.H.; Abriel, H. Four TRPM4 cation channel mutations found in cardiac conduction diseases lead to altered protein stability. *Front. Physiol.* **2018**, *9*, 177. [CrossRef]
58. Stallmeyer, B.; Zumhagen, S.; Denjoy, I.; Duthoit, G.; Hébert, J.L.; Ferrer, X.; Maugenre, S.; Schmitz, W.; Kirchhefer, U.; Schulze-Bahr, E. Mutational spectrum in the Ca^{2+}-activated cation channel gene TRPM4 in patients with cardiac conductance disturbances. *Hum. Mutat.* **2012**, *33*, 109–117. [CrossRef]
59. Zhao, Y.; Feng, M.; Shang, L.-X.; Sun, H.-x.; Zhou, X.-H.; Lu, Y.-M.; Zhang, L.; Xing, Q.; Li, Y.-d.; Tang, B.-P. KCNQ1 G219E and TRPM4 T160M polymorphisms are involved in the pathogenesis of long QT syndrome: A case report. *Medicine* **2021**, *100*, e24032. [CrossRef] [PubMed]
60. Saito, Y.; Nakamura, K.; Nishi, N.; Igawa, O.; Yoshida, M.; Miyoshi, T.; Watanabe, A.; Morita, H.; Ito, H. TRPM4 mutation in patients with ventricular noncompaction and cardiac conduction disease. *Circ. Genom. Precis. Med.* **2018**, *11*, e002103. [CrossRef] [PubMed]
61. Daumy, X.; Amarouch, M.-Y.; Lindenbaum, P.; Bonnaud, S.; Charpentier, E.; Bianchi, B.; Nafzger, S.; Baron, E.; Fouchard, S.; Thollet, A. Targeted resequencing identifies TRPM4 as a major gene predisposing to progressive familial heart block type I. *Int. J. Cardiol.* **2016**, *207*, 349–358. [CrossRef] [PubMed]
62. Janin, A.; Bessière, F.; Georgescu, T.; Chanavat, V.; Chevalier, P.; Millat, G. TRPM4 mutations to cause autosomal recessive and not autosomal dominant Brugada type 1 syndrome. *Eur. J. Med. Genet.* **2019**, *62*, 103527. [CrossRef] [PubMed]
63. Xian, W.; Wang, H.; Moretti, A.; Laugwitz, K.-L.; Flockerzi, V.; Lipp, P. Domain zipping and unzipping modulates TRPM4's properties in human cardiac conduction disease. *Biophys. J.* **2020**, *118*, 21a–22a. [CrossRef]
64. Xian, W.; Hui, X.; Tian, Q.; Wang, H.; Moretti, A.; Laugwitz, K.-L.; Flockerzi, V.; Ruppenthal, S.; Lipp, P. Aberrant deactivation-induced gain of function in TRPM4 mutant is associated with human cardiac conduction block. *Cell Rep.* **2018**, *24*, 724–731. [CrossRef]
65. Asatryan, B.; Medeiros-Domingo, A. Molecular and genetic insights into progressive cardiac conduction disease. *EP Eur.* **2019**, *21*, 1145–1158. [CrossRef]
66. Demion, M.; Thireau, J.; Gueffier, M.; Finan, A.; Khoueiry, Z.; Cassan, C.; Serafini, N.; Aimond, F.; Granier, M.; Pasquié, J.-L. Trpm4 gene invalidation leads to cardiac hypertrophy and electrophysiological alterations. *PLoS ONE* **2014**, *9*, e115256. [CrossRef]
67. Brugada, P.; Brugada, J. Right bundle branch block, persistent ST segment elevation and sudden cardiac death: A distinct clinical and electrocardiographic syndrome: A multicenter report. *J. Am. Coll. Cardiol.* **1992**, *20*, 1391–1396. [CrossRef]
68. Veerman, C.C.; Wilde, A.A.; Lodder, E.M. The cardiac sodium channel gene SCN5A and its gene product NaV1. 5: Role in physiology and pathophysiology. *Gene* **2015**, *573*, 177–187. [CrossRef]
69. Mizusawa, Y.; Horie, M.; Wilde, A.A. Genetic and clinical advances in congenital long QT syndrome. *Circ. J.* **2014**, *78*, 2827–2833. [CrossRef]
70. Tester, D.J.; Ackerman, M.J. Genetics of long QT syndrome. *Methodist DeBakey Cardiovasc. J.* **2014**, *10*, 29. [CrossRef]
71. Wang, F.; Han, Y.; Sang, W.; Wang, L.; Liang, X.; Wang, L.; Xing, Q.; Guo, Y.; Zhang, J.; Zhang, L. In Vitro Drug Screening Using iPSC-Derived Cardiomyocytes of a Long QT-Syndrome Patient Carrying KCNQ1 & TRPM4 Dual Mutation: An Experimental Personalized Treatment. *Cells* **2022**, *11*, 2495. [PubMed]
72. Bruegmann, T.; Smith, G.L.; Lehnart, S.E. Cardiac optogenetics: Using light to observe and excite the heart. *Front. Physiol.* **2022**, *13*, 1031062. [CrossRef] [PubMed]
73. Entcheva, E.; Kay, M.W. Cardiac optogenetics: A decade of enlightenment. *Nat. Rev. Cardiol.* **2021**, *18*, 349–367. [CrossRef] [PubMed]
74. Tucker, N.R.; Chaffin, M.; Fleming, S.J.; Hall, A.W.; Parsons, V.A.; Bedi Jr, K.C.; Akkad, A.-D.; Herndon, C.N.; Arduini, A.; Papangeli, I. Transcriptional and cellular diversity of the human heart. *Circulation* **2020**, *142*, 466–482. [CrossRef] [PubMed]

75. Zhao, D.; Lei, W.; Hu, S. Cardiac organoid—A promising perspective of preclinical model. *Stem Cell Res. Ther.* **2021**, *12*, 272. [CrossRef] [PubMed]
76. Hughes, A.; Shandhi, M.M.H.; Master, H.; Dunn, J.; Brittain, E. Wearable devices in cardiovascular medicine. *Circ. Res.* **2023**, *132*, 652–670. [CrossRef]
77. Jiang, D.; Zhang, J.; Xia, Z. Structural Advances in Voltage-Gated Sodium Channels. *Front. Pharmacol.* **2022**, *13*, 908867. [CrossRef]
78. Guinamard, R.; Hof, T.; Del Negro, C. The TRPM 4 channel inhibitor 9-phenanthrol. *Br. J. Pharmacol.* **2014**, *171*, 1600–1613. [CrossRef]
79. Arullampalam, P.; Preti, B.; Ross-Kaschitza, D.; Lochner, M.; Rougier, J.-S.; Abriel, H. Species-specific effects of cation channel TRPM4 small-molecule inhibitors. *Front. Pharmacol.* **2021**, *12*, 712354. [CrossRef]

International Journal of *Molecular Sciences*

MDPI

Article

Bioinformatic Identification of Potential RNA Alterations on the Atrial Fibrillation Remodeling from Human Pulmonary Veins

Wataru Igarashi [1,†], Daichi Takagi [1,†], Daigo Okada [2], Daiki Kobayashi [3], Miho Oka [4], Toshiro Io [4], Kuniaki Ishii [5], Kyoichi Ono [3], Hiroshi Yamamoto [1] and Yosuke Okamoto [3,*]

1 Department of Cardiovascular Surgery, Akita University Graduate School of Medicine, 1-1-1 Hondo, Akita 010-8543, Japan
2 Center for Genomic Medicine, Graduate School of Medicine, Kyoto University, Shogoinkawahara-cho, Kyoto 606-8507, Japan
3 Department of Cell Physiology, Akita University Graduate School of Medicine, 1-1-1 Hondo, Akita 010-8543, Japan
4 Research Department, Ono Pharmaceutical Co., Ltd., Kyutaromachi, Osaka 541-0056, Japan
5 Department of Pharmacology, Faculty of Medicine, Yamagata University, Iida-Nishi, Yamagata 990-9585, Japan
* Correspondence: okamoto@med.akita-u.ac.jp
† These authors contributed equally to this paper.

Abstract: Atrial fibrillation (AF) is the most frequent persistent arrhythmia. Many genes have been reported as a genetic background for AF. However, most transcriptome analyses of AF are limited to the atrial samples and have not been evaluated by multiple cardiac regions. In this study, we analyzed the expression levels of protein-coding and long noncoding RNAs (lncRNAs) in six cardiac regions by RNA-seq. Samples were donated from six subjects with or without persistent AF for left atria, left atrial appendages, right atria, sinoatrial nodes, left ventricles, right ventricles, and pulmonary veins (PVs), and additional four right atrial appendages samples were collected from patients undergoing mitral valve replacement. In total, 23 AF samples were compared to 23 non-AF samples. Surprisingly, the most influenced heart region in gene expression by AF was the PV, not the atria. The ion channel-related gene set was significantly enriched upon analysis of these significant genes. In addition, some significant genes are cancer-related lncRNAs in PV in AF. A co-expression network analysis could detect the functional gene clusters. In particular, the cancer-related lncRNA, such as *SAMMSON* and *FOXCUT*, belong to the gene network with the cancer-related transcription factor *FOXC1*. Thus, they may also play an aggravating role in the pathogenesis of AF, similar to carcinogenesis. In the least, this study suggests that (1) RNA alteration is most intense in PVs and (2) post-transcriptional gene regulation by lncRNA may contribute to the progression of AF. Through the screening analysis across the six cardiac regions, the possibility that the PV region can play a role other than paroxysmal triggering in the pathogenesis of AF was demonstrated for the first time. Future research with an increase in the number of PV samples will lead to a novel understanding of the pathophysiology of AF.

Keywords: atrial fibrillation; pulmonary vein; co-expression network analysis; long non-coding RNA; ion channels; inflammatory cytokines

Citation: Igarashi, W.; Takagi, D.; Okada, D.; Kobayashi, D.; Oka, M.; Io, T.; Ishii, K.; Ono, K.; Yamamoto, H.; Okamoto, Y. Bioinformatic Identification of Potential RNA Alterations on the Atrial Fibrillation Remodeling from Human Pulmonary Veins. *Int. J. Mol. Sci.* **2023**, *24*, 10501. https://doi.org/10.3390/ijms241310501

Academic Editors: Demetrios A. Arvanitis and Buddhadeb Dawn

Received: 18 March 2023
Revised: 16 May 2023
Accepted: 20 June 2023
Published: 22 June 2023

1. Introduction

Atrial fibrillation (AF) is the most common cardiac arrhythmia. Researchers worldwide have devoted much effort to finding its etiology. Although we currently know that AF has a complex pathology, investigation of the specific molecular mechanisms of atrial fibrillation began with identifying the causative gene for familial atrial fibrillation in the last century. From the ancestry of familial AF, linkage analysis identified causal loci on the genetic map [1]. In 2003, it was reported that one of the causal genes in familial AF was due to mutations in the K^+ channel KCNQ1 [2]. KCNQ1 abnormality was known to cause long QT syndrome [3], so this finding attracted much interest. Subsequently, many mutations in ion channel genes causing familial AF have been reported, and the genetic mutations leading to familial AF are well summarized in the review by Ragab et al. [4]. However, familial AF is a rare genetic disorder that hardly explains the pathogenesis of common AF.

For this reason, genome-wide association analysis (GWAS), a comprehensive analysis of single-gene polymorphisms (SNPs), has been repeatedly performed to explore the genetic background of the disease as a more common condition, i.e., non-familial AF. First 2007, an SNP at locus 4q25 was reported [5]. This locus locates between the *PITX2* gene and the translational region of the *C4orf32* gene, whose function is poorly understood, and its association with AF is overwhelmingly significant. Furthermore, the transcription factor *PITX2* has attracted attention in the understanding of the pathophysiology of AF [6], as it is involved in the development of the pulmonary veins (PVs) [7], the regional origin of AF. Subsequently, similar analyses have reported the association of the transcription factors *ZFHX3* [8,9] and the K^+ channel *KCNN3* [10], and now nearly 120 loci have been reported to be associated with AF [11]. The problem is that SNPs at each locus have a low relative risk (RR) for developing AF. Most loci have an RR between 1.0 and 1.1 [12]. In other words, AF-associated loci explain the inherited susceptibility to AF, but a single SNP alone does not lead to the development of the disease.

Therefore, the inherited combination of several SNPs or the involvement of environmental factors leads to the onset of the disease. Transcriptome analysis helps elucidate the direct pathophysiology. Hitherto, oxidative stresses [13], metabolic changes [14], and inflammations [11] have been suggested from transcriptome analysis. Recent advances in sequencing technology allow the transcriptome and non-coding RNAs to be analyzed together, allowing even post-transcriptional considerations [15,16]. However, transcriptome analysis generally has been limited to human patient samples from the atria and has yet to elucidate the pathogenesis by multi-cardiac regions. In the present study, we compared six different cardiac regions of donated samples. Surprisingly, the PV was the cardiac region most affected by atrial fibrillation in gene expressions. Moreover, our results have the potential to explain that AF is induced by an accumulation of acquired environmental stresses, similar to carcinogenesis.

2. Results

2.1. General Overview of Analysis Results

In this study, a total of 46 samples were obtained for RNA-seq; three samples each of left atrium (LA), left atrial appendage (LAA), left ventricle (LV), pulmonary vein (PV), right atrium (RA), right ventricle (RV) and sinoatrial node (SAN) were obtained from donors with or without persistent atrial fibrillation (AF); right atrial appendages (RAAs) were excised during mitral valve replacement surgery for two samples each from patients with or without AF. Patient information is summarized in Table 1. From 33535 RNAs detected by RNA-seq, statistical testing of expression differences between controls and AFs using the false discovery ratio (FDR) < 0.05 criterion resulted in 413 RNA hits, of which 201 were protein-coding, and 212 were long non-coding RNAs (lncRNAs). Using the R package, we applied the authorized method to compare two groups by computing differential expression genes (DEGs). The full results of the analysis can be viewed in Supplementary File S1. The hierarchical clustering of the DEGs in protein-coding and lncRNA are shown in Figure 1A,B, respectively. Besides the presence or absence of AF, differences in expression patterns were observed among cardiac regions. Therefore, each cardiac region was evaluated separately. Hereafter, LAA and RAA data were included in LA and RA, respectively.

Table 1. Patient information of donated or collected samples. SR and AF indicate sinus rhythm and persistent atrial fibrillation, respectively. For sex, M and F indicate male and female, respectively. For medical history: HD; heart disease, HT; hypertension, DM; diabetes, AF; atrial fibrillation, MR; mitral regurgitation, TR; tricuspid regurgitation, CAD; coronary artery disease, HCH; hypertensive cardiac hypertrophy. For heart samples: SAN; sinoatrial node, LA; left atrium, LAA; left atrial appendage, RA; right atrium, LV; the left ventricle, RV; the right ventricle, PV; pulmonary vein, RAA; right atrial appendage.

NO.	AF/SR	Age	Sex	Strain	Tabaco	Alcohol	HD	HT	DM	Cancer	Samples
1	SR	52	M	Caucasian	Yes	Yes	No	No	No	No	SAN, LA, LAA, RA, LV, RV, PV
2	SR	46	M	Filipino	Yes	Yes	No	medication	medication	No	SAN, LA, LAA, RA, LV, RV, PV
3	SR	23	F	Caucasian	Yes	No	No	No	No	No	SAN, LA, LAA, RA, LV, RV, PV
4	SR	50	M	Japanese	Yes	Yes	MR, TR	medication	No	No	RAA
5	SR	50	M	Japanese	Yes	Yes	MR	No	No	No	RAA
6	AF	58	M	Caucasian	Yes	Yes	CAD, AF	No	No	No	SAN, LA, LAA, RA, LV, RV, PV
7	AF	54	M	Caucasian	Yes	Yes	AF	No	No	No	SAN, LA, LAA, RA, LV, RV, PV
8	AF	58	F	Caucasian	Yes	Yes	AF	medication	No	No	SAN, LA, LAA, RA, LV, RV, PV
9	AF	59	M	Japanese	Yes	Yes	MR, TR, AF	medication	No	No	RAA
10	AF	70	F	Japanese	No	Yes	MR, HCH AF	medication	No	No	RAA

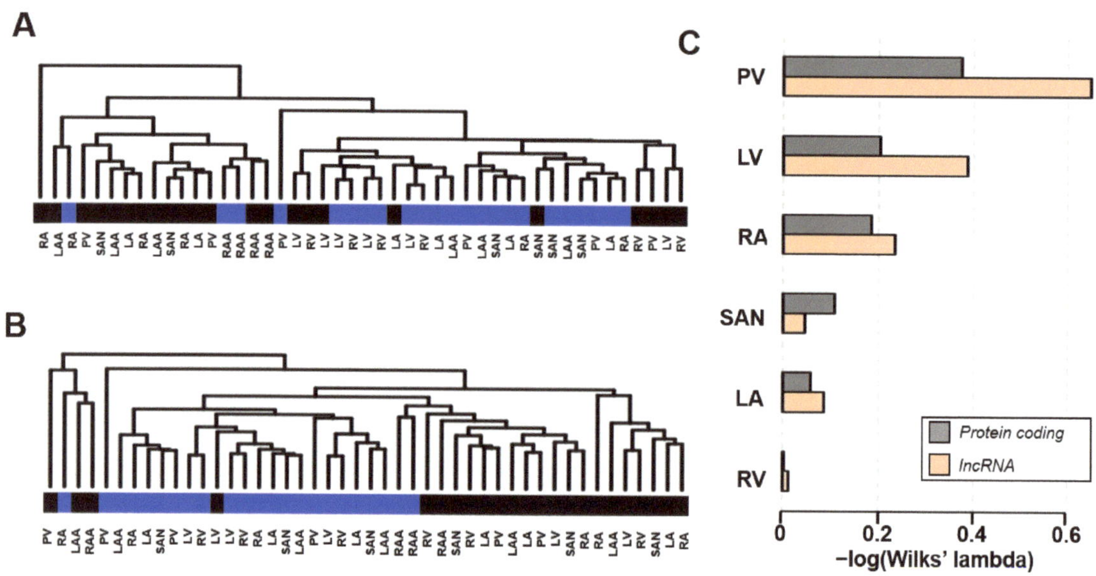

Figure 1. Overview of the obtained data. Twenty-three samples, each with or without AF, were compared, and (**A**) 201 coding RNA and (**B**) 212 long non-coding RNA (lncRNA) were detected as differential expression genes (DEGs). Samples were hierarchically clustered, and the origin of the

samples was color coded below the dendrogram. Black codes control and blue codes AF. (**C**) Degree of separation between control and AF samples in each cardiac region. The acquired data were analyzed using principal component analysis to construct a two-dimensional space with first and second principal components. Wilks' lambda calculates how much each cardiac region data is separated between control and AF. The smaller the value, the greater the degree of separation. The negative logarithm is scaled to present the degree of separation.

First, principal component analysis was applied to the gene expression matrices of each region. Wilks' lambda was calculated to indicate the distance between groups with or without AF in the two-dimensional space defined by the first and second principal components. Wilks' lambda is a statistic that measures how groups are separated in space; it is expressed from 0 to 1, with the closer to 0, the greater the degree of separation. Figure 1C displays the negative logarithm of Wilks' lambda for protein-coding (gray) and lncRNAs (orange). The most considerable difference between groups was demonstrated in PVs. Except for the SAN, differences between groups were more remarkable for lncRNAs than for protein-coding genes. It was suggested that the degree of association with the disease might differ from region to region. There is little difference in RV because the development of AF has little effect on gene expression in RV.

2.2. Differential Expression Genes (DEGs) in PV and LV

DEG analysis was performed by cardiac region (FDR < 0.05). The results showed that 145 DEGs were identified in the PV and 14 in the LV, but not in other regions. This supports the results from the analysis of Wilks' lambda. Four genes belonging to both PV and LV DEGs were *S100A1*, *GABARAP*, *CCDC144A*, and *ENSG00000289474* (Figure 2). Among the DEGs, the top 50 coding RNAs in PV, all lncRNAs in PV, and all DEGs in LV are listed in Figure 3A–C, according to statistical significance. The full results for DEGs in PV and LV can be viewed in Supplementary File S2. In addition, the Gene Ontology (GO) analysis was performed on the PV and LV gene sets to determine the functional annotation of protein-coding RNAs among the DEGs. Significantly enriched ontologies were identified in the PV DEGs, including "ion channel activity". The GO IDs, descriptions, and statistics of the identified GOs are graphed in Figure 4. The gene names of the DEGs belonging to each GO are listed in Table 2.

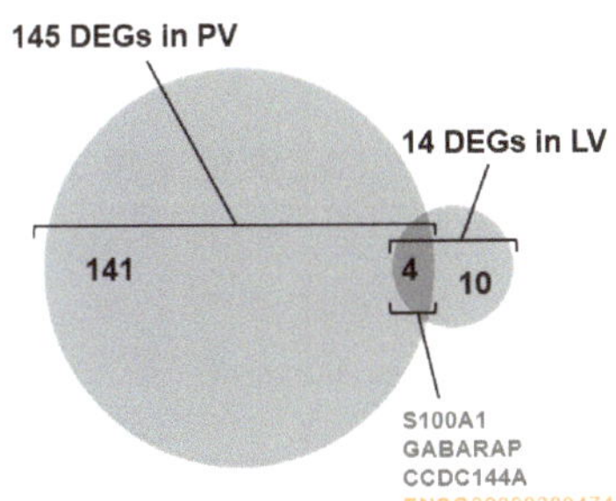

Figure 2. Populations of differential expression genes (DEGs) in the pulmonary vein (PV) and left ventricle (LV). 145 DEGs in PV and 14 in LV share four genes, *S100A1, GABARAP, CCDC144A,* and long non-coding *ENSG00000289474.*

In contrast, no significantly enriched Ontology was detected in DEGs of LV. Although most lncRNAs have poorly understood functions and cannot be generally annotated like coding RNAs, cancer-associated lncRNAs have been studied relatively well. In Table 3, lncRNAs corresponding to PV DEGs are addressed to cancer-associated lncRNAs in the database. Significant lncRNAs in PV in AF are also comparable in cancer.

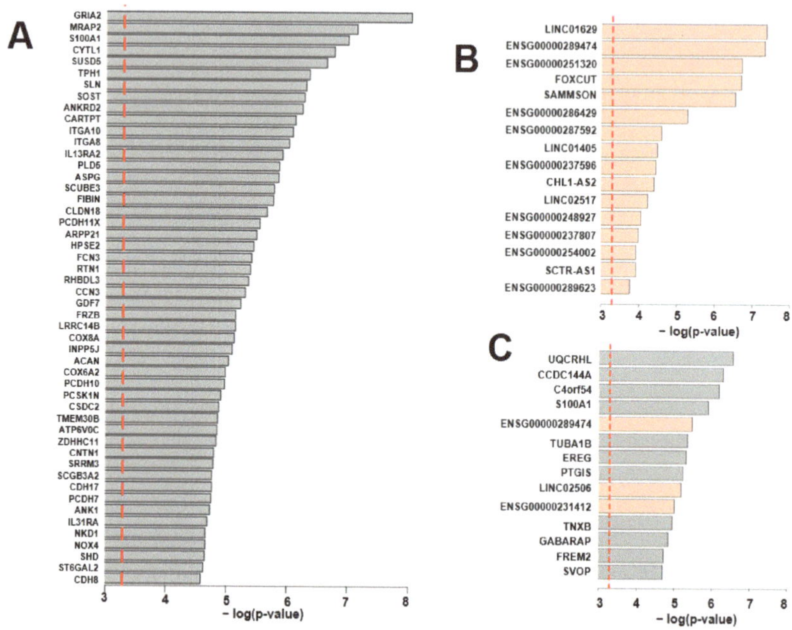

Figure 3. Differential expression genes (DEGs) in the pulmonary veins (PVs) and left ventricles (LVs). DEGs are presented in order of statistical significance. The red dotted line indicates a p-value = 0.0005. Protein-coding RNAs and long non-coding RNAs (lncRNAs) are color-coded by gray and orange, respectively. (**A**) Top 50 protein-coding DEGs in PV. (**B**) lncRNAs as DEGs in PV. (**C**) All DEGs in the left ventricle.

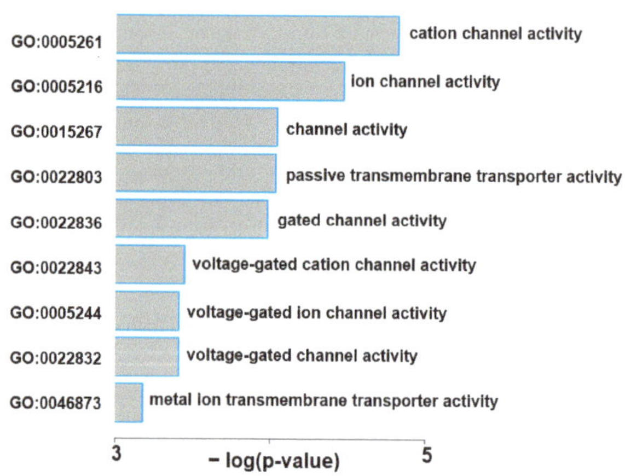

Figure 4. Functional annotation of differential expression genes (DEGs) in the pulmonary veins (PVs) by Gene Ontology (GO) analysis. Enriched GOs are presented in order of statistical significance. GO IDs are provided on the left. The corresponding negative logarithm of p-values and descriptions are in the middle and on the right, respectively. The corresponding genes are listed in Table 2.

Table 2. Enriched gene sets detected by gene ontology (GO) analysis in the pulmonary veins (PVs). The Detected GOs are on the left, and corresponding gene symbols are on the right. Most genes are significantly downregulated in PV of AF patients except GRIA2, GABRR1, and PKD2.

Descriptions	Gene Symbols
cation channel activity	*GRIA2, ATP5F1D, ATP6V0C, KCNH2, NALF2, LRRC38, CACNA1E, PKD2, SCN2B, KCNIP2, GRIN2C*
ion channel activity	*GRIA2, GABRR1, ATP5F1D, ATP6V0C, KCNH2, NALF2, LRRC38, CACNA1E, PKD2, SCN2B, KCNIP2, GRIN2C*
channel activity	*GRIA2, GABRR1, ATP5F1D, ATP6V0C, KCNH2, NALF2, LRRC38, CACNA1E, PKD2, SCN2B, KCNIP2, GRIN2C*
passive transmembrane transporter activity	*GRIA2, GABRR1, ATP5F1D, ATP6V0C, KCNH2, NALF2, LRRC38, CACNA1E, PKD2, SCN2B, KCNIP2, GRIN2C*
gated channel activity	*GRIA2, GABRR1, KCNH2, NALF2, LRRC38, CACNA1E, PKD2, SCN2B, KCNIP2, GRIN2C*
voltage-gated cation channel activity	*KCNH2, LRRC38, CACNA1E, PKD2, KCNIP2, GRIN2C*
voltage-gated ion channel activity	*KCNH2, LRRC38, CACNA1E, PKD2, SCN2B, KCNIP2, GRIN2C*
voltage-gated channel activity	*KCNH2, LRRC38, CACNA1E, PKD2, SCN2B, KCNIP2, GRIN2C*
metal ion transmembrane transporter activity	*SLC22A3, KCNH2, NALF2, LRRC38, CACNA1E, PKD2, SCN2B, KCNIP2, GRIN2C, SLC5A1*

Table 3. Long non-coding RNAs (lncRNAs) annotation of atrial fibrillation-induced alterations in pulmonary veins based on cancer-related lncRNAs. Sixteen significant lncRNAs in PV were expressed on a log2 scale. Related carcinomas are listed based on two databases. (http://www.bio-bigdata.net/lnc2cancer/index.html, and http://fcgportal.org/TCLA/index.php, accessed on 1 February 2023). Red or blue letters and numbers indicate the increase or decrease in expression in the related disorder, respectively. GBM, glioblastoma; HCC, hepatocellular carcinoma; MME, malignant melanoma; NCP, nasopharyngeal carcinoma; GAC, gastric adenocarcinoma; BLBC, basal-like breast cancer; ESCC, esophageal squamous cell carcinoma; OSCC, oral squamous cell carcinoma; KIRC, kidney renal clear cell carcinoma; LUSC, lung squamous cell carcinoma; HNSC, head and neck squamous cell carcinoma; PRAD, prostate adenocarcinoma; BRCA, breast invasive carcinoma; COAD, colon adenocarcinoma; NA, not applicable.

Unique Names/Ensembl IDs	Fold Changes (log2) in PV in AF	Expressional Changes in Extra-Cardiovascular Carcinoma
SAMMSON	9.29	GBM, HCC, MME
FOXCUT	8.41	NCP, GAC, BLBC, ESCC, OSCC
ENSG00000251320	5.99	KIRC, LUSC, HNSC
ENSG00000248927	4.16	KIRC, PRAD
ENSG00000237807	2.46	BRCA, COAD, PRAD
ENSG00000286429	−4.73	NA
SCTR-AS1	−3.60	LUSC
ENSG00000289474	−9.35	NA
CHL1-AS2	−3.41	NA
LINC02517	−3.55	NA
ENSG00000237596	−3.72	NA
ENSG00000289623	−6.12	NA
ENSG00000287592	−3.61	NA
ENSG00000254002	−3.82	NA
LINC01405	−6.77	HNSC
LINC01629	−6.76	NA

2.3. Co-Expression Network Analysis on DEGs

To explore functional relationships between DEGs, co-expression network analysis was performed on 155 DEGs detected in PV and LV. This method is a valuable data analysis technique in characterizing gene function globally [17] and is also used in investigations of lncRNA function [18]. By clustering Pearson correlation coefficients between gene expression levels, compartmentalized groups of genes with correlated expressions are visualized (Figure 5A). A gene network was constructed by linking pairs of these gene groups with correlation coefficients >0.8. In Figure 5A, the largest compartment (a), followed by compartment (b), and the negatively linked compartments (c1) and (c2) are constructed as subnetworks A, B, and C. Subnetwork A, B, and C are extracted into Figure 6A–C, respectively. Figure 5B displays the distribution of gene degree (number of intergenic connections). These networks have typical gene network properties, with a few genes having many connections and most genes having few connections, referred to as scale-free [19]. Genes with a degree above 30 are listed in Table 4 as hub genes, which play essential roles in the gene network. In Figure 6, subnetwork A is a population of up-regulated 62 genes, subnetwork B is a population of down-regulated 40 genes, and subnetwork C is mixed by up-regulated five and down-regulated ten genes. Subnetworks A and C have no known genes that reportedly induce AF and have several lncRNAs, which are thought to be involved in regulating post-acquired gene expression. In contrast, subnetwork B has known AF-related genes such as *NKX2-5*, *KCNH2*, and *SCN2B* and no lncRNAs.

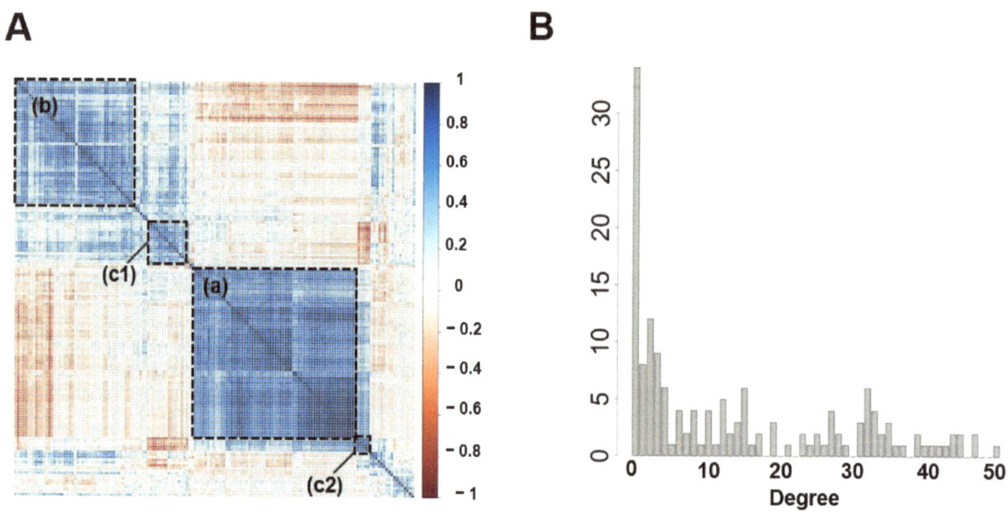

Figure 5. The fundamental structures of the co-expression network in differential expression genes (DEGs). (**A**) Co-expressed gene pairs are clustered and visualized as a heatmap. The color scale corresponding to the correlation coefficient is inserted on the right. Based on absolute values of a correlation coefficient of 0.8 or higher, there are three subnetworks: a compartment containing 63 genes (a) and a compartment containing 46 genes (b), as well as a compartment (c1) and (c2), which are composed of 10 and 5 genes, respectively. (c1) and (c2) are linked by negative correlation. The rest of the subnetworks are isolated genes with fewer than two nodes. Compartment (a) is represented as subnetwork A, compartment (b) as sub-network B, and compartments (c1) and (c2) as sub-network C, which are represented in Figure 6A–C, respectively. (**B**) The histogram of gene degree. The degree is proportional to the number of intergenic connections. The distribution of this histogram indicates that the network is scale free.

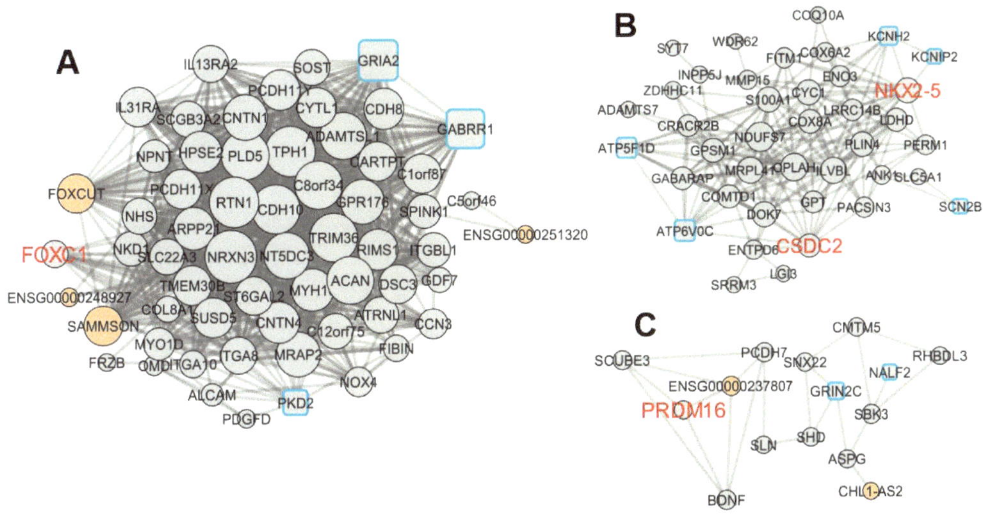

Figure 6. Subnetworks A, B, and C are expanded into (**A–C**) in this figure. The diameter of the nodes is proportional to the degree that the gene has. Coding RNAs are grey, lncRNAs are orange nodes, and square nodes indicate ion channel-related genes enriched in the GO analysis with a marine blue border. Transcription factors are highlighted in red letters. Subnetwork A consists of 62 genes and is a network of genes whose expression is up-regulated compared to controls. Subnetwork B is a network of 40 genes whose expression is down-regulated. Subnetwork C consists of five up-regulated and ten down-regulated genes. All nodes represent differential expression genes of the pulmonary vein except *PTGIS*, which is in subnetwork A.

Table 4. Hub genes detected in the co-expression analysis (degree > 30). Marine blue and orange text color indicates ion channel activity and long non-coding RNA, respectively.

Unique Name	Gene type	Degree
NRXN3	protein_coding	50
TPH1	protein_coding	47
RTN1	protein_coding	47
PLD5	protein_coding	45
PCDH10	protein_coding	45
TRIM36	protein_coding	44
C8orf34	protein_coding	44
ACAN	protein_coding	43
ADAMTSL1	protein_coding	42
NT5DC3	protein_coding	41
CNTN1	protein_coding	40
MRAP2	protein_coding	39
GPR176	protein_coding	39
MYH10	protein_coding	37
ARPP21	protein_coding	36
CNTN4	protein_coding	35

Table 4. *Cont.*

Unique Name	Gene type	Degree
GRIA2	protein_coding	35
HPSE2	protein_coding	35
RIMS1	protein_coding	34
CDH8	protein_coding	34
SUSD5	protein_coding	33
GABRR1	protein_coding	33
TMEM30B	protein_coding	33
PCDH11Y	protein_coding	33
C1orf87	protein_coding	32
CYTL1	protein_coding	32
CARTPT	protein_coding	32
IL31RA	protein_coding	32
PCDH11X	protein_coding	32
SAMMSON	lncRNA	32
ST6GAL2	protein_coding	31
DSC3	protein_coding	31
FOXCUT	lncRNA	31

3. Discussion

Although there have been many comprehensive RNA expression analyses on atrial fibrillation (AF), most studies have limited their analysis to atrial samples on the assumption that pathological RNA alterations occur in the atria [13–16]. We are the first to perform comparisons in six different heart regions. Surprisingly, the most robust RNA alterations occurred in the pulmonary veins (PVs) among the cardiac regions (Figure 1B). Of these, coding RNAs related to ion channels are particularly enriched as RNAs influenced by AF (Figure 4). This may illustrate the basis of electrical remodeling in AF patients. In addition, the expression of the transcription factor FOXC1 was strongly up-regulated with long non-coding RNA (lncRNA) *FOXCUT* in AF patients (Figure 6A), which may be involved in structural remodeling by post-acquired gene regulation as discussed below.

3.1. The Priority of PV among Cardiac Regions on AF Remodeling

AF is known to be mainly triggered by PV sleeves [20]. Nevertheless, AF itself is defined as an atrial arrhythmia, and it has long been considered that substrate changes in the atria are responsible for the persistence of the arrhythmia [21]. Thus, most AF researchers believe that the initiation and maintenance of AF result from different molecular mechanisms occurring in different cardiac regions [22]. However, intense fibrosis, representing the substrate changes in AF patients, also develops in PV [23,24]. If substrate changes provide the basis for AF persistence, the atria would not be the only cardiac region contributing to AF persistence. Instead, the current analytic results indicate that biological change is more potent in PVs (Figure 1B). In our analysis, the cardiac regions with statistically significant differences between AF and sinus rhythm patients were PV and LV (Figure 2). There were 145 significant hits in the PV samples, including coding RNAs and long non-coding RNAs (lncRNAs), compared to 14 gene hits in the LV. Furthermore, samples from donors with coronary artery disease are largely unaffected in atrial and PV samples, while ventricular samples may be affected by ischemic changes in the data.

3.2. The Possible Basis of the Electrical Remodeling in PV in AF

Gene ontology (GO) analysis detected several ion transport-related genes from the enriched gene sets on PV, such as a subunit of mitochondrial ATPase (ATP5FD1), endoplasmic reticulum proton pump (ATP6V0C), hERG channel (KCNH2), an auxiliary subunit of BK channel (LRRC38), voltage-gated Ca^{2+} channel (CACNA1E), ion channel scaffold molecule (PKD2), β subunit of voltage-gated Na^{+} channel (SCN2B), β subunit of voltage-gated K^{+} channel (KCNIP2), a part of ion channel-type glutamate receptors (GRIA2, GRIN2c) and GABA receptors (GABRR1).

ATP5FD1, *ATP6V0C*, *KCNH2*, *SCN2B*, and *KCNIP2* are essential to myocardial functions and significantly downregulated in AF samples. In addition, these cardiac function-related proteins are all on subnetwork B in Figure 6B. As noted, the co-expression network analysis was calculated from n = 46. Subnetwork B contains the transcription factor *NKX2-5*. *NKX2-5* is a well-established transcription factor that determines morphological and functional features of the myocardium [25] and is also down-regulated in AF samples. It is, thus, estimated that cardiomyocyte-like characteristics might be lost in the PVs of AF patients. Among these genes, *NKX2-5*, *KCNH2*, and *SCN2B* have already been reported as single nucleotide polymorphisms for AF [26–28]. Unlike mutations, polymorphisms by themselves do not induce disease and are therefore explained as a genetic background rather than a causative gene. In the present analysis, the simultaneous finding of the down-regulation of these genes is meaningful in considering the cause of the proarrhythmic state. Namely, the down-regulation of *NKX2-5* indicates a deviation from cardiac tissue, accompanied by electrical remodelings, i.e., the down-regulation of *KCNH2* and *SCN2B*, resulting in the formation of atrial fibrillation. Also, the present analysis, which does not assume genetic mutations, shares results with the analysis of congenital causes, making it reasonable to assume that the polymorphisms identified in the previous reports are responsible for the proarrhythmic state.

Other ion transport proteins belonging to subnetworks A and C are tentatively ion channel-related molecules expressed in the smooth muscle layer and the pulmonary veins' nerve plexus.

3.3. Possible Classifying Post-Acquired Disturbances in AF by the Network Analysis

In the co-expression network analysis in Figure 6, subnetwork A is the extensive network with the highest number of correlated RNAs. It exhibits up-regulated expressions of the cancer-associated lncRNAs *FOXCUT*, *SAMMSON*, *ENSG00000251320*, and *ENSG00000248927*, as well as ion channels in nerve fibers, such as *GRIA2* and *GABARR1*. In addition, the expression of *FOXC1*, a *FOXCUT*-related transcript, is also up-regulated. Furthermore, prothrombotic factors, such as *PDGFD* and *ALCAM*, and molecules in the inflammatory response, such as receptors for IL-13 and 31, are also up-regulated. On the other hand, unlike subnetwork B, none of the known AF-causing genes are listed.

Given the above, it is likely that subnetwork A reflects post-acquired pathological disturbances due to the disease rather than inherited conditions. LncRNAs are often influential in regulating acquired gene expression through epigenomic regulation and post-transcriptional and post-translational modifications [29], e.g., *FOXCUT* binds to FOXC1 mRNA and promotes protein translation [30]. The transcription factor FOXC1 is known to contribute to the formation of the heart tube outflow tract early in development in the cardiovascular research field [31]. The PV is the cardiac region that forms a part of the outflow tract [32]. However, considering the age of the donors (Table 1), the increased expression in AF may represent a pathological rather than a developmental change. As for pathological changes, FOXC1 is up-regulated in many carcinomas. It plays a unique role in promoting the epithelial–mesenchymal transition (EMT) [33]; If EMT occurs in the pulmonary veins, it is assumed that endothelial cells will be converted to fibroblasts. Some reports have already found that EMT occurs in the atria of AF patients [34,35]. Then, fibroblasts secrete cytokines and adhesion molecules that may influence the substrate changes in the patients. Indeed, *PDGE* and *ALCAM*, prothrombotic factors in subnetwork

A, can be secreted by fibroblasts. Thus, fibroblasts would be expected to proliferate and the substrate alteration to progress irreversibly. Our analysis indicates that the focus of the substrate changes is the PV rather than the atria, which may be related to thrombi being more likely to occur in the left atrium (~90%) than in the right atrium [36].

SAMMSON is also a well-studied lncRNA initially described as up-regulated in malignant melanoma [37]. The *SAMMSON* is located 30 kilobase pairs downstream of *MITF*, a melanoma-specific transcriptional regulator, whose expression correlates with *MITF*. However, unlike the relationship between *FOXCUT* and FOXC1, *SAMMSON* does not directly contact MITF but epigenetically regulates transcription factors influencing MITF expression. While MITF is a melanoma-specific factor, *SAMMSON* has also been observed to be up-regulated in thyroid cancer [38] and glioblastoma [39]. What genes *SAMMSON* couples to in the pulmonary veins of patients with AF is a subject for future research.

Furthermore, lncRNAs can imply regulation of gene expression by environmental factors, but the specific environmental factors themselves are often unknown. On the other hand, increased cytokine receptor expression has been detected in sub-network A, indicating that one of the environmental factors is inflammation or oxidative stress. Both *IL13RA2* and *IL31RA* are receptor genes for inflammatory cytokines, IL13 and IL31. These cytokines play essential roles in acquired immunity [40,41]. In recent years, *IL13RA2* has also been reported as a marker for several cancer cells [42–45].

Although few studies have investigated the analogies between AF and cancer, both diseases are characterized by a high prevalence in the elderly and an increased incidence with age. Previous studies have revealed that the transcriptome profile of age-related changes is characterized by increased inflammation-related genes, among other features [46]. Therefore, age-related changes in molecular profiles may contribute to the risk of age-related diseases through increased inflammation. On the other hand, we do not know of any other transcriptomic studies of human cardiac PV, so we do not know the relationship between the molecules identified in this study and aging. Further profiling of pulmonary veins will lead to a better understanding of age-related diseases.

In this way, it has been explained that sub-network A represents environmental factors and sub-network B myocardial functional factors. In contrast, sub-network C is a mixture of up-and-down-regulated genes and a few genes with a low degree, so biological considerations are challenging. Increasing the number of significant genes would be necessary to make sense of these complementary networks by increasing the sample size.

3.4. Limitations

The primary limitation of the current study is the limited size of sample numbers. The present analysis has been performed on three AF samples and three normal heartbeat samples. It is not guaranteed that the results would be replicated if the sample size were huge or if the property of the domination were aligned by sex, age, or race. However, our study focuses on comparing multiple cardiac regions, and it is not easy to include more replicates. In the RNA-seq research community, a sample size of n = 3 is often considered a minimum requirement. This is even clearly stated in the submission guidelines of specialized omics journals such as Genomics (https://www.sciencedirect.com/journal/genomics, accessed on 3 May 2023). Thus, our study design is considered acceptable. Besides, we took advantage of the properties of omics data to make our analysis more robust with the GO and co-expression gene network analyses. GO analysis uses genome-wide information and provides more reliable biological findings even with small sample sizes. The gene network analysis identifies essential genes using genome-wide information from all n = 46 samples.

Second, we are unable to analyze molecules regulated by micro RNAs because our analysis depends on detecting poly A tails. Some connexin genes, one of the AF-related gene family [47–49], are regulated by micro RNAs [50,51], but unfortunately, post-transcriptional modifications through micro RNAs are outside the focus of this study. Moreover, our hypothesis discussed above should also be confirmed in vitro to see how endothelial cells transform into fibroblasts in the future.

4. Materials and Methods

4.1. RNA Extractions from Donated Hearts and Right Atrial Appendage Collections

Human heart samples from subjects with or without persistent atrial fibrillation (AF) for left atrium (LA), left atrial appendage (LAA), right atrium (RA), sinoatrial node (SAN), left ventricle (LV), right ventricle (RV), and pulmonary vein (PV) were purchased from AnaBios Corporation (San Diego, CA, USA), which provides heart organs through the organ procurement organization of the USA in compliance with the Health Insurance Portability and Accountability Act. All personal information of donors is protected. Donated hearts were separated into SAN, LA, LAA, RA, SAN, LV, RV, and PV. These samples were frozen in RNAlater (Thermo Fisher Science, Waltham, MA, USA) at −20 °C. Our laboratory extracted RNAs using QIAGEN RNeasy mini columns (QIAGEN, Venlo, Netherland). Right atrial appendage (RAA) samples were obtained from four patients undergoing mitral valve surgery. Two of these patients had permanent atrial fibrillation; the other two were in sinus rhythm with no history of atrial fibrillation. After excision, all tissue specimens were frozen in liquid nitrogen and stored at −80 °C for later use. RNAs were extracted using ISOGEN II (NIPPON GENE CO., LTD. Kandashiki-machi, Tokyo, Japan).

4.2. RNA-Seq and Data Processing on the Samples

The high-throughput sequence was demonstrated with NovaSeq 6000 (Illumina, San Diego, CA, USA). First, extracted RNAs were purified by poly(A) capture. Resultant mRNAs were then fragmented and reverse-transcribed into single-stranded complementary DNAs (cDNAs). Subsequently, cDNAs were double-stranded by a DNA polymerase. Deoxy UTPs (dUTPs) were mixed in nucleotide materials during the polymerase reactions. Both double-stranded DNA (ds DNA) ends were ligated to a 13 bp adapter sequence. Both double-stranded DNA (ds DNA) ends were ligated to a 13 bp adapter sequence. Next, the ds DNAs were subjected to PCR amplification to prepare the multi-sized DNA library. NovaSeq Control software v1.4.0 analyzed the sequencing runs and tag sequences classified each read in the raw sequencing data. fastp software (version 0.12.4) was used for read quality control and adapter removal [52]. Reads were aligned using STAR software (version 2.7.0a) [53]. The fasta and gtf files of GRCh38/release105 obtained from the Ensemble database were used as reference genome and gene annotation information. The count values of each gene were quantified from the alignment results by featureCounts software (see [54]). We calculated the Transcripts Per Kilobase Million (TPM) values based on the calculated gene lengths and gene counts [55,56]. The current study targeted eight heart regions: left atrium (LA), left atrial appendage (LAA), left ventricle (LV), pulmonary vein (PV), right atrium (RA), right atrial appendage (RAA), right ventricle (RV), and sinoatrial node (SAN). Twenty-three samples from sinus rhythm (SR) and 23 samples from AF patients, a total of 46 samples of RNA-seq data, were used for analysis. Of these, the 21 control samples other than RAA are the same as in the previous report [57] and have been validated by a different experiment (GSE112339 in Gene Expression Omnibus (GEO) database). The TPM values were transformed by log2(intensity + 1) and used in subsequent analyses as gene expression intensity. Genes with zero TPM values in all 46 samples were excluded from the analysis. Protein coding genes and long non-coding RNA (lncRNA) genes were extracted based on gene feature format files provided by the Ensemble database in the same version used for alignment. Finally, 33,535 genes were included in this analysis, including 18,899 protein-coding genes and 14,636 lncRNA genes. The obtained data were subsequently analyzed in the following order: (1) differential expression gene (DEG) analysis, (2) multivariate analysis-of-covariance by manova function in R to calculate Wilks' lambda, (3) coding RNA annotation by gene ontology (GO) analysis, (4) cancer-related lncRNA annotation, (5) co-expression network analysis.

4.3. *Differential Expression Gene (DEG) Analysis*

DEG analysis between SR and AF was performed using TCC (an acronym for Tag Count Comparison) in the R package. Here, the trimmed mean of M values (TMM) and edgeR methods were employed for normalization and testing, respectively [58]; genes with FDR < 0.05 were identified as significant DEGs. Gene Ontology (GO) analysis was performed using R's clusterProfiler package to determine the GO-associated gene set (Benjamini-Hochberg adjusted *p*-value < 0.05) [59]. In the GO analysis, all genes included in the analysis were set to background. The R package's org.Hs.eg.db (version 3.13.0) was used as the GO and gene annotation database. The list of transcription factors in humans was obtained from AnimalTFDB4.0 [60]. A gene co-expression network analysis was performed to perform a functional analysis of the identified DEGs. Pearson correlation coefficient matrices were calculated for DEGs. A gene network was constructed by connecting gene pairs with absolute values of correlation coefficients >0.8. The Cytoscape software (version 3.9.1, https://cytoscape.org/) visualizes the network.

4.4. *Clustering and Heatmap Visualization*

Clustering and heatmap drawings of the samples were applied to the gene expression intensity matrix. Hierarchical clustering of samples was obtained by applying the dist and hclust functions in R with default settings. The heatmap.2 function and dendextend packages [61] in the gplots package of the R were used for heatmap and clustering drawing (https://CRAN.R-project.org/package=gplots). Each expression data was standardized before the heatmap drawing.

5. Conclusions

At least, the current study suggests two findings: (1) the pulmonary veins (PVs) were the regions with the most intensive gene expression alterations by the atrial fibrillation (AF) incidence. Therefore, the PV region should be focused more on researching the etiology of both paroxysmal and chronic AF in the future. (2) the AF remodeling is attributable to a post-transcriptional genetic regulation similar to carcinogenesis, e.g., the FOXCUT-FOXC1 axis, SAMMSON, and IL13RA2.

Thus, our analytic results may suggest that rather than searching the inborn genetic backgrounds, a healthy lifestyle, such as smoking cessation and exercise, is a more protective strategy against AF progression in analogy with cancer prevention. This is because that familial AF is a sporadic disease, and the relative risk of AF-associated SNPs is small enough. Future research with an increase in the number of PV samples will lead to a novel understanding of the pathophysiology of AF.

Supplementary Materials: The supporting information can be downloaded at: https://www.mdpi.com/article/10.3390/ijms241310501/s1.

Author Contributions: W.I. wrote the manuscript and performed experiments and performed the data analysis. D.T. funded and supervised the study. D.O. is the main analyzer of this project. D.K. performed the experiment. T.I. designed the experiment and funding. M.O. processed RNA-seq data. K.I. and H.Y. supervised the study. K.O. funded and supervised the study. Y.O. wrote the manuscript and designed the study and performed experiments and analyzed the data and supervised the study. All authors have read and agreed to the published version of the manuscript.

Funding: This research was funded by JSPS KAKENHI (Grant-in-Aid for Young Scientists) grant number 20K16116 to Y.O., JSPS KAKENHI (Grant-in-Aid for Young Scientists) Grant number 21K16045 to D.T., JSPS KAKENHI (Grant-in-Aid for Scientific Research C) Grant number 22K08093 to K.O., and by collaboration with Ono Pharmaceutical Cooperation, Ltd. (Japan).

Institutional Review Board Statement: The protocols used in this study were approved by the Ethics Review Committees of the Yamagata University Faculty of Medicine (#110;June 2nd 2017), the Ethics Committees of Kyoto University Graduate School and Faculty of Medicine (R3118;September 17th 2021).Akita University Certified Clinical Research Revies Board (#1982;May 21st 2021, and #2496; April 25th 2022).

Informed Consent Statement: For all samples except right atrial appendage (RAA), informed consent was obtained appropriately from all subjects involved in the study by AnaBios Corporation, along with the Health Insurance Portability and Accountability Act. For the RAA, patients gave written informed consent, and these samples were obtained with the approval of Akita University's Ethics Review Board, as mentioned above.

Data Availability Statement: The analytic results in this study are available on a request basis. All RNA-seq data we analyzed in the current study are uploaded into the Gene Expression Omnibus (GEO) database. GSE203367 and GSE226283 are accession numbers for RNA-seq on seven cardiac regions from sinus rhythm control and atrial fibrillation patients, respectively. GSE226282 is the accession number for RNA-seq on cut-off right atrial appendages.

Acknowledgments: We thank Steve Wilhite for curating the GEO submissions.

Conflicts of Interest: The authors declare no conflict of interest.

References

1. Brugada, R.; Tapscott, T.; Czernuszewicz, G.Z.; Marian, A.; Iglesias, A.; Mont, L.; Brugada, J.; Girona, J.; Domingo, A.; Bachinski, L.L.; et al. Identification of a Genetic Locus for Familial Atrial Fibrillation. *N. Engl. J. Med.* **1997**, *336*, 905–911. [CrossRef] [PubMed]
2. Chen, Y.-H.; Xu, S.-J.; Bendahhou, S.; Wang, X.-L.; Wang, Y.; Xu, W.-Y.; Jin, H.-W.; Sun, H.; Su, X.-Y.; Zhuang, Q.-N.; et al. KCNQ1 Gain-of-Function Mutation in Familial Atrial Fibrillation. *Science* **2003**, *299*, 251–254. [CrossRef] [PubMed]
3. Wang, Q.; Curran, M.E.; Splawski, I.; Burn, T.C.; Millholland, J.M.; VanRaay, T.J.; Shen, J.; Timothy, K.W.; Vincent, G.M.; De Jager, T.; et al. Positional cloning of a novel potassium channel gene: KVLQT1 mutations cause cardiac arrhythmias. *Nat. Genet.* **1996**, *12*, 17–23. [CrossRef] [PubMed]
4. Ragab, A.A.Y.; Sitorus, G.D.S.; Brundel, B.B.; Groot, N.M.S.D. The Genetic Puzzle of Familial Atrial Fibrillation. *Front. Cardiovasc. Med.* **2020**, *7*, 14. [CrossRef]
5. Gudbjartsson, D.F.; Arnar, D.O.; Helgadottir, A.; Gretarsdottir, S.; Holm, H.; Sigurdsson, A.; Jonasdottir, A.; Baker, A.; Thorleifsson, G.; Kristjansson, K.; et al. Variants conferring risk of atrial fibrillation on chromosome 4q25. *Nature* **2007**, *448*, 353–357. [CrossRef]
6. Wang, J.; Klysik, E.; Sood, S.; Johnson, R.L.; Wehrens, X.H.T.; Martin, J.F. Pitx2 prevents susceptibility to atrial arrhythmias by inhibiting left-sided pacemaker specification. *Proc. Natl. Acad. Sci. USA* **2010**, *107*, 9753–9758. [CrossRef]
7. Mommersteeg, M.T.M.; Brown, N.A.; Prall, O.W.; de Gier-de Vries, C.; Harvey, R.P.; Moorman, A.F.M.; Christoffels, V.M. *Pitx2c* and *Nkx2-5* are required for the formation and identity of the pulmonary myocardium. *Circ. Res.* **2007**, *101*, 902–909. [CrossRef]
8. Gudbjartsson, D.F.; Holm, H.; Gretarsdottir, S.; Thorleifsson, G.; Walters, G.B.; Thorgeirsson, G.; Gulcher, J.; Mathiesen, E.B.; Njølstad, I.; Nyrnes, A.; et al. A sequence variant in *ZFHX3* on *16q22* associates with a trial fibrillation and ischemic stroke. *Nat. Genet.* **2009**, *41*, 876–878. [CrossRef]
9. Benjamin, E.J.; Rice, K.M.; Arking, D.E.; Pfeufer, A.; van Noord, C.; Smith, A.V.; Schnabel, R.B.; Bis, J.C.; Boerwinkle, E.; Sinner, M.F.; et al. Variants in *ZFHX3* are associated with atrial fibrillation in individuals of European ancestry. *Nat. Genet.* **2009**, *41*, 879–881. [CrossRef]
10. Ellinor, P.T.; Lunetta, K.L.; Glazer, N.L.; Pfeufer, A.; Alonso, A.; Chung, M.; Sinner, M.; De Bakker, P.; Mueller, M.; Lubitz, S.; et al. Common variants in *KCNN3* are associated with lone atrial fibrillation. *Nat. Genet.* **2010**, *42*, 240–244. [CrossRef]
11. Miyazawa, K.; Ito, K.; Ito, M.; Zou, Z.; Kubota, M.; Nomura, S.; Matsunaga, H.; Koyama, S.; Ieki, H.; Akiyama, M.; et al. Cross-ancestry genome-wide analysis of atrial fibrillation unveils disease biology and enables cardioembolic risk prediction. *Nat. Genet.* **2023**, *55*, 187–197. [CrossRef]
12. Ellinor, P.T.; Lunetta, K.L.; Albert, C.M.; Glazer, N.L.; Ritchie, M.D.; Smith, A.V.; Arking, D.E.; Müller-Nurasyid, M.; Krijthe, B.P.; Lubitz, S.A.; et al. Meta-analysis identifies six new susceptibility loci for atrial fibrillation. *Nat. Genet.* **2012**, *44*, 670–675. [CrossRef]
13. Kim, Y.H.; Lee, J.H.; Lim, D.S.; Shim, W.J.; Ro, Y.M.; Park, G.H.; Becker, K.G.; Cho-Chung, Y.S.; Kim, M.-K. Gene expression profiling of oxidative stress on atrial fibrillation in humans. *Exp. Mol. Med.* **2003**, *35*, 336–349. [CrossRef]
14. Barth, A.S.; Merk, S.; Arnoldi, E.; Zwermann, L.; Kloos, P.; Gebauer, M.; Steinmeyer, K.; Bleich, M.; Kääb, S.; Hinterseer, M.; et al. Reprogramming of the human atrial transcriptome in permanent atrial fibrillation: Expression of a ventricular-like genomic signature. *Circ. Res.* **2005**, *96*, 1022–1029. [CrossRef]
15. Zhang, J.; Huang, X.; Wang, X.; Gao, Y.; Liu, L.; Li, Z.; Chen, X.; Zeng, J.; Ye, Z.; Li, G. Identification of potential crucial genes in atrial fibrillation: A bioinformatic analysis. *BMC Med. Genom.* **2020**, *13*, 1–8. [CrossRef]
16. Xie, L.; Huang, G.; Gao, M.; Huang, J.; Li, H.; Xia, H.; Xiang, X.; Wu, S.; Ruan, Y. Identification of Atrial Fibrillation-Related lncRNA Based on Bioinformatic Analysis. *Dis. Markers* **2022**, *2022*, 8307975. [CrossRef]
17. Stuart, J.M.; Segal, E.; Koller, D.; Kim, S.K. A Gene-Coexpression Network for Global Discovery of Conserved Genetic Modules. *Science* **2003**, *302*, 249–255. [CrossRef]
18. Li, J.; Li, C.; Wang, J.; Song, G.; Zhao, Z.; Wang, H.; Wang, W.; Li, H.; Li, Z.; Miao, Y.; et al. Genome-wide analysis of differentially expressed lncRNAs and mRNAs in primary gonadotrophin adenomas by RNA-seq. *Oncotarget* **2017**, *8*, 4595–4606. [CrossRef]

19. Jordan, I.K.; Mariño-Ramírez, L.; Wolf, Y.I.; Koonin, E.V. Conservation and Coevolution in the Scale-Free Human Gene Coexpression Network. *Mol. Biol. Evol.* **2004**, *21*, 2058–2070. [CrossRef]
20. Haïssaguerre, M.; Jaïs, P.; Shah, D.C.; Takahashi, A.; Hocini, M.; Quiniou, G.; Garrigue, S.; Le Mouroux, A.; Le Métayer, P. Clémenty, J. Spontaneous initiation of atrial fibrillation by ectopic beats originating in the pulmonary veins. *N. Engl. J. Med.* **1998** *339*, 659–666. [CrossRef]
21. Nattel, S.; Heijman, J.; Zhou, L.; Dobrev, D. Molecular Basis of Atrial Fibrillation Pathophysiology and Therapy: A Translational Perspective. *Circ. Res.* **2020**, *127*, 51–72. [CrossRef] [PubMed]
22. Michaud, G.F.; Stevenson, W.G. Atrial Fibrillation. *N. Engl. J. Med.* **2021**, *384*, 353–361. [CrossRef] [PubMed]
23. Roney, C.H.; Bayer, J.D.; Cochet, H.; Meo, M.; Dubois, R.; Jaïs, P.; Vigmond, E.J. Variability in pulmonary vein electrophysiology and fibrosis determines arrhythmia susceptibility and dynamics. *PLoS Comput. Biol.* **2018**, *14*, e1006166. [CrossRef] [PubMed]
24. Steiner, I.; Hájková, P.; Kvasnička, J.; Kholová, I. Myocardial sleeves of pulmonary veins and atrial fibrillation: A postmortem histopathological study of 100 subjects. *Virchows Arch.* **2006**, *449*, 88–95. [CrossRef] [PubMed]
25. Akazawa, H.; Komuro, I. Cardiac transcription factor Csx/*Nkx2-5*: Its role in cardiac development and diseases. *Pharmacol. Ther.* **2005**, *107*, 252–268. [CrossRef] [PubMed]
26. Roselli, C.; Chaffin, M.D.; Weng, L.C.; Aeschbacher, S.; Ahlberg, G.; Albert, C.M.; Almgren, P.; Alonso, A.; Anderson, C.D.; Aragam, K.G.; et al. Multi-ethnic genome-wide association study for atrial fibrillation. *Nat. Genet.* **2018**, *50*, 1225–1233. [CrossRef]
27. Sinner, M.F.; Pfeufer, A.; Akyol, M.; Beckmann, B.-M.; Hinterseer, M.; Wacker, A.; Perz, S.; Sauter, W.; Illig, T.; Näbauer, M.; et al. The non-synonymous coding IKr-channel variant KCNH2-K897T is associated with atrial fibrillation: Results from a systematic candidate gene-based analysis of KCNH2 (HERG). *Eur. Heart J.* **2008**, *29*, 907–914. [CrossRef]
28. Watanabe, H.; Darbar, D.; Kaiser, D.W.; Jiramongkolchai, K.; Chopra, S.; Donahue, B.S.; Kannankeril, P.J.; Roden, D.M. Mutations in Sodium Channel β1- and β2-Subunits Associated With Atrial Fibrillation. *Circ. Arrhythmia Electrophysiol.* **2009**, *2*, 268–275. [CrossRef]
29. Statello, L.; Guo, C.-J.; Chen, L.-L.; Huarte, M. Gene regulation by long non-coding RNAs and its biological functions. *Nat. Rev. Mol. Cell Biol.* **2021**, *22*, 96–118, Correction in *Nat. Rev. Mol. Cell Biol.* **2021**, *22*, 159. [CrossRef]
30. Yang, Z.; Jiang, S.; Cheng, Y.; Li, T.; Hu, W.; Ma, Z.; Chen, F.; Yang, Y. FOXC1 in cancer development and therapy: Deciphering its emerging and divergent roles. *Ther. Adv. Med. Oncol.* **2017**, *9*, 797–816. [CrossRef]
31. Seo, S.; Kume, T. Forkhead transcription factors, Foxc1 and Foxc2, are required for the morphogenesis of the cardiac outflow tract. *Dev. Biol.* **2006**, *296*, 421–436. [CrossRef]
32. Peng, T.; Tian, Y.; Boogerd, C.J.; Lu, M.M.; Kadzik, R.S.; Stewart, K.M.; Evans, S.M.; Morrisey, E.E. Coordination of heart and lung co-development by a multipotent cardiopulmonary progenitor. *Nature* **2013**, *500*, 589–592. [CrossRef]
33. Debnath, P.; Huirem, R.S.; Dutta, P.; Palchaudhuri, S. Epithelial–mesenchymal transition and its transcription factors. *Biosci. Rep.* **2022**, *42*, 1–29. [CrossRef]
34. Kato, T.; Sekiguchi, A.; Sagara, K.; Tanabe, H.; Takamura, M.; Kaneko, S.; Aizawa, T.; Fu, L.-T.; Yamashita, T. Endothelial–mesenchymal transition in human atrial fibrillation. *J. Cardiol.* **2017**, *69*, 706–711. [CrossRef]
35. Lai, Y.J.; Tsai, F.C.; Chang, G.J.; Chang, S.H.; Huang, C.C.; Chen, W.J.; Yeh, Y.H. miR-181b targets semaphorin 3A to mediate TGF-β–induced endothelial-mesenchymal transition related to atrial fibrillation. *J. Clin. Investig.* **2022**, *132*, e142548. [CrossRef]
36. Klein, A.L.; Grimm, R.A.; Murray, R.D.; Apperson-Hansen, C.; Asinger, R.W.; Black, I.W.; Davidoff, R.; Erbel, R.; Halperin, J.L.; Orsinelli, D.A.; et al. Use of Transesophageal Echocardiography to Guide Cardioversion in Patients with Atrial Fibrillation. *N. Engl. J. Med.* **2001**, *344*, 1411–1420. [CrossRef]
37. Leucci, E.; Vendramin, R.; Spinazzi, M.; Laurette, P.; Fiers, M.; Wouters, J.; Radaelli, E.; Eyckerman, S.; Leonelli, C.; Vanderheyden, K.; et al. Melanoma addiction to the long non-coding RNA SAMMSON. *Nature* **2016**, *531*, 518–522. [CrossRef]
38. Shao, L.; Sun, W.; Wang, Z.; Dong, W.; Qin, Y. Long noncoding RNA SAMMSON promotes papillary thyroid carcinoma progression through p300/Sp1 axis and serves as a novel diagnostic and prognostic biomarker. *IUBMB Life* **2020**, *72*, 237–246. [CrossRef]
39. Ni, H.; Wang, K.; Xie, P.; Zuo, J.; Liu, W.; Liu, C. LncRNA SAMMSON Knockdown Inhibits the Malignancy of Glioblastoma Cells by Inactivation of the PI3K/Akt Pathway. *Cell. Mol. Neurobiol.* **2021**, *41*, 79–90. [CrossRef]
40. Khurana Hershey, G.K. IL-13 receptors and signaling pathways: An evolving web. *J. Allergy Clin. Immunol.* **2003**, *111*, 677–690. [CrossRef]
41. Cornelissen, C.; Lüscher-Firzlaff, J.; Baron, J.M.; Lüscher, B. Signaling by IL-31 and functional consequences. *Eur. J. Cell Biol.* **2012**, *91*, 552–566. [CrossRef] [PubMed]
42. Zeng, J.; Zhang, J.; Yang, Y.Z.; Wang, F.; Jiang, H.; Chen, H.-D.; Wu, H.-Y.; Sai, K.; Hu, W.-M. IL13RA2 is overexpressed in malignant gliomas and related to clinical outcome of patients. *Am. J. Transl. Res.* **2020**, *12*, 4702–4714. [PubMed]
43. Sai, K.K.S.; Sattiraju, A.; Almaguel, F.G.; Xuan, A.; Rideout, S.; Krishnaswamy, R.S.; Zhang, J.; Herpai, D.M.; Debinski, W.; Mintz, A. Peptide-based PET imaging of the tumor restricted IL13RA2 biomarker. *Oncotarget* **2017**, *8*, 50997–51007. [CrossRef] [PubMed]
44. Chong, S.T.; Tan, K.M.; Kok, C.Y.L.; Guan, S.P.; Lai, S.H.; Lim, C.; Hu, J.; Sturgis, C.; Eng, C.; Lam, P.Y.P.; et al. IL13RA2 Is Differentially Regulated in Papillary Thyroid Carcinoma vs. Follicular Thyroid Carcinoma. *J. Clin. Endocrinol. Metab.* **2019**, *104*, 5573–5584. [CrossRef]
45. Shibasaki, N.; Yamasaki, T.; Kanno, T.; Arakaki, R.; Sakamoto, H.; Utsunomiya, N.; Inoue, T.; Tsuruyama, T.; Nakamura, E.; Ogawa, O.; et al. Role of IL13RA2 in sunitinib resistance in clear cell renal cell carcinoma. *PLoS ONE* **2015**, *10*, e0130980. [CrossRef]

46. de Magalhães, J.P.; Curado, J.; Church, G.M. Meta-analysis of age-related gene expression profiles identifies common signatures of aging. *Bioinformatics* **2009**, *25*, 875–881. [CrossRef]
47. Kato, T.; Iwasaki, Y.K.; Nattel, S. Connexins and atrial fibrillation: Filling in the gaps. *Circulation* **2012**, *125*, 203–206. [CrossRef]
48. Gollob, M.H.; Jones, D.L.; Krahn, A.D.; Danis, L.; Gong, X.-Q.; Shao, Q.; Liu, X.; Veinot, J.P.; Tang, A.S.; Stewart, A.F.; et al. Somatic Mutations in the Connexin 40 Gene (GJA5) in atrial fibrillation. *N. Engl. J. Med.* **2006**, *354*, 2677–2688. [CrossRef]
49. Tribulova, N.; Egan Benova, T.; Szeiffova Bacova, B.; Viczenczova, C.; Barancik, M. New aspects of pathogenesis of atrial fibrillation: Remodelling of intercalated discs. *J. Physiol. Pharmacol.* **2015**, *66*, 625–634.
50. Xu, H.F.; Ding, Y.J.; Shen, Y.W.; Xue, A.-M.; Xu, H.-M.; Luo, C.-L.; Li, B.-X.; Liu, Y.-L.; Zhao, Z.-Q. MicroRNA-1 represses Cx43 expression in viral myocarditis. *Mol. Cell. Biochem.* **2012**, *362*, 141–148. [CrossRef]
51. Yang, B.; Lin, H.; Xiao, J.; Lu, Y.; Luo, X.; Li, B.; Zhang, Y.; Xu, C.; Bai, Y.; Wang, H.; et al. The muscle-specific microRNA miR-1 regulates cardiac arrhythmogenic potential by targeting GJA1 and KCNJ2. *Nat. Med.* **2007**, *13*, 486–491. [CrossRef]
52. Chen, S.; Zhou, Y.; Chen, Y.; Gu, J. Fastp: An ultra-fast all-in-one FASTQ preprocessor. *Bioinformatics* **2018**, *34*, i884–i890. [CrossRef]
53. Dobin, A.; Davis, C.A.; Schlesinger, F.; Drenkow, J.; Zaleski, C.; Jha, S.; Batut, P.; Chaisson, M.; Gingeras, T.R. STAR: Ultrafast universal RNA-seq aligner. *Bioinformatics* **2013**, *29*, 15–21. [CrossRef]
54. Liao, Y.; Smyth, G.K.; Shi, W. featureCounts: An efficient general purpose program for assigning sequence reads to genomic features. *Bioinformatics* **2014**, *30*, 923–930. [CrossRef]
55. Li, B.; Ruotti, V.; Stewart, R.M.; Thomson, J.A.; Dewey, C.N. RNA-Seq gene expression estimation with read mapping uncertainty. *Bioinformatics* **2010**, *26*, 493–500. [CrossRef]
56. Wagner, G.P.; Kin, K.; Lynch, V.J. Measurement of mRNA abundance using RNA-seq data: RPKM measure is inconsistent among samples. *Theory Biosci.* **2012**, *131*, 281–285. [CrossRef]
57. Okada, D.; Okamoto, Y.; Io, T.; Oka, M.; Kobayashi, D.; Ito, S.; Yamada, R.; Ishii, K.; Ono, K. Comparative Study of Transcriptome in the Hearts Isolated from Mice, Rats, and Humans. *Biomolecules* **2022**, *12*, 859. [CrossRef]
58. Sun, J.; Nishiyama, T.; Shimizu, K.; Kadota, K. TCC: An R package for comparing tag count data with robust normalization strategies. *BMC Bioinform.* **2013**, *14*, 1–14. [CrossRef]
59. Wu, T.; Hu, E.; Xu, S.; Chen, M.; Guo, P.; Dai, Z.; Feng, T.; Zhou, L.; Tang, W.; Zhan, L.; et al. clusterProfiler 4.0: A universal enrichment tool for interpreting omics data clusterProfiler4.0: A universal enrichment tool for interpreting omics data. *Innovation* **2021**, *2*, 100141.
60. Shen, W.; Chen, S.-Y.; Gan, Z.; Zhang, Y.-Z.; Yue, T.; Chen, M.-M.; Xue, Y.; Hu, H.; Guo, A.-Y. AnimalTFDB 4.0: A comprehensive animal transcription factor database updated with variation and expression annotations. *Nucleic Acids Res.* **2023**, *51*, D39–D45. [CrossRef]
61. Galili, T. Dendextend: An R package for visualizing, adjusting and comparing trees of hierarchical clustering. *Bioinformatics* **2015**, *31*, 3718–3720. [CrossRef] [PubMed]

MDPI AG
Grosspeteranlage 5
4052 Basel
Switzerland
Tel.: +41 61 683 77 34

International Journal of Molecular Sciences Editorial Office
E-mail: ijms@mdpi.com
www.mdpi.com/journal/ijms

www.ingramcontent.com/pod-product-compliance
Lightning Source LLC
Chambersburg PA
CBHW041346210926
51850CB00013B/3004